BIOTECHNOLOGY OPERATIONS

Principles and Practices

Michael J. Roy

CRC Press
Taylor & Francis Group
Boca Raton London New York

CRC Press is an imprint of the
Taylor & Francis Group, an **informa** business

CRC Press
Taylor & Francis Group
6000 Broken Sound Parkway NW, Suite 300
Boca Raton, FL 33487-2742

Printed in the United States of America on acid-free paper
10 9 8 7 6 5 4 3 2 1

International Standard Book Number: 978-1-4398-3027-7 (Hardback)

Library of Congress Cataloging-in-Publication Data

Roy, Michael J. (Michael Joseph), author.
Biotechnology operations : principles and practices / Michael J. Roy.
p. ; cm.
"CRC title."
Includes bibliographical references and index.
Summary: "This book resulted from experiences gained working in biotechnology and while teaching a graduate course entitled Biotechnology Operations and offered to graduate students in the Master of Science (M.S.) in Biotechnology Program at the University of Wisconsin-Madison (http://www.ms-biotech.wisc.edu/). In this course we examine the undertaking of developing biotechnology products, focusing on the scientific and management skills of biomanufacturing, clinical trials, non-clinical studies, project management, quality assurance, quality control, and regulatory affairs. The course emphasizes both operational planning for success and integration of plans and efforts in these seven functional areas. The instructors realized from their experience in the biotechnology industry the great need to carefully plan and fully integrate biotechnology development projects. The course is taught in that manner and this book reflects that philosophy with the outcome is a practical guide for students and for those employed or interested in biotechnology"--Provided by publisher.
ISBN 978-1-4398-3027-7 (hardcover : alk. paper)
1. Biotechnology. 2. Biotechnology industries. 3. Biotechnological process control. 4. Project management. I. Title.
[DNLM: 1. Biomedical Technology--organization & administration. 2. Biotechnology--organization & administration. 3. Planning Techniques. 4. Program Development--methods. 5. Total Quality Management--methods. W 82]

TP248.2.R69 2011
660.6--dc22 2010048796

Visit the Taylor & Francis Web site at
http://www.taylorandfrancis.com

and the CRC Press Web site at
http://www.crcpress.com

Contents

Preface

This book resulted from experiences gained working in biotechnology and while teaching a graduate course entitled Biotechnology Operations and offered to graduate students in the Master of Science (MS) in the Biotechnology Program at the University of Wisconsin–Madison (http://www.ms-biotech.wisc.edu/). In this course, we examine the undertaking of developing biotechnology products, focusing on the scientific and management skills of biomanufacturing, clinical trials, nonclinical studies, project management, quality assurance, quality control and regulatory affairs. The course emphasizes both operational planning for success and integration of plans and efforts in these seven functional areas. The instructors realized from their experience in the biotechnology industry the great need to carefully plan and fully integrate biotechnology development projects. The course is taught in that manner and this book reflects that philosophy with the outcome a practical guide for students and for those employed or interested in biotechnology.

Biotechnology Operations: Principles and Practices hopefully meets a need and fills a gap. Despite the wealth of experience with operations in the biotechnology industry, there was no single comprehensive and practical, yet fundamental guide available. Many books and most individual scientific or trade publications are highly technical and focused on a specific aspect of biotechnology. And, they do not emphasize the themes of planning and integrating the seven operational endeavors. This book is written with the objective of presenting a roadmap and reference for biotechnology operations, integrating these functional areas through the processes of product planning and design and the practice of project management. It also applies lessons learned in the biotechnology industry over past decades as novel products have been developed from emerging scientific discoveries. The lessons highlight development principles that could help the industry to bring to market more efficiently and quickly the safe and effective biotechnology products of the future. While focused largely on biopharmaceuticals, the book also reflects development of other biotechnology products. It is anticipated that this book will provide the reader a clear understanding of basic principles and practices, and assist in reducing risks and in resolving problems as future biotechnology discoveries are developed into products.

Michael J. Roy
University of Wisconsin–Madison

Acknowledgments

I would like to acknowledge the aid and encouragement of many professional colleagues and friends and the wonderful discussions with students in the Master of Science Biotechnology Program, especially those who have made helpful suggestions. Special thanks goes to Anthony (Tony) Clemento, who along with Derek Hei and me developed the course, Biotechnology Operations, and have, under Tony's direction, instructed over 150 students in this subject over the past nine years. Tony has been especially supportive by encouraging the preparation of this book, providing timely information to me on all aspects of biotechnology operations, and directing the course and its content.

I am grateful to the reviewers who took time from their busy lives to review the book and make helpful suggestions. Thank you Lori Barr, Natalie Betz, Anthony Clemento, Derek Hei, Florence Kaltovich, Laurence Lemiale and Trish MacDonald for your reviews and insightful comments. Tracy Ulderich expertly prepared and edited the figures included in this book. And thanks to Kurt Zimmerman, Program Director of the MS Biotechnology Program at the University of Wisconsin–Madison, for providing a program in which students are trained and encouraged to become industry leaders. He has set a professional, progressive and positive tone for instructors and students and encouraged us to take on the many challenges related to biotechnology education. I am very grateful to Phyllis Dubé and Kirsten Roy for their understanding, encouragement and support.

The Author

Michael J. Roy, PhD, RAC, is an adjunct professor in the Master of Science Biotechnology Program in the School of Medicine and Public Health at the University of Wisconsin–Madison. He teaches Biotechnology Operations and Project Management and Leadership.

He has successfully developed biopharmaceutical and medical device products for more than 22 years, serving as a consultant in biotechnology since 2001. Much of his work has focused on early development of novel biotechnology products and medical devices with emphasis on product development planning, regulatory affairs, quality systems and project management. Dr. Roy has worked with many biopharmaceutical products and biologicals with an emphasis on vaccines and therapeutic proteins.

At PowderJect Vaccines, Inc., he was vice president of regulatory affairs, responsible for national and international regulatory strategies and programs and managed clinical and regulatory efforts for the firm's vaccine development programs. Prior to its merger with PowderJect, he served as director of regulatory affairs at Agracetus, Inc. In these roles, he established and managed manufacturing, quality assurance, quality control and regulatory affairs, and was involved in development of vaccines, genetic therapies, medical delivery devices and biopharmaceutical products raised in plants. Prior to this, Dr. Roy was director of regulatory affairs at MedImmune, Inc. (Gaithersburg, Maryland), where he was responsible for regulatory submissions for vaccines, monoclonal antibodies and intravenous immunoglobulin products. Earlier in his career, he spent 10 years as a civilian scientist at Walter Reed Army Institute of Research (Silver Spring, Maryland) and the U.S. Army Medical Materiel Development Activity (Fort Detrick, Frederick, Maryland) where he managed and led biopharmaceutical development projects aimed at developing over 20 biopharmaceuticals.

He is a graduate of the University of Wisconsin–Madison with a PhD in Pathology, of Louisiana State University Medical Center with an MS in Tropical Medicine and Medical Parasitology, and of the University of Wisconsin–Platteville with a BS in Biology. He holds Regulatory Affairs Certification (RAC).

Colonel Roy is retired from the U.S. Army Reserve. While serving in the army, he was involved in the development of *in vitro* diagnostics and vaccines and in establishing ISO (International Organization of Standards) quality systems at the U.S. Army Medical Research Institute of Infectious Diseases, Ft. Detrick.

He also enjoys raising hardwood trees on land in southwestern Wisconsin, hiking and fishing.

1

Introduction to Biotechnology Operations: Planning for Success

Biotechnology Operations

Biotechnology encompasses a wide variety of scientific, business and operational endeavors in the life sciences. It is applied across a broad range of specific disciplines—plant, animal, medical, microbiological, biopharmaceutical, agricultural, environmental, to name just a few. Biotechnology is practiced worldwide and at many institutions such as small private firms, large public corporations, nonprofit organizations, universities and research institutes. Those practicing biotechnology include individuals with diverse skills and backgrounds, namely, entrepreneurs, scientist, business persons, managers, product developers and other highly educated and motivated specialists. As seen by the uninitiated and at the macro level, biotechnology appears to be a vast, three-dimensional matrix, broad and oftentimes baffling in scope and operation. However, to those experienced in biotechnology, there is organization and rationale. The keys to successfully managing a biotechnology firm are to focus on carefully crafted plans and efforts to accomplish a specific objective and integration of operational activities within the operational matrix. This is especially true for biotechnology product development operations where the objective is to increase the value of specific products by moving them through sequential phases and to the marketplace.

Virtually every aspect of biotechnology has two common themes: (1) to extend our knowledge of life sciences and (2) to produce a product or service that someday will improve the condition of mankind. In the commercial sector of biotechnology, there also is the objective to profit financially. There are subplots to every biotechnology endeavor as well. Developing a novel biotechnology product, especially a biopharmaceutical, is an extremely technical, highly regulated, complex, expensive and long process. Biopharmaceuticals are in development for over five years and it is not unusual for schedules to extend, from research to market approval, beyond 10 years. The risks associated with biotechnology are tremendous because most biopharmaceuticals fail at some point in development. And, yet, there are compelling reasons to undertake biotechnology product development. The profits can be substantial, if investors provide support and the firm markets useful products. For some individuals, it is not financial incentives but altruistic purposes or the challenge of pursuing an ambition and life-long dream. This provides a stream of bright individuals willing to labor at bringing biotechnology products to market. So, biotechnology development continues to grow in importance, size and scope, and is highly regarded by the public.

This book focuses on biotechnology product development, specifically the scientific skills commonly applied worldwide to move, in an ordered manner, a product from concept at the laboratory bench to the marketplace. It emphasizes product design, development planning, project management and elements of each major operational function applied to the development process. We refer to these combined activities as biotechnology operations. The seven major functional areas of biotechnology operations, identified in Box 1.1 with a brief description of each, are further described in individual chapters of this book. Additional functional areas, such as business development and finance, also directly impact biotechnology operations and these

BOX 1.1 SEVEN MAJOR AREAS OF BIOTECHNOLOGY OPERATIONS

Operational Area	Definition	Chapter
Project Management	Lead the planning, organization and management of the overall development project and associated resources	2
Regulatory Affairs	Advise on regulatory aspects and climate for product development, coordinate activities with regulatory agencies, and ensure regulatory compliance	3, 4
Quality Assurance	Provide support to ensure that all efforts and the product are of highest quality through quality management, audits, documentation and other quality functions	5
Biomanufacture	Produce the highest quality product through phased manufacturing development and final commercial production	6
Quality Control	Ensure quality product through laboratory testing	7
Nonclinical	Develop pharmacology and toxicology laboratory and animal studies and reports to ensure the safe and proper use of the product	8
Clinical	Determine the safety and effectiveness of product when used to treat human volunteers	9

are recognized because they are keys to the success of various functions and efforts.

The focal point of a biotechnology operation is the product and at the heart of product development are the user and intended use. The operational team of professionals works together to add value and, eventually, to bring this product to market—actually to the intended user. Hence, a key to building a successful biotechnology operation is to maintain this focus on the product and its intended use and the user. In biopharmaceutical development, the intended use is the product indication, a word that will be used repeatedly in this book. In medicine, an indication is defined as the reason a product is used to diagnose, prevent or treat a specific disease or condition. An indication also identifies, to a great extent, the intended user of a biotechnology product. This is especially true for biopharmaceuticals. In addition to having an indication, biotechnology development also is based on an understanding of the molecular or cellular nature of a product and how its safety, strength, purity and potency impact the user.

The seven functional areas listed in Box 1.1 form the backbone of a biotechnology product development operation and, thus, provide the basic skills and resources to accomplish objectives. An additional key element is integration and coordination of these skills in an effective and timely manner and focused on making headway, moving the product toward market approval. Given the complexity of biotechnology operations, the need for careful planning is intuitive. A plan establishes the objective and also maps out a means of integrating the skills and events that lead to success. Indeed, a product development plan allows a development program to be successful. Without a carefully crafted and functionally integrated plan, biotechnology operations typically fail.

Biotechnology has its own jargon as evidenced by terms used in this book and other references listed in the Additional Reading at the back of the book. A great amount of operational information, notably regulatory, is available at Web sites and some of these are identified in the text.

Words, some considered jargon, have developed to describe certain aspects of the biotechnology operational trades and these can be confusing, even counterintuitive, to the uninitiated. To assist the reader with definitions as used in this text, a glossary has been provided.

To begin our journey through biotechnology operations, this chapter introduces the planning process for product development. Think of the plan as a "skeleton" and each element of the plan a bone that gives structure to the overall operation. Chapter 2 describes project management, the operational function that serves as the "neural system" to a biotechnology operation, coordinating movement of operational elements according to the plan. Chapter 3 through Chapter 9 describe individual functional areas that execute or flesh-out the plan and provide operational activities (Box 1.1). The functional areas do the heavy lifting, so to speak, in an operation and six of them are considered the muscles of an operation.

Marketing, Financial and Business Considerations for Development

Biotechnology products in general and biopharmaceutical products in particular, with their stringent regulatory guidelines and strict need for a high benefit to risk ratio, are particularly expensive to develop. So expensive, in fact, that investment capital and public funding often provide insufficient resources to support the complete product development cycle. Today, the total development cycle costs for a biopharmaceutical can reach or exceed $1 billion. While somewhat less expensive to develop, other types of biotechnology products, such as those in the agricultural or environmental sectors, might still cost in excess of $500 million. Indeed, some biotechnology firms never even enter the development arena due to high cost and inability to raise capital to meet projected expenses.

Biotechnology firms rely upon both public or private financing and partnerships with traditional pharmaceutical firms to provide capital needed to reach their development goals. Of course, money always comes with tradeoffs and an investor or partner may project definite ideas and opinions regarding how the biotechnology firm should develop the product. In the end, some biotechnology firms are acquired by the partner during the development cycle and well before a product comes to market. Raising capital is not a subject of this book, but one must consider expenses and budgets during development planning and again at every milestone.

Once a project has begun, financing and budgets continue to have an impact on decisions made both in planning and in executing a project. Indeed, they are often the *primary* consideration regarding a decision on whether or not to continue a product development project. There are tradeoffs for the biotechnology firm. Development of a specific product may necessitate the sacrifice of other endeavors, such as pursuing promising lines of research. The nature of a company may have to be changed to pursue development, with hiring of development staff offset by the loss of research scientists. Facilities inevitably must be added or modified to suit development efforts and, as will be noted later, this can be extremely resource intensive. Once these resources have been committed, there is no turning back without incurring significant loss of time and money. No wonder biotechnology executives typically refer to the decision to embark on development as "betting the farm" or "entering the valley of death."

Given all these warnings, what *is* the prudent way for a biotechnology firm to enter product development? The answer is simple: one step at a time, with a market analysis, a carefully defined product and indication and a well-considered product development plan.

Earlier in this chapter, a metaphor (skeletal, muscular and neural systems) was used to introduce the concepts of biotechnology development plans, operational elements and integration by project management, respectively. This metaphor is further explained and developed in Box 1.2. Adding to

BOX 1.2 A BIOLOGICAL METAPHOR FOR PLANNING BIOTECHNOLOGY OPERATIONS

A metaphor to planning a biotechnology operation is taken from the organized development of the mammalian neural, muscular and skeletal systems. This metaphor seems relevant, given the biological nature of our professional work.

An organism is composed of individual organs and tissues, and as they develop and function they work together in harmony and allow the animal to function and survive. The skeletal, muscular and neural tissues provide functions, respectively of support, movement and perception of or reaction to stimuli. Each tissue arises in an exact manner, shaped according to a plan, this programmed in the genetic code. The developing skeletal system is composed of bones, logically arranged and able to provide the outline of a unique organism.

To begin the metaphor a biotechnology operation functions, or should function, in the manner of a healthy organism, with the individual organs and tissues coordinated and working in harmony. An operational plan is the skeleton of that operation. The organism's skeletal structure represents the product development plan. It provides shape to the overall project. While the individual bones of an animal form a strong framework, they must move in an integrated and coordinated manner. For this to happen in an organism, muscle is the organ system holding bones in a particular manner yet moving them so they are useful structural elements. In a like manner, the product development plan is motivated by the various functional areas of biotechnology development—clinical, manufacture, nonclinical, quality assurance, quality control and regulatory affairs—that implement the plan, providing outcomes yet allowing movement and flexibility of operation.

Returning to the organism in this metaphor, a neural system signals the bones and muscle to work together in a timely and effective manner. The neural system makes bone and muscle useful to the body by coordinating endeavors, both as affecters and effectors. Thus, the bones and muscles achieve specific objectives. In biotechnology product development, the neural system is represented by project management, a key function that ensures the various elements work together in harmony, sensing the operating environment and reacting accordingly.

Perhaps the most important part of this metaphor is to imagine an organism deficient in one of these three elements: skeleton, muscle or neural. Indeed, there are diseases for which this is the case. The result is illness and eventual death. Here the metaphor carries to the biotechnology operation because without each of the functional elements, a product development plan to bring them all together and a system to integrate and manage their operation, the product development program does not function properly and eventually it will die. Alternatively, if skeletal, muscle and neural systems are healthy and carry their weight, then the organism, and by analogy the biotechnology operation, prospers.

this metaphor, consider that these three organ systems would not function properly in any animal without support provided by other organs: the heart, liver and kidneys, for example. So, it is in biotechnology development where support from research, marketing, business development, management and other areas is essential to the life of the operation. An important element of any good development program is the need to consider the advice, expertise and support of individuals with skills that do not apply directly to the technical agenda of an operation, but have great impact nonetheless. We have mentioned financing and now consider input from the business and marketing professionals. While these professionals might seem at times to perceive situations and issues differently from operational staff, their skills and judgment are indeed important throughout the development process and their input is especially critical to success at the planning stage.

They sit on the product development team, advising and planning from business, finance and marketing standpoints. The team members should pose to them critical questions from the outset of the planning process. Is there a market for the biotechnology product as it is currently designed and, if so, is the market large and extensive enough to generate a profit and is it open to introduction of this new or improved product? Or, should another product design be chosen? Is there competition and, if so, is it prohibitive to the intended market? Will it be necessary for the firm to develop or further develop the market and, if so, how long might this take? Are there advantages and disadvantages to the market due to regulatory pressure, not just the U.S. Food and Drug Administration (FDA), but any regulatory agencies? How might we price this product in the current market? Here, the business and finance elements of the entity become especially important and a well considered business plan provides valuable information for development planning purposes. At this time, it may be difficult to exactly identify business advantages of a particular product, but certain elements can be considered. At a strategic level, several questions are posed. In theory, is money available to develop products in this market sector and, if so, is there precedent? What are the potential sources of funding and are partnerships with other firms possible? Alternatively, might competitors try to impede our progress in an effort to retain their market share? At this early stage of predevelopment, it is impossible for even the most seasoned business experts to have all the answers; indeed, meetings at this time may generate more questions than answers.

Product Development Planning

Rationale for Product Development Planning

Biotechnology operations have borrowed many concepts and operating principles from the drug industry. Indeed, both drug and biopharmaceutical development projects often focus on preventive and therapeutic biopharmaceuticals intended for use in humans. Drug development, a phased or stepwise process well established by the drug industry and regulatory authorities, is commonly applied to biopharmaceutical development. Figure 1.1 outlines functional elements involved in a phased scheme for a biopharmaceutical or drug development project and the approximate schedule for each. It represents a project beginning with discovery or engineering of a novel biologic and ending with a product entering the marketplace. It is an idealized and simplified cartoon, but, in reality, the process is much more complex than depicted and functions presented here may abbreviate or lengthen the overall process. Nonetheless, such schemes are developed and applied as planning and operational management tools, thus providing visual representation of the major events, processes and milestones, and facilitating communication and understanding by project teams and upper management.

Discovery research is the foundation upon which most biotechnology products are based. Some refer to it as Phase 0 in the development process because discovery must happen before Phase 1 or early development may begin. It is scientists in laboratories who discover, sometimes serendipitously and, in other instances, by plan, the information upon which biotechnology product development is based. Gene cloning, propagation of stem cells, engineering a drought-resistant trait into plants and a monoclonal antibody directed against a tumor protein are but a few of the thousands of proven discoveries that have been the foundations for important and useful products. In most instances, these discoveries are patented, a legal means of ensuring that the discoverer, or the affiliated institution, receives proper credit for any worthwhile product that might be developed from their invention. Patents ensure that the patent holder, the discoverer, reaps a monetary reward if he or she reaches the marketplace with a product upon which the patent is based. Because product development requires substantial resources, typically tens to hundreds of millions in U.S. dollars, only a few biologic discoveries are taken through the development life cycle to become a product. Most biotechnology products, therefore, are based upon a unique discovery that either has a patent or is patentable. However, few discoveries or inventions in biotechnology are themselves marketable products; they must first be developed.

What then can we do with an exciting, patented biological innovation that holds potential value to mankind and in the marketplace? We can develop it into a useful product. But, what must we do to develop that product? First, we carefully and exactly define the product. While this may sound simple,

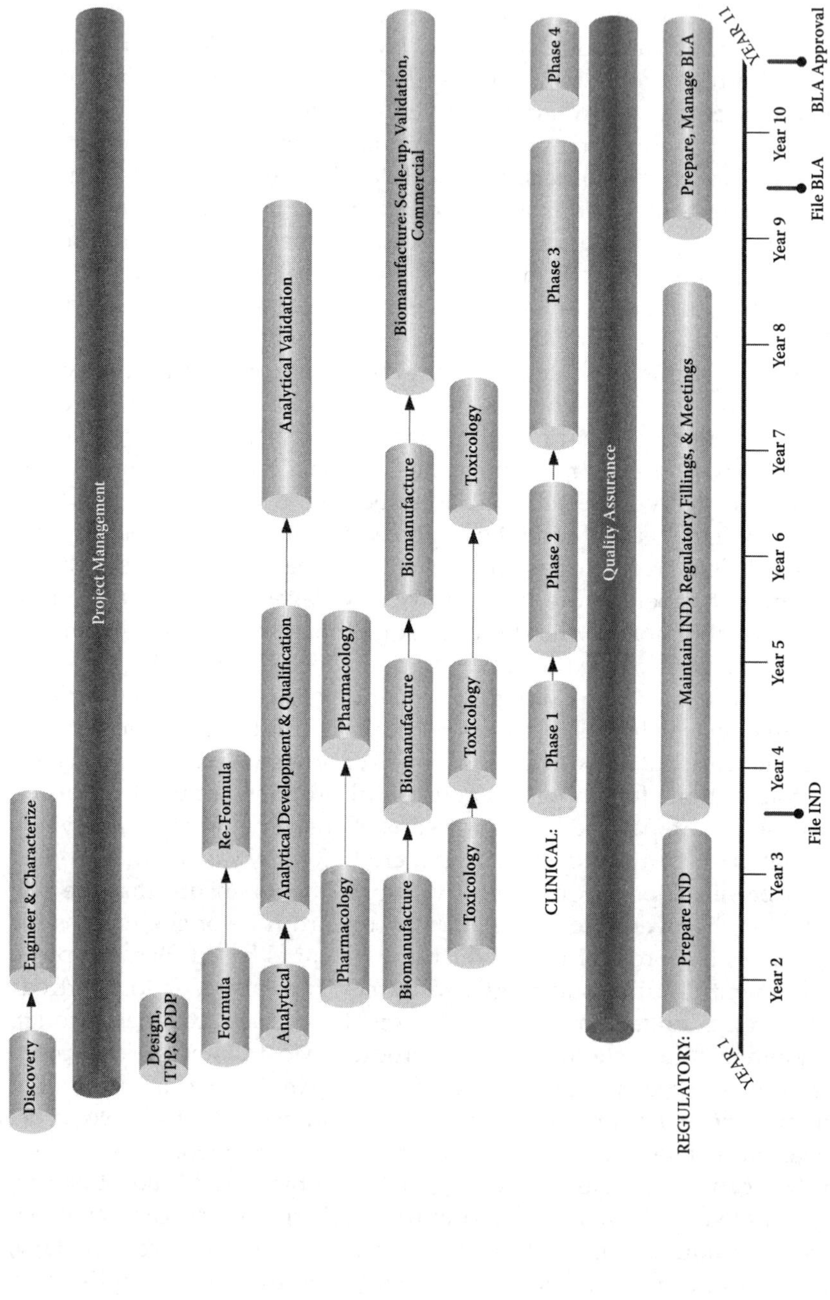

FIGURE 1.1
Example of a development pathway, highlighting functional areas and general schedule of events.

Continued on facing page.

reaching a definition is no easy task and, unfortunately, many discoveries enter product development without an exact definition of what the intended product is or what it is expected to do to be useful to mankind and to the patent holder. In such cases, the product development plan, and the execution of that plan, is unfocused, wasteful and, far too often, unsuccessful. A biotechnology product, and hence the developmental pathway, must be planned, it cannot simply evolve. There is not enough time and money to take any product development route other than execution of a well-considered product development plan.

How does planning work? Let us consider another metaphor to explore targeted product planning. Suppose we had the power to design and then develop a new animal. Our first step would be to define, in various ways, the purpose of this desired mammal. Specifically, we begin by asking how it would meet needs of the user. Let's say that we wish our new animal to pick fruit from trees in orchards. Then our design, based on this user need, would be bipedal and tall, with long arms and dexterous fingers. It should have the strength to stand for hours and muscles that allow it both to stretch and to rapidly pluck fruit from a tree. It should have intellect, an ability to differentiate oranges from apples and to discern ripe oranges from immature fruit, and a brain to signal the muscles and skeleton to pick that fruit. Hence, we have defined a creature intended to pick fruit from trees. In planning the bone structure of this animal, would we borrow the design of a dog or a meadow vole? Certainly not, instead we would shape our plan, the bones if you will, around a bipedal creature, perhaps a primate. However, we would design especially long bones in the arms and legs, a vertical or erect and strong vertebral column and lengthy arm bones with many digits. Indeed, we might include bones for four arms, one to grasp the tree branch, one to pluck the fruit, one to catch the fruit and yet another to transfer it to a basket. Hence, our bone structure forms a framework for the intended creature.

The plan for the muscular system would make these bones useful to the creature's intended purpose. Would we link these bones with muscles that allow our creature to run fast, like a cheetah? Probably not. We would instead give it muscles that allow these bones to stand all day, to continually grasp limbs and to carefully pick, grasp and transfer fruit.

We would plan a neural system that coordinates these musculoskeletal functions, one that achieves the primary objective of picking fruit, but also

---

FIGURE 1.1 (*Continued*)
A development pathway, here defined over 11 years, begins with discovery of a novel and potential product. The research laboratory remains involved in the early years, assisting in characterization and, perhaps, reengineering or refinement of the product to meet design criteria. The targeted product profile (TPP) and product development plan (PDP) are prepared early under the guidance of formal project management and is often revised in the first year. Formulation, biomanufacture and analytical development efforts begin early in development and continue through two or more phases as additional information becomes available from nonclinical, laboratory and clinical studies. Quality assurance efforts are continuous from beginning to end and regulatory input is consistent and focused on compliance, meetings and filings.

allows the grasp to rapidly change as the creature reaches for a new branch, to pluck fruit with one arm even as the grasp is changed with another arm and fruit is transferred with a third arm, and to discriminate ripe from unripe fruit, immediately before it is plucked.

Biotechnology product development, to be successful, follows a specific planning process in much the same way as we designed the fruit-picking creature. However, in biotechnology, the long process of discovery research and economic realities do not allow us the luxury of millennia, the time needed for evolutionary processes in nature. In biotechnology operations, products are developed rapidly and efficiently from innovations. We begin with a clear understanding of what the product is and what it must do and how the product will be used. This is written in a targeted product profile (TPP).

Targeted Product Profile

Product development planning is said to happen in a backward manner, this because the process begins with generation of a TPP, which in fact is a draft of the product label with product claims. The planning process is outlined in Figure 1.2. In biopharmaceutical development, the TPP has in the past been referred to as draft or concept product labeling. The FDA strongly encourages

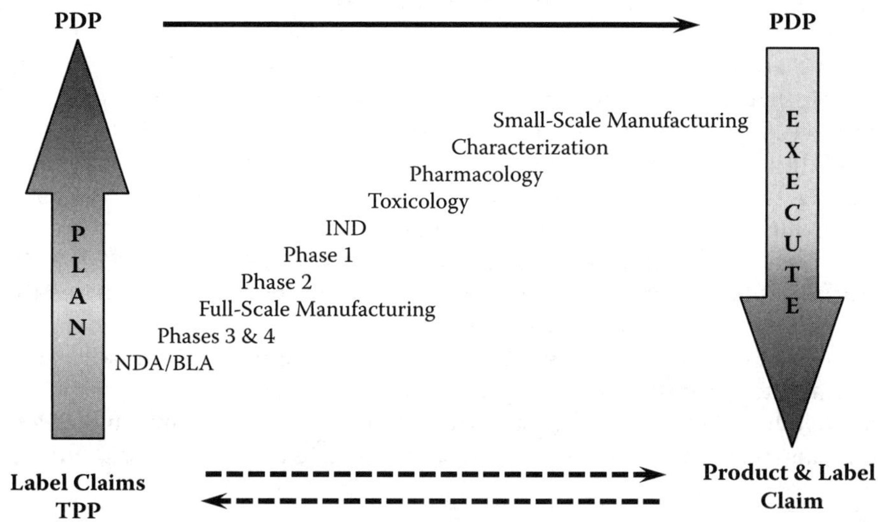

FIGURE 1.2
Planning backwards: Development of a targeted product profile and product development plan. Beginning in the lower left, a targeted product profile (TPP) defines label claims in the first step of the planning process. Then, working backwards through product development events and phases, a product development plan (PDP) is established, as seen on the left side proceeding from bottom to top. The plan is implemented (right side of the figure) in a forward manner through each stage of development to produce a final product with approved label claims. (Graphic courtesy of Anthony Clemento, 2008.)

sponsors, defined as the entity responsible for developing a biopharmaceutical, to prepare and use a TPP to support communication with regulatory authorities. More recently, investors in biotechnology have asked petitioners to provide them with a TPP along with the business and technical plans.

Simply stated, the TPP is a clear and detailed description of what a product should be, how it will appear and, most importantly, what it must do. The term TPP says it all, establishing a target of or focus on the product and profiling or summarizing characteristics of the product. Box 1.3 lists elements of a typical TPP for a biopharmaceutical product. Production of a TPP is truly the first step in managed product development. A TPP is written by a lead author, someone familiar with both the product and with various aspects of biopharmaceutical development. Teamwork is essential for a successful start and this means appointing a product development team and project manager and holding team meetings at regular intervals (see Chapter 2). Team members review and recommend changes to the draft TPP. Members of the product development team, each representing a functional area, investors and upper management are involved in this process with a final TPP as the team's first goal. These early interactions set the context and tone for later discussions, team members bond, agree or disagree, and leadership skills become evident. The need for additional professional skills is recognized and so teams are filled out to meet management and technical objectives and, early on, thought is given to the nature, scope and possible general schedule of the development project. Thus, preparation of a TPP provides a critical guidance document and solidifies the new product development team.

Of all elements attributed to a TPP, the first, most important and, often times, the most contentious, is establishing *the* indication, also referred to as the *label claim* or simply the *"claim"* (see Figure 1.2). Note similarities between information in a TPP and in an actual product label. The differences are largely due to the fact that a TPP is an expectation in nature and tone, while the actual labeling of a marketed product is FDA approved, the real thing, based upon data. For a biopharmaceutical, an indication might be defined as a treatment or prevention for a disease or condition that has a specific cause and symptoms. Let's begin by demonstrating poorly worded indications. One, for a peptide therapeutic product, is to "lower blood pressure in benign hypertension." Another vague example, this for a vaccine composed of a recombinant protein, is to "prevent malaria." It is critical that an indication be as specific as possible and that it be matched with a biopharmaceutical well suited and most likely to succeed in clinical trials. Here the biotechnology development team must set aside bias and grand or long-range projections of safety or efficacy (e.g., "this biopharmaceutical is so great it will cure every type of cancer and never result in a side effect") and instead focus on the research data, settling on an indication for which the product would likely reach the market in short order. Having said this and returning to our examples, the indication for the peptide might be more rationally stated as "lowering blood pressure in individuals with uncomplicated benign

BOX 1.3 ELEMENTS OF A TARGETED PRODUCT PROFILE

Trade name and chemical name. A draft trade name or interim designation, such as a compound number, is developed for the product.

Warnings. Warnings might be added for this product based upon its class of product and previous experience. Messages to the prescriber or user for this product are noted.

Description. The product's nature and classification are developed. The formulation in the final container, with excipients, is included.

Clinical pharmacology. The mechanism of action, pharmacodynamics and pharmacokinetics that are known to date or should be explored are presented. Drug interactions are given.

Clinical studies. Identified are pivotal clinical studies to include patient populations, endpoints and outcomes.

Indications and usage. The expected indication is given as is the intended patient or user population.

Contraindications. Situations in which the product should not be used (e.g., pregnancy or congestive heart failure) are stated.

Warnings/precautions. The users or physicians become aware of events or reactions to the product and the more serious or common of these are given in a warning. They are written for physician and user, or composed as Information for Patients, specifically written for the user. Instructions for special situations also are placed here and specific items are highlighted as paragraph headings. Recommendations may be given to stop using a product, for example, if a disease progresses or if certain symptoms are noted. Drug interactions, use in nursing mothers or in pregnancy, pediatric and geriatric use, or use in other special populations are generally included in this section.

Adverse reactions. Types of adverse reactions that might be acceptable, given the intended use and user profile, are identified.

Overdosage. This describes reactions or remedies, should a patient take more than the prescribed amount.

Dosage and administration. This provides a statement on how the product will be provided as dosage form to a patient in final format. The intended final container or delivery system is described.

How supplied. This describes the packaging format that will be produced and marketed.

> **References.** A few key scientific publications regarding the product and indication are included.
> **Patient information.** This expands special instructions that might be required for proper handling, storage or use by the patient.

hypertension and between the ages of 60 and 82 years, where blood pressure has remained elevated above 140/90 despite the use of other common drugs and where there are no symptoms of congestive heart failure." The example of the malaria vaccine might better be stated as "indicated for use in infants and children between the ages of 6 months and 5 years for prevention of serious disease and death from falciparum malaria in endemic regions of Africa, Asia and South America." The TPP also makes claims for safety parameters and, thus it is important to consider the safety profile that would be acceptable for the product and include this in the final profile.

Biotechnology products are all accompanied by printed labeling and, in this, the claims on product safety and efficacy are made. Claims are reflected in advertisements and labeling, not just for medicines, but for all types of products. Consumers read these claims (or should read them) when making purchases and before use. A TPP is draft labeling with predicted claims and, as such, the TPP is used to guide the planning and actual development of all biotechnology products, not only biopharmaceuticals. Some examples follow. A recombinant bacterium is indicated for remediation of crude oil spills in salt water where the air and water temperatures are above 40° F, wave action is not severe and the spill is contained to a geographic region under 100 square kilometers. A genetically engineered soybean has the indication to increase yields 20% over other varieties when grown in zones 3 or 4 and where rainfall averages between 12 and 20 inches per year, there is no irrigation and the soil is slightly acidic or neutral.

Returning to biopharmaceutical development, a TPP discusses the other objectives and these are listed in Box 1.3. It is worth noting that results from research or early development completed to date, market drivers and, perhaps, the experiences and ingenuity of individual team members are the basis for deciding on each claim. The target population is identified as part of the claim. For example, a product is to be used only in adults over the age of 50 years and in otherwise good health. Often, specialists, such as physicians highly regarded in a specialty area, are consulted before the team reaches a consensus on a target indication and population.

The next step in developing a TPP for a biopharmaceutical is to determine the target dosage and route of administration best suited to the product and the population identified in the indication. The peptide example might be best administered by the parental route, such as subcutaneous, because peptides might not be amenable to the hostile environment of the gut. The vaccine might be preferred as an intramuscular injection from a disposable syringe. In each example, the dosage may need to be 1 milliliter (mL) or less. Dosage forms and strength refer to the formulation of the product and how it might be presented in a marketed or "final" container, such as a vial or syringe. The peptide might only be stable in a buffer of low pH. In the case of the malaria vaccine, the product profile includes a preservative, so it can be used in the tropics. This, in turn, necessitates a formulation that allows it to be shipped and stored with breaks in refrigeration. For the microbes used to remediate an oil spill of over 100 square miles, it might be necessary to consider a product that could be disbursed from large mechanical sprayers on aircraft. The seeds of a drought-resistant soybean plant might need to be planted farther apart from each other, as compared to current soybeans. This type of information is agreed by the product development team and included in the TPP.

The TPP also considers dosage form and strength. For medical products, there will be limitations as to the mass of product, peptide or recombinant protein that can be held in 1 mL of solution. The optimal formulation, one that is simplest and least expensive, may not be feasible and the product team could decide a special formulation was necessary, e.g., instructions to keep a protein product from aggregating and, thus, preventing loss of activity. In these examples, the TPP prompts the team to consider manufacture, formulation and quality control issues, and highlights the need for additional steps in development and, perhaps most importantly, identifies both complexities and costs of actually developing the individual product.

Contraindication refers to those times when the team recommends that the product simply should not be used, when it might be unsafe, for example. Basic contraindications should be considered and here again it might be helpful to use a medical consultant with experience treating the disease in the indicated patient population. Warnings and precautions, on the other hand, are more difficult to define at this very early stage of development and in the absence of any safety information on the product. However, warnings and precautions from products similar in nature, treatment indication and target patient populations may be instructive as to what may or may not be acceptable for this product. The contraindications, warnings and precautions often narrow the indication and this is important information to consider in product development. For the peptide used to treat hypertension, it might be contraindicated to use the drug in patients with certain other cardiovascular diseases as known from experience in cardiovascular medicine and pharmacology. The malaria vaccine might be contraindicated when the patient was already infected with the parasite. The remedial bacterium might be contraindicated when other petrochemicals, such as gasoline or diesel fuel, were present. The drought-resistant soybean plant might not be used within a kilometer of other soybean fields. The main point is that a knowledgeable product development team confronts these issues during the process of developing a TPP and well before development begins. This facilitates early planning to resolve, if possible, each potential problem or issue.

Identification of undesirable and product-related adverse reactions, risk of overdosage, or interactions of the biopharmaceutical with other drugs are, to a great extent, items that must be addressed during clinical studies (see Chapter 9). However, it is possible during TPP preparation to consider the limits of adverse events or precautions the team might allow for a product. With the peptide antihypertensive, serious illness or death resulting from therapeutic doses, no matter how infrequent, might pass the acceptable threshold for adverse events. The malaria vaccine for children should not cause local reactions and discomfort that can be of great concern to the child or a parent. In the case of the remedial bacteria or the soybean plant, one might respectively establish limits regarding how extensively the bacteria could multiply in the environment in the absence of crude oil or how far the soybean could spread to neighboring fields. The process does identify, to the development team, certain limits that might be applied to the development program.

Use in special populations further defines when and how one might use the product; it extends the indication by considering individuals of certain age groups, such as adolescents or the elderly, or of physiologic status, such as nursing mothers or pregnant women. Drug abuse and dependence is typically not an issue with biopharmaceuticals, but can be important with certain types of drugs.

Adding a description of the product to the TPP would seem, on the surface, to be a simple task but product development teams often find it to be a challenge, especially in regard to describing all intended physical, chemical and biological characteristics. This is discussed in greater length in Chapter 7. Any biotechnology product that will be used in man, animals or the environment will need to be very well characterized in all respects but, at this juncture, product characteristics are unknown. Preparation of the TPP forces the team to consider what types or classes of characteristics must be examined in the product during development. For any of the examples we have used, biological characteristics should include potential toxicity or half-life and description of any living cells. Chemical characteristics might be the chemical nature of a molecule and also any impurities or contaminants. Physical characteristics are size or shape or the ability to withstand adverse conditions of an acceptable molecule or organism.

Clinical pharmacology also may be unknown at this early stage of development. The term takes into consideration the distribution of the biopharmaceutical in the body, the kinetics of distribution from the time of dosing through the time of clearance and the dynamic properties while it is in tissues (see Chapter 8). However, there should be some information, from laboratory or animal studies, upon which the team can develop desirable parameters or acceptable limits. The antihypertensive peptide should clear itself from the body prior to another dose being given and the malaria vaccine should not remain in a subcutaneous tissue indefinitely. The remedial bacterium should be cleared from the environment and not be present long after the crude oil has been eliminated and there should be limits on how long the soybeans can self-reproduce under field conditions.

Nonclinical toxicology testing (see Chapter 8) is very important because the toxicity profile of a product in animals is often a predictor of toxicity in humans. Clinical studies follow nonclinical toxicology and, as covered in Chapter 9, they are designed with the results of nonclinical studies in mind. The nonclinical and clinical toxicology profiles are certainly unknowns at this early stage, but the development team does have the opportunity and obligation to set limits for safety and efficacy parameters, even if they are general, for each product in its TPP. Here the history and labels of competitive products or good medical judgment come into play, along with scientific and medical experience. Would one consider developing and marketing the example peptide antihypertensive if it consistently caused rats to die of hypotensive shock at the intended human dose? Might the malaria vaccine be advanced to clinical trials in children and infants or would it even be marketable if it caused severe local reactions in both rabbits and nonhuman primates? And, for the remedial bacterium, would one wish to put it into a field trial in an ocean lagoon if the bacterium itself led to the death of fish or invertebrates in an aquarium setting? Would it be prudent to place the drought-tolerant soybeans into field trials if they were found, in greenhouse studies, to spread the resistance gene to other species of legumes? Limits for nonclinical and clinical toxicity can and must be established.

The preparation of a TPP not only motivates the team to discuss potential issues early on, even before a product development plan (PDP) is written, it also forces members of the team to consider limits to the technology well before a major investment is made in developing the product. Hence, the value of a TPP goes far beyond internal use by the sponsor. Once completed, a TPP often becomes the technological extension of a business plan and is invaluable for business development and helps to raise working capital from investors. It is a foundation for communicating the technology, potential benefits and possible risks, to the public and to regulatory agencies, simply because it clearly demonstrates that the sponsor has considered implications, good and bad and known and unknown, of the technology.

Product Development Plan

A PDP (also called a product development strategy (PDS)) extends the TPP, providing a roadmap to reach the stated goals. Further, it defines how issues and unknowns identified in the TPP will be addressed and, thus, becomes the basis for scheduling activities and budgeting resources over that schedule. And the PDP is shared by everyone on the product development team as a common narrative understanding of what has to be done and how it will be accomplished. Project managers use the PDP for exact task integration, scheduling and tracking. The PDP may be shared, in confidence, with potential investors or partners and regulatory agencies so as to demonstrate that the sponsor has the will and a valid strategy to take the product to market and, thus, make a return on investment. Consider again that planning is a process and a plan is a written document. Typically, the planning process is managed by a project manager while the plan itself is written by individual members of the project team. There is no established order to preparing the individual chapters, discussed below, but most organizations find it is quite helpful to develop a draft or at least an outline of three sections—clinical, regulatory and project management—before beginning the others. The draft project management plan establishes the project team and provides guidelines, early on, as to the planning process itself. The planning process requires much discussion and this comes at team meetings or teleconferences. Therefore, having a project management approach established early on facilitates communication and preparation of each section of the PDP. Because the planning process works backwards in a development scheme (see Figure 1.2) and because a Phase 3 clinical study is critical to achieving market approval of a biopharmaceutical, it is very helpful to draft a clinical plan before other sections are prepared. Also, having regulatory guidance upfront provides important guidelines, especially for biopharmaceutical products.

Once the project management, regulatory and clinical plans are in outline or early draft format, then a designate from each functional area drafts the appropriate section of the plan. The planners each apply the method of working backwards to prepare at least a solid outline. They can then expand the plan, adding detail while working forward through early, mid and late phases. Integration of the elements is important and the process is facilitated by effective project management, frequent meetings and cooperation on the part of every team member.

The contents of any one PDP are difficult to predict because every product is unique. Yet, experience provides suggestions to ensure any PDP is understandable. The PDP has a clearly stated purpose and objective, focused on the product as described in the TPP. It considers every one of the seven functional areas. It identifies significant risks and foreseeable difficulties and makes arrangements in the plan to address them. The risk-to-benefit discussion is real and not overly optimistic. The plan is comprehensive by providing precise technical descriptions. Important steps or stages are not avoided or omitted and adequate resources are committed for each functional area. Finally, the PDP offers a realistic schedule.

Provided below are elements found in most PDPs, written as statements that should, along with the writings in this book and other publications, stimulate thought and focus on planning your product. The order of sequence includes

1. Clinical
2. Project management

3. Regulatory
4. Nonclinical
5. Manufacturing
6. Quality control
7. Quality systems and quality assurance

Additional planning considerations are discussed in a separate section.

Clinical Development Planning

Overall Clinical Development Planning

- Prepare a broad overview plan for clinical development and confirm intended label claims and intended medical outcomes following treatment.
- List, by phase of development, all clinical trials that are anticipated.
- Define the most challenging aspects of clinical development for this product and indication.
- Identify safety, tolerability or toxicity factors that are of concern for the investigational product.
- Describe what has been learned from previous clinical studies with products of this type and for similar indications.
- Define pharmacoeconomic and marketing issues related to the product should it be approved for the stated indication and patient population.
- Identify key decision points in the clinical development scheme.

Clinical Development Planning by Phase

- Describe the Phase 3 clinical trial design and elements of a Phase 3 concept protocol to include hypothesis, objectives, outcomes, endpoints and measurements.
- State the regulatory guidance or precedent needed to develop the Phase 3 clinical approach.
- Given the intended design and outcomes of Phase 3, identify a study or studies to be performed in Phase 2.
- Describe how Phase 2 studies are to be temporally staggered.
- Provide a brief concept design for each Phase 2 study, indicating the outcomes, endpoints, measurements and number and nature of subjects tested.
- List Phase 1 studies to be completed prior to beginning Phase 2.

- Provide a brief concept design for each Phase 1 study, indicating the objectives, endpoints, measurements, number and nature of subjects tested and most likely outcomes.

General Clinical Development Plans at Each Phase or Study

- Identify criteria for patients or subjects enrolled.
- Describe unique designs, such as adaptive or crossover, contemplated for any study.
- List the resources required to perform the study and describe requirements for clinical study centers or sites.
- Identify multicenter studies to be performed in the late stage and include foreign clinical trials considered at early- and midstage studies.
- Describe logistical considerations and management of multicenter trials.
- Name the most likely opportunities for clinical study sites and identify studies to be outsourced and the sponsor's roles and responsibilities to be delegated to outside consultants or contractors.
- Identify analytical or medical tools or procedures that will be developed to measure clinical endpoints. Describe how and when each is to be developed.
- Name the internal staff requirements at each stage of development.
- Provide the general statistical approach and list requirements for data handling, statistical analysis and report preparation.
- Outline the clinical study's monitoring and auditing plans and describe how clinical quality will be ensured for each study.
- Identify clinical trial material (product) requirements at each phase of development based upon the concept protocols and number of subjects and doses per protocol.

Project Management Planning

- Define the overall objectives and scope of the project.
- Identify the overall policy for project management applied to the development project.
- Define requirements for support from upper management.
- Perform a general work breakdown structure of the major areas of effort known to date. Provide an estimated schedule for the project; illustrate this in a chart (e.g., Gantt or PERT).
- Define roles, responsibilities and authority of the project manager.

- Define team composition in all areas over the course of the project.
- Define team communication methods along with anticipated frequency of each type of communication. Identify special communication requirements due to distances or international participation on the project team.
- Identify methods to involve contractors, consultants or vendors with the team.
- Identify responsibilities of the team and of the project manager for risk assessment and risk management.
- Identify methods to be used by the team to solve problems.
- Identify methods for the team's decision-making process.
- Define responsibilities and processes for risk analysis, mitigation and management.
- Define tracking and metrics procedures to be applied and indicate their frequency of use.
- Discuss budget and human resource responsibilities of the project manager.
- Develop a project schedule.
- Provide, in general, the objectives and schedule for project closure.

Regulatory Planning

Planning the regulatory approach and operational elements requires several skills. First, regulatory intelligence is conducted. Next, a draft plan is formulated. Finally, all other sections of the PDP are reviewed to ensure that each is consistent and compliant with the current regulatory environment.

Regulatory Intelligence

- Describe what is known about this product or a similar product (predicate) from the regulatory literature.
- Describe how predicate products were designed, mention their origin and history, and identify the methods and technologies used in their discovery and development.
- Identify potential regulatory routes of approval, both United States and ex-United States, used to develop similar or predicate products.
- List the technical (e.g., manufacture, control, nonclinical or clinical) and regulatory successes and failures for each of these products and explain why each succeeded or failed to gain market approval.
- Discuss the impact this technical and regulatory intelligence might have on the intended PDP.

- Discuss how the national political environment may or may not be supportive of this product. List state, local, or even cultural practices or laws that might be unfavorable to such a product or indication.
- Discuss, during the investigational phases, how the public might perceive the relative benefits and risks of this product. Will public opinion matter, one way or another, to regulatory agencies in regards to this product and indication? Mention outstanding safety issues that might concern regulatory agencies and regulatory precedent for handling these issues.
- Identify how regulations might be expected to change prior to approval in any given market.

Regulatory Planning

- Identify regulatory objectives, such as Investigational New Drug Applications and Biologics License Application.
- Define any special regulatory pathways, activities or options that will or might be considered.
- Prepare a regulatory risk-to-benefit analysis for this product.
- Provide one or more possible regulatory outlines or roadmaps with proposals to overcome perceived or real regulatory hurdles.
- Identify and propose means to manage regulatory risks in the United States.
- Define global, i.e., ex-United States, regulatory strategies, primary and alternative. Consider each major market separately and explain unique regulatory guidance and country-specific regulatory hurdles.
- Propose responses to some possible regulatory changes that could occur prior to market approval.
- Identify likely postmarketing regulatory activities and anticipated advertising and promotion guidelines and restraints for the product and the labeling claims.
- Identify methods that would most effectively facilitate regulatory communication with each agency or office within an agency. Define each means of communication with a regulatory agency and at each phase—early, mid and late—of development. Discuss the most challenging aspects of the regulatory communication plan.
- Provide an estimate of the number of investigational documents and market applications that must be filed and the temporal relationships of each.
- What are alternative regulatory routes to approval, such as orphan drug or fast-track status, that might apply to this product and

indication? Have any of these routes been tried with this class of product and, if so, what were the outcomes?

- How will compliance be accomplished under current Good Manufacturing Practices, current Good Clinical Practices and current Good Laboratory Practices and at which phase of development will they be needed?
- If compliance activities are managed in-house, what are the internal programs and guidelines for handling FDA inspections? Based on risks to the user associated with the product, is an FDA inspection likely during early investigational phases of development?

Nonclinical Planning

- Identify precedence and regulatory guidance for pharmacokinetic and pharmacodynamic studies performed at each phase of development for this class of biopharmaceutical and any predicate products.
- Identify safety, tolerability or toxicity factors that are of concern for the investigational product.
- Refer to intended human dose, dosing regimen, length of dosing, route and method of administration in the clinical plan.
- Define the most challenging aspects of nonclinical development for this product and indication.
- Outline objectives, concept study design and relative schedule for all intended studies:
  - Pharmacokinetic and ADME
  - Pharmacodynamic
  - *In vitro* toxicology
  - Acute toxicology
  - Subchronic toxicology
  - Chronic toxicology
  - Specialty toxicology in animals
- Identify analytical or clinical evaluation tools or procedures that will be developed or used to measure endpoints in animals. Describe how and when each is to be developed.
- Name the internal staff requirements at each stage of development.
- Provide the general statistical approach for these nonclinical studies and give requirements for data handling, statistical analysis and report preparation.
- Outline the nonclinical study's monitoring and auditing plans and describe how clinical quality will be ensured for each study.

- Identify nonclinical study materials (product) requirements, which are based on the concept protocols and number of animals and treatment doses per protocol.
- Once a clinical plan has been drafted, define each concept study design, putting them into perspective with the overall development scheme, schedule, precedence and guidance. Examine all requirements to achieve the objective: scientific, material, time and monetary limitations. Propose budget and schedule for each.

Biomanufacturing Planning

- Having considered the TPP and research results on the product, identify and describe the product's type or class and summarize information on the biomanufacture of predicate or similar products.
- Referring to the draft product design, outline a biomanufacturing design, including overall objectives and goals for each phase of development: early, late and middle. Consider product risks and hurdles for the biomanufacturing plan.
- Draft or outline a biomanufacturing plan based upon this design and consider product quality attributes both from the standpoint of process control and for quality control testing. In drafting the plan, consider the ultimate objective, biomanufacture of commercial product upon market approval, and work backwards, i.e., begin with commercial manufacture of the product and proceed in planning each phase in reverse order.
- Provide a plan for scale up of biomanufacture to produce required amounts of product at each phase. Consider also purity and potency requirements at each phase.
- Define plans for application of current Good Manufacturing Practices at each phase of biomanufacturing development.
- Consider each raw material or component that will be used in production. Identify potential quality criteria and the sources. Identify any regulatory guidelines on the quality of proposed raw materials.
- Identify and review the history of any expression system or host cell line that will be used. Determine if there is precedent for using the proposed production system and, if so, consider issues revealed in previous biomanufacturing efforts.
- Having reviewed the research background on the product and its current status in research or early development, identify any genetic engineering or other biological manipulations that might be required of the product or a host cell line before the product enters biomanufacture. For example, the need to develop or modify an expression

vector, to evaluate a construct for a particular trait, or to characterize or do further research on a gene, a vector or a host cell line.

- Plan the production and in-process and quality testing of any cell banks.
- Define early, middle and late stage development production schemes for this product focusing on quality specification and quantity requirements and the chosen processes. Consider upstream production and downstream purification processes for bulk substance and formulation, fill and finish for final product. Apply objectives and criteria for quality and quantity, yield and scale up at each stage.
- Having defined the processes, identify requirements for in-process testing.
- Once again consider risks associated with the chosen processes, raw material requirements and unique aspects of production.
- Identify special requirements for formulation, fill and finish and labeling of the final product.
- Define the containers or delivery devices to be used and storage conditions and requirements.
- Define facility requirements for each stage of biomanufacture. Consider both quantity and quality. Discuss approaches for meeting these facility requirements or for utilizing contract manufacturers.
- Discuss the need to provide aseptic manufacturing environments. Discuss requirements for clean work areas with classified air supply, segregation of product, potential for campaign manufacturing or shared manufacture and flow of product within a facility.
- Identify equipment and utility requirements at each stage of biomanufacturing development and consider special environmental issues that are relevant to production of this product.
- Provide an overview of validation requirements and plans for the biomanufacturing facilities, utilities, equipment and processes proposed in the manufacturing scheme and at each stage of development.

Quality Control Planning

- Understand from the TPP and draft manufacturing plan any requirements for quality control testing of product for both product release and stability.
- Identify product, bulk substance and final product attributes (e.g., safety, purity, potency) of the product as they will be considered for testing. Identify one or more analytical requirements for each attribute.

- Design a quality control assay for each analytical requirement and consider a hypothetical specification for each.
- For each quality control assay, design the remainder of the assay development life cycle and harmonize this with phases of manufacturing, nonclinical and clinical development.
- For each assay, provide a plan to identify how and where the assay will be performed and estimate resource requirements both for assay development and to perform the assay on expected samples, release and stability.
- Identify analytical methods that will be developed or used to measure the quality of each cell bank that is to be tested under quality control. For each method, describe how and where the assay will be performed and estimate resource requirements both for assay development and to perform the assay on each sample. Harmonize this with the manufacturing plan.
- For each assay, identify analytical controls and reference standards and describe how and when they will be developed or otherwise obtained and, in general, give qualitative and quantitative requirements.
- Outline the initial (early phase) stability test requirements for bulk substance, final product and cell banks or other intermediates. Describe the attributes that will be tested and identify one or more tests for each attribute. Outline the stability test criteria that will be applied at later phases of development. Describe any stability indicating assays that must be developed beyond those considered and planned for release of bulk substance and final product. Outline the frequency of testing under stability protocol.
- Describe any requirements for quality control to measure the quality of the manufacturing environment or output of utilities.
- Identify quality control tests that will be verified to ensure compliance with compendial methods and identify the phase of development for each verification.
- Identify quality control tests that are candidates for qualification and state the requirements, expected outputs and phase of development for each assay qualification.
- Identify quality control tests that should be validated and give expected outputs and phase of development for each assay qualification. Describe the resources that might be required for assay validation and harmonize the analytical validation with manufacturing development and the manufacturing process validation plan.
- Describe the program that will be developed to investigate test failures or out of specification test results and investigations. Harmonize this with quality assurance plans.

- Describe efforts planned to ensure quality control is in compliance with current Good Manufacturing Procedures (cGMP) and harmonize this with plans of Regulatory Affairs and Quality Assurance.

Quality Systems and Quality Assurance Planning

- Consider the appropriate quality systems (e.g., cGMP, current Good Clinical Practices, current Good Laboratory Practices) that must be in place for compliance.
- How and at which stage of development will each quality system be developed and instituted? Will they be performed in-house or by a contractor or partner?
- Identify the "Hallmarks of Quality" that must be established for each quality system that will be instituted in-house or at a contract site.
- Identify applicable U.S. and ex-U.S. regulations that drive the requirement for each quality system.
- Describe the requirement for in-house quality activities, and discuss requirements for a quality assurance unit, quality policy, quality manual and quality plan.
- Outline the elements of the quality manual.
- Outline the elements of the quality plan.
- Describe the roles for quality in "Quality by Design" and at each stage of design control.
- Identify requirements for the quality assurance unit and specifically refer to needs for quality management, document control, auditing and training.
- Identify needs for quality agreements with contractors or collaborators.
- Identify any special quality management requirements, such as Total Quality Management, Six Sigma or risk-based approaches.
- Identify any requirements for quality assurance support in research activities.

Additional Elements of Product Planning

In addition to these seven functional area plans, three additional elements of project planning and implementation deserve mention. These are product design, project risk management and the risk-to-benefit ratio of the product itself. Failure to apply these concepts and practices can result in delay or failure of product development; hence, they are considered during the product development planning process and identified in the product development plan.

Product design immediately brings to mind an engineering endeavor, something that is applied to medical devices but not to a biological product. In fact, product design is an important aspect of the planning process for a biotechnology product and, in this book, the concepts and practices of design, design control and quality by design are discussed at length in Chapter 5. Design is a process that focuses on the product itself and, as one might expect, design begins with product criteria and attributes listed in the TPP. Design focuses on critical quality attributes that are often realized only after a certain amount of product development planning has taken place and the product team has had an opportunity to look at the candidate product and the proposed development plan in some detail. In a practical sense, this means that design activities, and these are described in Chapter 5, often interrupt the planning process and require the team to revisit the nature of the product. This can even mean returning to the research laboratory bench and changing or "tweaking" the candidate product to improve it before it enters, or reenters, the development arena. Quality by design goes one step farther, building quality into the product and, hence, into the product's design process. In effect, this means that, in addition to user and performance requirements, there is a conscious effort to design quality criteria into the product. Of course, this means the quality criteria must be available as product development planning begins and suggests they be included in the TPP. Design also introduces the idea of design controls, steps in a formal design process, as discussed further in Chapter 5. Design is also applied to development processes and, in Chapter 6, an example is given with the design of the manufacturing scheme.

Project risk assessment, mitigation and management strategies involve application of procedures and practices to identify potential or actual risks and to reduce their chance of occurring or, should they happen, their impact on a project. Risk management has a significant impact on improving product quality, safety and effectiveness and, thus, is of direct consequence to the user. It should be part of every product development plan, considering both the product and the development processes. Because risk management activities are often the purview of the project team, this subject is discussed in Chapter 2, Project Management.

Risk-to-benefit evaluations are related to risk management and bring with them other connotations. The term and its concept were developed in the health products industries and by regulatory agencies as a means to convey a specific idea: Any product must deliver more benefit than it poses risk to the user. Immediately, one realizes this concept carries with it philosophical as well as practical and technical implications. Specifically, we ask how on earth do we weigh risk versus benefit for any given biotechnology product. The simple answer is that somehow we do this for every biopharmaceutical before it reaches the marketplace. A biopharmaceutical intended to treat cancer as a terminal disease is allowed to have significant associated risk, usually seen as side effect. A vaccine intended to prevent a nonlife threatening infection in infants is allowed to have a low incidence of risks and these side effects must be considered mild. We make these choices, relying upon the judgment of experts with input from the public because, as the user, they are the object of risk or benefit. We are not always correct in these analyses, but overall the record is excellent. A PDP always considers risk-to-benefit of the product and, because it is driven in large part by regulatory authorities, it is also discussed in Chapter 3, Regulatory Affairs.

Summary of Planning for Success

The primary theme of this book, one that is ingrained into each chapter, is to carefully plan biotechnology operations. This demands that, once a product vision, or Targeted Product Profile, has been established, a long-range plan, the PDP, is produced to guide development, manage resources and reveal to upper management and investors the progress (or lack thereof) of development during this long period. It also means that a project must be well managed and this is especially true for development operations, where at least seven distinct functional areas and several quality practices are brought to bear in an orchestrated manner. We discuss early in this book the principles and practices of project management in the biotechnology industry because this function must be in place before any technical, regulatory and quality efforts begin.

2

Project Management

Project Management in Biotechnology

Discovery research is the heart and soul of a biotechnology organization. Here, the innovative genius of the scientific team works to create new products vital to the success of our industry. The new cellular and molecular entities represent a vast array of functional materials that are changing our lifestyles and improving health, leading to new and expanding markets.

Market demand for products encourages entrepreneurial small firms and large, well-established companies to move biotechnology innovations from the discovery laboratory through the development cycle and to the marketplace. Successful development itself adds value to a product and, if successful, the product ultimately yields profit. The process of transitioning any new product out of discovery research and through development is important to the success of the biotechnology firm because investors closely watch progress of development projects. Yet, the transition from discovery research to the market is beset with failures. Some new and exciting biotechnology products do not survive development and, thus, fail to reach the marketplace or ever yield a profit. Examples of biopharmaceutical firms mismanaging product development are noted regularly in trade and business publications. Some products that fail in early development are acquired, often at low cost, by firms that then take them through to marketing approval.

What is it that these rescue specialists know and from this how can we learn what a biotechnology firm might do to increase the chances of successful product development? One answer is simple and inexpensive: Apply the principles and practices of project management to every product development endeavor and throughout the product development cycle. Indeed, today this is the case in most biotechnology organizations where managed project teams facilitate the transfer of new biotechnology products from research though development and to the marketplace. Nowhere is this more evident than in the field of biopharmaceutical product development, a complex undertaking where project teams simultaneously and sequentially apply project management to integrate and coordinate the six functional areas that include: manufacture, quality control, regulatory affairs, quality assurance, nonclinical studies and clinical research. The project management professional, a critical member of any biopharmaceutical development team, takes scientific, marketing and business ideas and intentions and, through project management, planning and monitoring, converts these resources into a feasible, coherent project, When done correctly, it results in a successful outcome.

Project management is the discipline of applying tools, techniques and skills to plan, organize and manage resources through the various phases of a project to accomplish project goals. Project management has both strategic planning and operational phases. In the planning phase, a group, composed of functional area managers, the project manager and corporate executives, define the objectives and scope of a project and develop a long-term schedule. This product development planning process was described in Chapter 1. A project management plan is an important part of any product plan and no biotechnology development project should be initiated without a distinct project management plan. Elements of a project management plan are given in Chapter 1 as well.

During the operational phases of project management, a project team, led by a project manager, follows the project plan, always reflecting upon the technical tasks, milestones and schedule of resources and activities provided in the overall product development plan. Project management planning maps this process out and provides a foundation and charter for the project team. The project management plan applies concepts of project management, such things as team composition, communication, risk analysis and mitigation management, tracking, human elements, project completion, project management tools and resources, both human and monetary. These concepts are discussed later in this chapter. Project management planning greatly increases the chances of project success, i.e., meeting the objective of getting a product developed and to market, on budget and on schedule.

A history of success is the reason that project management has proved so popular and effective in the biotechnology industry and why it is almost universally applied to biotechnology product development. There are other reasons. First, project management is very malleable and it allows a firm to customize project teams and a management structure for development of each product, no matter how unique the product or the project. Second, it relies on team leadership and, today, well-led biotechnology product development teams meet objectives, keep schedules and move products to the marketplace. The cost of a single day's delay in the biopharmaceutical industry can be $1 million. Third, project management is goal oriented. By its very nature and definition, a project team has clear goals, enabling the team to focus on larger product development objectives. Fourth, projects are structured and this structure is in a written product development plan (PDP), described in Chapter 1. Structure assists a project team to consistently achieve tangible and profitable results. Fifth, a project team applies management principles. Each project has a defined beginning, an end, a schedule for completion and tools, such as task lists and schedules, that assist the project manager and the team. Project teams are consistent and resilient. A team stays with the program from start to finish pursuing the objective no matter if individuals leave or new persons join the team. Another tool is the shared budget for resources: human, fiscal and capital. Resources are allocated per plan and according to schedule as project managers strive to maintain a balance in resources, expended toward a common goal. Sixth, a team is diverse and professional. People with various skills coming from all backgrounds—contractors, employees, consultants and clients—are intimately involved and work together on the team. Finally, there is synchronization, as phases and activities of the project are sequenced to balance resources, time and performance against the objectives and the plan. This sounds a bit idealistic, but for many well-led teams, it is a reality.

In this chapter, we review the field of project management as practiced by biotechnology operations, notably in biopharmaceutical firms. It discusses the history of project management, the construction of a project team and selection of a project manager, reviews operational principles and practices of leading a team, and provides information on project management tools and techniques. It is not intended to be a complete review of the subject, but rather to serve as an introduction and to provide some idea of how a biotechnology operation can be managed to succeed. Hopefully, the reader will appreciate the value of project management and also understand how he or she could apply these skills in his or her work environment.

Background of Project Management

Project management evolved within the engineering industry. Specifically, it was first used on large, high-cost and complex projects that applied cutting edge technology. Examples are projects to build the first atomic weapons, to construct large bridges, to put a man on the moon and to build a major defense system. Advances in technology drove the need for project management. Projects became larger and more complex. Consider a feat like construction of the Panama canal over 100 years ago. It was so technically complex and grand for that time (and perhaps even for our generation) that the project begged for organized management. Further, individual workers brought to the workplace special skills and these individuals, and their work, had to be integrated and scheduled. Choreographed might be a better word. Two hundred years ago shipbuilding required woodworkers, blacksmiths, sail makers and perhaps a few other skills. Today, designing and building a new aircraft carrier depends on the integration of individuals with thousands of skills and subskills. Costs of shipbuilding are managed in part by careful scheduling of parts and labor. Project management appeared because it was needed in a technological society and it has evolved to meet demands of cost, quality and schedule.

In the 1960s and 1970s, as pharmaceutical development technologies became more complex and regulation of the drug industry further complicated this endeavor, the largest pharmaceutical firms began to adapt, from other industries, the principles of project management. At first, these principles were applied to pharmaceutical manufacture, as engineers, trained and practiced in project management, brought skills to increasingly more complex pharmaceutical plants. They were successful in managing teams and complex technologies and this was noted by upper management. By the 1980s, project management was being applied to the full scope of pharmaceutical development, from discovery to postlicensure activities. Also at this time, the new industry called *biotechnology* was just beginning to emerge. Not surprisingly, as scientists and engineers migrated from pharmaceutical to biotechnology firms, they transferred project management knowledge and skills to biopharmaceutical companies. Today, project management has been adopted by pharmaceutical, medical device and biotechnology firms worldwide.

Project Management in Biotechnology

Once an objective has been established for a development program and the targeted product profile (TPP) and the PDP have been drafted (see Chapter 1), the process of planning project management itself may begin. A project may be a three-month process to produce a recombinant DNA molecule for sale as a laboratory reagent or it can be a complex, 10-year biopharmaceutical effort to develop a monoclonal antibody to treat a life-threatening disease of children. No matter the complexity or length of a project, the management portion must be carefully planned as well as the technical aspects.

Hence, a project management plan establishes goals or objectives for the life-cycle management of a project, recognizing hurdles and providing a long-range framework to minimize risks and to achieve goals and rewards. It also puts into place procedures and processes for management. Project management planning simply takes elements of good planning practices—planning for success, looking at the ultimate objective and defining goals along the way, incorporating quality systems—and formalizes them into a document or set of documents that can be shared by all team members throughout the life cycle of the product. If this is achieved and the objective is clear and shared by all team members, then transition from strategic project and management planning to operational project management is easy for the project manager and the biotechnology product development team. Indeed, many would argue that a comprehensive and well-written TPP and

PDP along with a realistic, long-range schedule of critical events, tasks and milestones constitute the core of a project management plan.

A project management plan for biotechnology product development may be considered as having five basic elements. Each reflects a phase in the life cycle of project management and indeed in the life of a technical project. These are

1. Initiation: Starting the project in a positive manner and formation of a team.
2. Planning: The subject of much of this chapter.
3. Executing: The technical and management aspects according to the plan.
4. Monitoring and controlling: Functions that ensure the project is meeting objectives.
5. Closing.

Project Management Environment

The environment in which a biotechnology product is developed matters to project management planning almost as much as the TPP and PDS. Biotechnology firms come in all sizes and with various types of structure or organization. These factors matter to effective project planning and management. For this discussion, firms are stratified and considered based on size and complexity, with virtual biotechnology firms at one end of the spectrum and large, experienced companies at the other.

Virtual firms have few full-time employees. A project management team at a virtual firm might be comprised of from one to a very few employees and, in addition, includes outside partners, consultants or contractors. A key representative, perhaps the titular CEO or a key investor, could lead this small team. Although there is little formal training or experience on the part of the project manager or project team members, and despite the fact that each team member may be responsible for two or more functional areas, small teams at virtual firms often outperform their counterparts at much larger biotechnology or pharmaceutical companies.

Small biotechnology firms normally have little project management infrastructure at the time their first product enters development. To establish project management at a small firm, the technology should have reached a level of maturity and the pathway forward must be clear. Specifically, five elements must be in place because they form the foundation for successful product management:

1. **Management decision and support:** A business decision to move forward, which is made by executive management or a board of directors based on project benefit or attributes, risk (technical and

commercial) and resources. There is the intention to apply project management to product development and there are or will be resources available.

2. **Planning:** A written PDP and TPP, or their equivalent, provide adequate information, stating objectives and spelling out a clear route forward.
3. **Feasibility to begin to move forward:** All elements of the plan are feasible in the current financial, technical and regulatory environments.
4. **Estimate of completion:** A realistic schedule based on estimates of experienced professionals.
5. **Decision points or milestones:** Milestones and decision points are evident in the plan. For example, Go/No Go criteria to advance the project to the next phase are established at the beginning of the project and revisited at the start of each phase.

Once a decision has been reached to apply project management to a product development pathway, management intent is best demonstrated by appointing the initial or core project management team and a project manager and, most importantly, relegating authority and responsibility for product development to this manager and team. Supporting a development project can be particularly difficult for the product discoverer, company founder or executive of an entrepreneurial firm that has, for years, focused on discovery research or business development. To many founders, a seemingly easier route to success may be continuation of discovery research, where perceived risks are lower than those in the route of product development. For some executives, it is difficult to let a project team take control of functions considered essential to the success of the firm. These are emotional decisions that must be made for the small biotechnology firm.

Established biotechnology firms have it a bit easier when they begin a new development project because they have the experience and infrastructure. Indeed, their mature project management programs provide experienced and highly trained staff dedicated strictly to building and managing teams. Many issues related to start-up operations—building the first team and introducing employees to the principles and practices of project management—may not apply to the larger biotechnology firm. However, the established firm has other hurdles to productive project management. A few are listed below.

- Complex organizational structure and rules confound efforts to complete any one project on time. For example, a merger or acquisition, a common occurrence in larger firms, results in changes to a major contractor, disrupting continuity of operations and schedules.

- Priorities change frequently and without clear direction from upper management. For example, the clinical indication for a product changes radically due to revised market objectives.
- Upper management is far removed from project teams.
- A large organization may be slow to respond to opportunity or to change when these are necessary or desirable.
- Projects are abandoned in midcourse and without explanation to the development team.
- Problems, incurred in one project, spill over to another project.
- Elements of a project must be "reworked" because they are considered unsatisfactory to someone outside the team.
- Communication breaks down due to change in mode of communication. For example, a new videoconferencing system is required for all project team meetings, but it does not work properly.
- Corporate politics impact project managers.
- Team membership changes during corporate reorganization.

These examples can complicate efforts toward successful product development in the larger biotechnology firm and they have led some project managers to wish they were employed by a smaller organization.

Project Objectives and Schedules

Biotechnology firms, especially small and midsized companies, often suffer from optimism and fail to recognize that few, if any, other firms have successfully completed projects under highly optimistic schedules. The project team is responsible for ensuring that a realistic schedule is composed and communicated to the team and to management. History suggests this is a difficult task because the typical story in the biotechnology industry goes something like this:

- Project development team is formed.
- Senior management provides ultimate objective for product development.
- Project development team prepares product development strategy with schedule.
- Senior management demands that work be completed in one-half the time allotted by the team.
- Project begins under the accelerated schedule.
- Within one year, project is off track, management is angry and team members are discouraged.

- Pessimism or outright failure in the face of a promising technology.

What is the solution to artificial compression of schedules, an issue that constantly plagues small and midsized biotechnology firms? First, upper management must recognize that the product management team is composed of an experienced group of individuals who, together, have years of experience in estimating development times. Also, these individuals have been responsible for meeting schedules composed by themselves and others. Secondly, members of the product development team must recognize the need to expedite development, but not at great risk to delaying development. Moving quickly along the development pathway is a hallmark of the industry and provides the biotechnology firm with competitive advantage. While it is inappropriate for upper management to establish unrealistic schedules or to diminish resources below a certain level, project managers must consider three project constraints: scope, time and budget, and periodically weigh these constraints while advancing the project through each phase. Balancing these constraints is a major challenge. Also quality and performance must be taken into account in the planning process. And communication is key to ensuring every team member is aware of the constraints.

Project scope and complexity are important considerations to planning. Simply stated, there are simple and easy projects and then there are complex and difficult projects. Yet others are somewhere in between these extremes. Project difficulty and complexity, if they exist, become apparent on reviewing the PDP and have a great impact on the project management plan. Complex projects call for more involved and extensive project management.

Sociotechnical Considerations

To be effective, project management focuses on two critical areas, one technical and the other social. First, it must apply project management skills to the plan with due consideration to implementation. Examples are establishing objectives, developing work breakdown structure and monitoring resources. Project management also must influence individuals whose cooperation and help is needed to complete the project successfully; for example, establishing buy-in from the supervisor of a key team member. This need to use both technical and social skills for effective project management has resulted in the realization that this trade is a "sociotechnical" endeavor. Executives of small biotechnology firms, though successful at influencing outsiders, such as investors or the scientific community, often lack the ability to influence technical aspects of product development. Hence, they call upon project managers to play this role. Project managers, therefore, must have strong interpersonal skills, notably the ability to influence others to achieve project goals. It is critical for upper management to understand this and to follow through by identifying and retaining experienced individuals to manage product development teams who possess both technical background and social skills.

Participants in Project Management

A biotechnology project team is composed of many individuals and they are led in this regard by an appointed project manager. Teams vary greatly in size and scope, depending on the complexity of the project and the size of the biotechnology firm. Each individual on a project team has a vested interest in reaching the same objective no matter what his/her technical skills, employment rank or title. Team members possibly aren't employees, but might be from outside the corporation, as consultants, contractors or investors. Individual skills to consider for inclusion on a biotechnology product development team are given in Box 2.1. Individuals who serve on teams have roles, both professional and managerial, and some have, shall we say, special status. It would be nice to think that everyone involved in a biotechnology project is equal in the eyes of the project and upper management. Unfortunately, this is seldom the case. A project manager and the team recognize key participants in a project, referred to as *stakeholders*. These individuals have a significant vested interest in the project, even though they are often not, from a product development standing, the most active members of the team. Indeed, some stakeholders, such as major investors or executive level management, seldom if ever participate in routine team functions. Nonetheless, their interests are held above those of others on the team and the project manager pays special attention to their opinions and desires. Even though they typically sit apart from the team, stakeholders

BOX 2.1 COMPOSITION OF A BIOTECHNOLOGY DEVELOPMENT PROJECT TEAM

Project manager
Project leader or project champion
Finance
Legal and contracts
Research
Business development
Marketing
Quality control
Quality assurance
Clinical studies
Nonclinical studies
Manufacture
Regulatory affairs

have great influence on team activities and each stakeholder expects regular direct communication from the team, usually by way of the project manager. From this, it can be inferred that the project manager, in addition to managing the team, is responsible for communicating with, indeed for influencing, stakeholders. This can be a stressful and time-consuming task in itself. Experience suggests that stakeholders often hold the positions described below.

- **Project champion:** This person, sometimes referred to as the project leader (as distinguished from the project manager), is capable of influencing biotechnology projects based on scientific expertise, organizational power, or responsibility for a critical resource (e.g., a patent). In other firms, the project champion is an executive manager. In either case, they may be a figurehead (e.g., the founder or discoverer) and may or may not serve on the project team or even have a specific technical role. However, they also may consult directly with project team members. They are often only accountable to upper management, such as board of directors or president, and not to a project manager or to the team.
- **Major investor:** The golden rule is sometimes stated as: "He who has the gold makes the rules." This has great meaning to a biotechnology firm where cash flow is always an issue. Today investors are very proactive. Few attend project meetings, but most major investors expect to be frequently informed by the project team manager on technical successes or failures, news about reaching or missing milestones and updates on expenditures.
- **Chief executive or board member:** Executive officers in small biotechnology firms are very hands-on with project teams. Most do not micromanage their teams, but instead stay in constant contact with the team leader and key team members. They are often scientifically astute, interested and inquisitive. Keep in mind that they are an important bridge for your firm, communicating the good, bad and ugly to analysts, investors and the public.

Individuals actually serving on the biotechnology development project team (Box 2.1) may or may not be considered stakeholders but, as a general rule, the project manager and functional area managers or directors are not stakeholders.

- **Functional area manager or director:** These individuals, and there may be many, are key architects of the project, responsible for decisions about strategy, plan, resource requirements, and determining status. While they may not have authority to allocate resources, their influence looms large in other ways. They might direct key technical or administrative aspects for project support and frequently maintain a commanding presence in the smaller biotechnology firm. They are accountable in two directions: (1) to corporate executives on project matters, but (2) to line management for functional responsibilities.

- **Project manager:** Individuals responsible for leading a team are influential as well. In some cases, they soon become stakeholders themselves, even though they may be subordinate to executives and directors. Project managers often have great responsibility, but without direct authority. We have referred to roles of a project manager throughout this chapter. The way a project is managed and executed is key to a project's success or failure. Hence, it stands to reason that selection of the appropriate project manager is an important decision. The manager should be experienced with a project of this scope and nature, although it is certainly not necessary for a candidate to have great technical knowledge in that area. Ideas for correctly matching a project manager with a project are listed in Box 2.2. Attributes of excellent project managers are given in Box 2.3.

BOX 2.2 CONSIDERATIONS FOR SELECTION OF A PROJECT MANAGER

- What are the objectives and what is the anticipated length of the project?
- What is the scope of the project management function and, hence, the project manager? Is it an individual project, a nested project, an integrated project or a series of projects?
- Has the project team been previously led by someone and, if so, what was the outcome and what are the lessons learned from that leadership?
- Is strategic and operational planning involved?
- Will he or she allocate resources, human or monetary, and make priority decisions?
- Is the project at more than one location or in more than one country?
- To whom will the project manager report and at what level within the organization?
- On what criteria will project management staff and team members be selected?
- How will their performance be evaluated?
- What are the roles and responsibilities of project champions?

BOX 2.3 ATTRIBUTES OF EFFECTIVE PROJECT MANAGERS

General management	Conflict resolution	Leadership
Team building	Planning and scheduling	Resource allocation
Anticipation of change	Acceptance of change	Adaption to change
Execution of change	Effective communication	Team building
Negotiation	Leading decisions	Risk analysis
Risk mitigation	Risk management	Organization
Technical knowledge	People and team skills	Critical thinking
Facilitation	Begging, nagging, playing devil's advocate	

But, where do we find a project manager with these attributes? In selecting a project manager, the small biotechnology firm, with a staff of perhaps 20 to 100 individuals, some of whom have previously served on project management teams, may have qualified applicants already on staff. While the firm may not have a seasoned and full-time project manager, there could be an employee who, through experience at another firm, has basic skills and thus qualifies to lead a product development team. The midsized biotechnology firm will have experienced project management processes and, like a large firm, have extensive project management staffing with individuals willing and available to move to a new project. Selection must be rigorous no matter the source. Individuals designated to the project team should have an opportunity to interview candidates. Final selection is influenced by those who understand and, preferably, have practiced project management as a profession in the biotechnology industry.

Project Management in Biotechnology Operations

Establishing Project Management

It is important to decide exactly when to begin the formal process of project management. Some guidelines and common practices are instructive in this regard. Project management, as described in this chapter, is seldom used in discovery research and, so, the concept is often foreign to the management of a small company. Certain scientists further argue that formal project management inhibits good research because a highly structured environment is not conducive to discovery. Others suggest that it inhibits direct management of projects by executives. However, most also would agree that an organization developing a product through application of more than one functional operational area must institute, at an early stage of product development, some method to coordinate and integrate activities and participants. In the end, most biotechnology firms elect to apply project management principles and practices to their projects.

Given that projects, and thus project management, have a defined beginning, when should the biotechnology firm make the transition and establish formal project management? The best answer may be whenever planning, coordination and scheduling activities will, in some way, help the team and the stakeholders achieve a common goal most quickly and efficiently.

The process is not difficult at experienced biotechnology or pharmaceutical firms typically large or medium sized. They begin project management at the outset of technical efforts, immediately upon approving an operational concept and even before a PDP is prepared. These firms have professional project management staff to draw from or the resources to hire new staff. Larger firms have significant infrastructure in project management, this headed by a vice president dedicated to the task. Also, institution of a project team usually follows internal guidance, instructions from upper management and established corporate guidelines.

This formal process is not often seen at the inexperienced and smaller biotechnology firms. Despite any recommendation to make a conscious decision and begin project management at a defined point, the fact is that project management usually evolves into smaller biotechnology operations, with little conscious effort on the part of executive management. Executives may realize (perhaps after witnessing a failure or setback) that, to develop a product within the allotted time, a team leader is needed to manage the project, lead the team, ensure a smooth and timely sequence of events and care for mundane items, such as setting an agenda, preparing minutes, communicating with stakeholders and preparing formal project management tools, such as Gantt charts and reports. This process is project management by evolution and it is characteristic of small biotechnology firms. Often in this situation, executives draw the project manager-designate from within the ranks of the project team, even if that person has other duties.

Another challenge to project management planning in small firms is open wariness of any managed development process or hesitation to appoint a project manager. At the start, team members may voice many and varied ideas and opinions concerning the scope, purposes and strategies for the project and disagree on management guidelines and styles. This is often the first sign that a new team is embarking on a sociotechnical endeavor and, in the absence of a full-time project manager, early conflicts must be handled gently so as to avoid delay or disruption to the new project. This is even more reason for executive management to complete project management planning and appoint a project manager and organize a project team as soon as the requirements are identified.

Work Breakdown Structure

In planning a new biotechnology project and preparing a project management plan, it is important to devise a work breakdown structure (WBS). This document or chart is based on an understanding of the deliverables, the materials, service or product that the project is intended to produce. To begin, it is necessary to have the TPP, a PDP and the intended scope of work laid out. At this stage of planning, these documents are often rough draft documents. The WBS simply breaks the intended work down into components that are manageable. In a WBS for development of a biopharmaceutical product, the organization looks somewhat like an organization chart with branches representing the functional areas, such as clinical, regulatory affairs, etc. In practice, this is done first on a large sheet of paper or it can be organized using project management software, a tool that will be discussed later in this chapter. Work is broken down first by deliverables or milestones, then into subheadings referred to as tasks, then into subtasks, etc. Examples are shown in Figure 2.1 to Figure 2.3. Each figure shows work breakdown of the same project, but using a different project management tool.

Forming a Project Team and Hands-On Project Management

However, or whenever, formal project management enters the picture, project management planning will begin with a meeting of project management team members, this led by the designated project manager. Please note that this first activity is *project management* planning, not operational or product development planning. To be successful, upper management must wholeheartedly and visibly support the institution of formal project management whether or not there is dissention regarding the need for a project team. Indeed, a way to open the first meeting is to have the president or CEO review for team members the development strategy and the initial or draft project management plan. This empowers the team and its leader and clearly demonstrates the intended objectives, deliverables and measurement criteria of the project; it gets everyone "on the same page" as a team. Network building, as team members form interpersonal relationships, is another important objective of the first meeting, as networks form links and bonds between individual team members. These first steps have been referred to as the "forming" stage of a project team.

Once the strategic objective has been reviewed, the project manager might focus the team on critical elements of a project management plan to ensure that all team members share the same vision. This stage has been referred to as the "storming" phase, perhaps because such discussions can seem quite unsettled. One of the most important tasks is development of the team charter and this is expanded upon in Box 2.4. If nothing else, this first meeting establishes, through the outline of a team charter, team identity both internally, among team members, and, externally, to senior management or to investors. Finally, the team agrees to certain attributes that have been

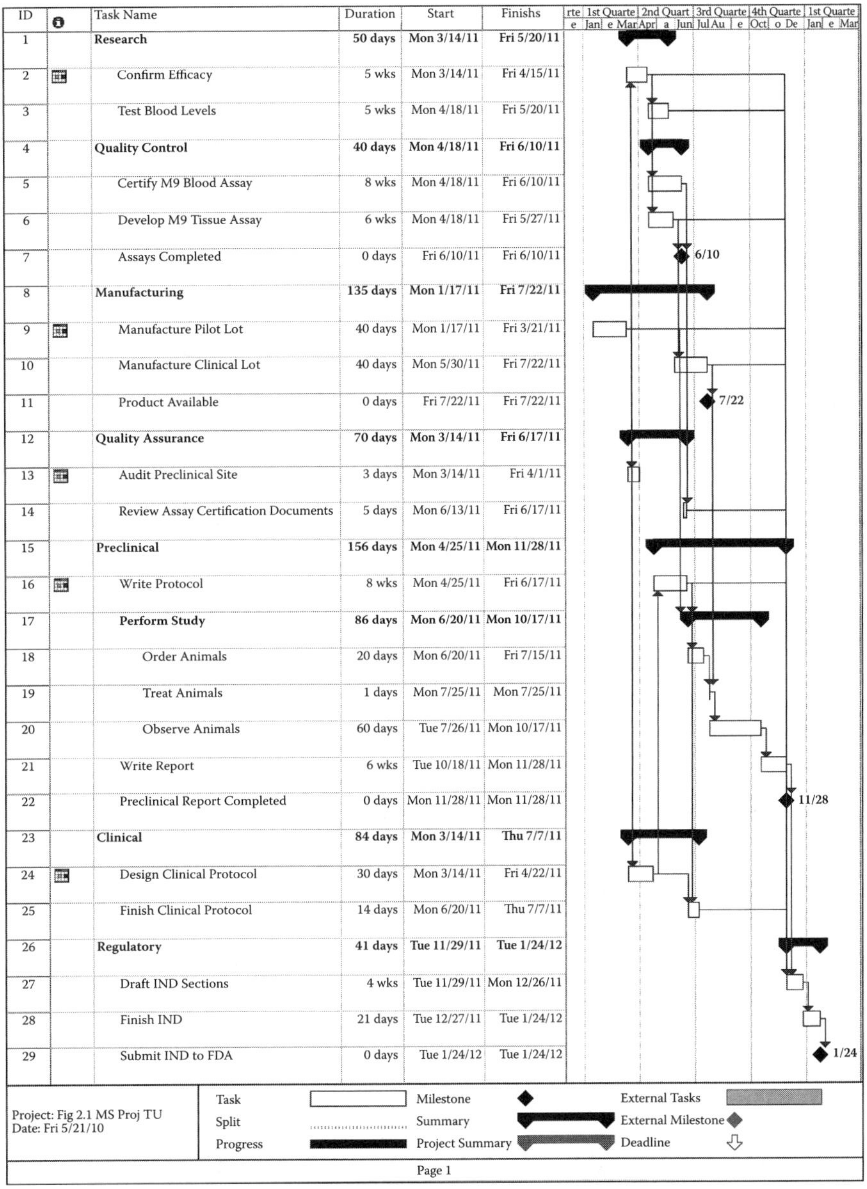

ID	Task Name	Duration	Start	Finishs
1	**Research**	**50 days**	**Mon 3/14/11**	**Fri 5/20/11**
2	Confirm Efficacy	5 wks	Mon 3/14/11	Fri 4/15/11
3	Test Blood Levels	5 wks	Mon 4/18/11	Fri 5/20/11
4	**Quality Control**	**40 days**	**Mon 4/18/11**	**Fri 6/10/11**
5	Certify M9 Blood Assay	8 wks	Mon 4/18/11	Fri 6/10/11
6	Develop M9 Tissue Assay	6 wks	Mon 4/18/11	Fri 5/27/11
7	Assays Completed	0 days	Fri 6/10/11	Fri 6/10/11
8	**Manufacturing**	**135 days**	**Mon 1/17/11**	**Fri 7/22/11**
9	Manufacture Pilot Lot	40 days	Mon 1/17/11	Fri 3/21/11
10	Manufacture Clinical Lot	40 days	Mon 5/30/11	Fri 7/22/11
11	Product Available	0 days	Fri 7/22/11	Fri 7/22/11
12	**Quality Assurance**	**70 days**	**Mon 3/14/11**	**Fri 6/17/11**
13	Audit Preclinical Site	3 days	Mon 3/14/11	Fri 4/1/11
14	Review Assay Certification Documents	5 days	Mon 6/13/11	Fri 6/17/11
15	**Preclinical**	**156 days**	**Mon 4/25/11**	**Mon 11/28/11**
16	Write Protocol	8 wks	Mon 4/25/11	Fri 6/17/11
17	**Perform Study**	**86 days**	**Mon 6/20/11**	**Mon 10/17/11**
18	Order Animals	20 days	Mon 6/20/11	Fri 7/15/11
19	Treat Animals	1 days	Mon 7/25/11	Mon 7/25/11
20	Observe Animals	60 days	Tue 7/26/11	Mon 10/17/11
21	Write Report	6 wks	Tue 10/18/11	Mon 11/28/11
22	Preclinical Report Completed	0 days	Mon 11/28/11	Mon 11/28/11
23	**Clinical**	**84 days**	**Mon 3/14/11**	**Thu 7/7/11**
24	Design Clinical Protocol	30 days	Mon 3/14/11	Fri 4/22/11
25	Finish Clinical Protocol	14 days	Mon 6/20/11	Thu 7/7/11
26	**Regulatory**	**41 days**	**Tue 11/29/11**	**Tue 1/24/12**
27	Draft IND Sections	4 wks	Tue 11/29/11	Mon 12/26/11
28	Finish IND	21 days	Tue 12/27/11	Tue 1/24/12
29	Submit IND to FDA	0 days	Tue 1/24/12	Tue 1/24/12

FIGURE 2.1
Gantt chart format for biopharmaceutical development project. Schedule of events for a project shown in Gantt chart format including the work breakdown by task and task identification (ID) number. The start and finish dates and the duration of each task are given in the narrative listing. The right-hand panel depicts the project in chart format using solid bars to summarize the duration of a set of tasks (e.g., Research) and shaded bars to represent individual tasks. Diamonds represent milestones and arrows interconnect tasks to reveal dependencies. Special computer software is used to compose complex Gantt charts.

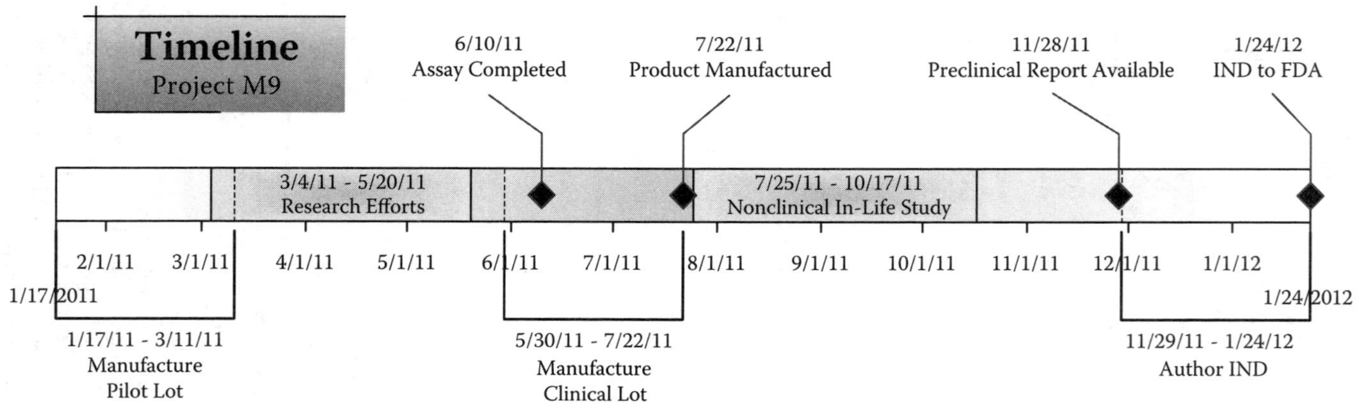

FIGURE 2.2
Timeline format for biopharmaceutical development project. This timeline format demonstrates in a linear pattern the various tasks, bracketed with labels or represented in the bars, and milestones seen as solid diamonds. This is done in software used to draw and prepare various organization charts.

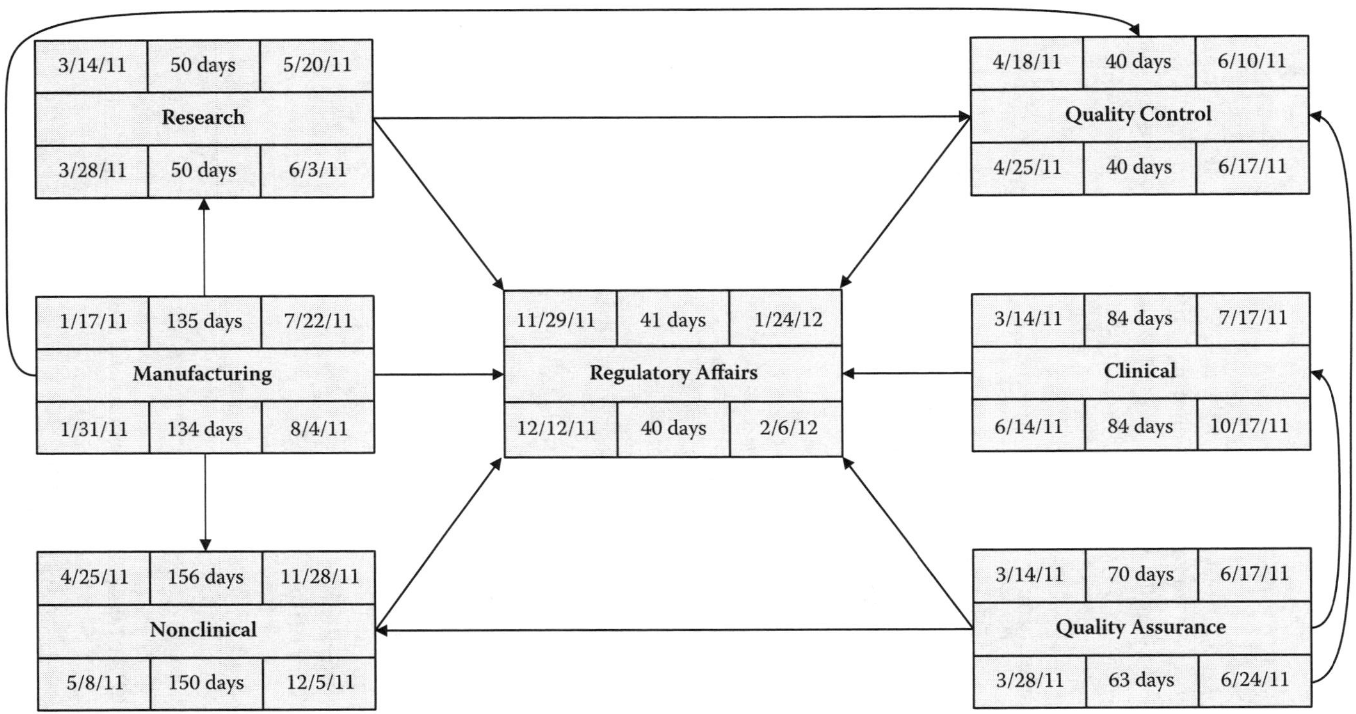

FIGURE 2.3
PERT chart format for biopharmaceutical development project. This is one of several ways of depicting a Program Evaluation and Review Technique (PERT) chart of a biotechnology project. Interrelationships are clearly delineated in this representation of the chart.

BOX 2.4 GUIDELINES AND ELEMENTS OF A CHARTER FOR A PROJECT TEAM

Establish team work rules:

- Identify means of selecting milestones and deliverables because these are key to project development and tracking.
- Agree on means to identify and track tasks. These basic building blocks, often technical, must be pursued in sequence.
- Discuss means of team communication.

Establish milestones:

- Points in the project, actually major events, at which time progress can be measured against an objective.
- It may be necessary to initiate planning in various functional areas before this begins.

Define required skills:

- Who must sit on the team and when will their skills be needed?

Reveal constraints:

- Identify boundary conditions, notably in resources.

Outline a team charter and include key elements:

- Our team goals
- Team's members and what each represents
- Roles of each team member
- Our project team manager, leaders and stakeholders
- Team quantity (size)
- Team's qualities (skills)
- Team's identity

Set the network:

- Draw out an organization chart for the team.

Share important technical information:

- Even use the first meeting to begin technical cross-training and discussions.

stressed by or seem important to upper management. Some examples (and these may be very different depending on project objectives, the firm and the team) are that the team must meet expectations of stakeholders, be on or under budget, and keep on schedule. Note that the *team*, not specific individuals, now meets goals and objectives. This has been referred to as the "norming" stage as team members agree to norms or standards. For some members new to project management teams or the concept of teamwork, this may be a new concept. Thus, it may require explanation and patience on the part of the project manager.

With this accomplished, the fourth, longest and final stage of a project team may begin and this is referred to as "performing."

Communication and Feedback

Once a team has been established, the charter written and the project manager appointed to lead the team, it is necessary to consider communication issues. Communication for the project team requires interactions between team members, the project sponsor (typically senior management), functional area managers, executive management (e.g., president, chief financial officer [CFO]) and business users, the likes of sales, marketing, legal, contracts and business development. While these individuals may not attend each project management team meeting, they are by nature team members. Indeed, team activities and decisions will affect them directly. Their membership and participation is encouraged and effective two-way communication, in whatever format, is essential to success. The project manager is kept informed of all current events by each team member.

The project manager outlines a communication plan explaining the purpose, frequency and format for communication between team members and with executive management and stakeholders. Examples of project messages are a corporatewide or public announcement, upon reaching a milestone, or the financial information, first reviewed in a budgetary meeting with finance prior to the close of a fiscal quarter. Today there are many methods of communication from which to choose, but some that are proven methods within the biotechnology industry are listed in Box 2.5.

Team Dynamics

Team dynamics refers to various team activities and interactions. Of greatest concern to the project manager are team identification of risks and resolution of problems and the decision process, three highly related and very dynamic responsibilities. Decisions are often required when problems or issues arise and must be addressed by the team. In a high technology industry such as biotechnology, we frequently encounter problems. These may be technical in nature, but often they are based in regulatory, quality or management

BOX 2.5 MEANS OF TEAM COMMUNICATION

Formal, Scheduled Meetings. Today, teleconferences or video conferences are often a regular venue for meetings. However, team members at a single location usually gather in a single meeting room, even if others join by video or telephone. If the team is scattered across several states or countries, there is still a need to meet, face-to-face, at regular intervals. Formal meetings always have an agenda and result in formal minutes.

Informal Meetings. Yes, there are times when some, but not all, members of the team must meet. Informal meetings are best summarized by e-mail messages or memoranda that identify the purposes and outcomes for the meetings and are shared with all members of the team. The project manager may be responsible for gathering these messages and providing them to the larger team.

E-mail, Telephone and FAX. Today, e-mail is a primary means of day-to-day communication, especially when individuals are separated, even by short distances. Telephone conversations remain very important to the professional. How and when are these communications copied to the project manager or to other members of the team? Often this is done periodically, perhaps at meetings, where important elements of previous conversations or messages are revealed to members of the larger team. In other cases, important communications between two team members are related to the project manager who ensures they are disseminated to other team members on a "need to know" basis.

Project Management Tools. Today the project manager has at his or her disposal computer-based tools, such as Microsoft® Project and Share-Point. These may be used to identify and outline tasks, define resources, track goals and tasks, and support preparation of reports.

and administrative areas. Our problems also are derived from the fact that this industry is competitive and, therefore, fast paced and dynamic. Recommendations for addressing problems typically encountered by project teams are listed in Box 2.6, where problems are classified as one of two types: adaptive or technical.

Risks are related to both technical problems and decision making and require skills to *solve* problems and to *make* decisions. Risks impact product development, adding to the already complex nature of projects and their management. Project managers are often called upon (or volunteer) to take

BOX 2.6 STEPWISE APPROACHES TO RESOLVING PROBLEMS

1. Identify the problem. Be very specific.
2. Classify the problem. Is it adaptive or technical?
3. Assess the problem in relation to the team and team members. Is it the team's or the project's problem or does it belong to another entity? If it does not belong to this team, then refer it to the proper authority.
4. Identify the importance of the problem to operations and objectives of the team.
5. A project manager ensures all team members are aware of the issue and that the team remains engaged and informed throughout the remainder of the problem-solving process. Keep the team informed, focused and involved, and maintain an urgency to finding a cause and resolution.
6. Look for causes and identify the nature and source of each possible cause.
7. Identify approaches to resolve the problem.
   a. For technical problems consider technical solutions. Engage technical experts and apply scientific methods.
   b. Adaptive problems are often more difficult to address or resolve. They may be complex, occur in a changing environment, and lack predictability. It may be necessary for a team composed largely of individuals with technical or scientific backgrounds to resolve an adaptive problem. Seek help of experts.
8. Once resolved, document the problem, the process and the resolution. Implement the solution.

the lead in risk identification and mitigation, or at least lead the team in these functions. Risk management is discussed later in this chapter.

Project managers cannot and should not attempt to measure risk or resolve problems by themselves. Instead their role is to motivate team members to find the best resolution in the shortest period of time and to lead the team in the problem-solving process. The project manager also communicates the problem, and its resolution, to team members, upper management and stakeholders.

The most important and challenging team dynamic for which project managers provide a leadership role is the decision process. Decisions face the team at virtually every meeting. Someone once said that the only reasons for having a team were to solve problems and to make decisions. While this might not be completely true, it makes an important point. An agenda would seem barren without the need to make at least one decision.

Experienced project managers have likened the process of leading the decision process with a biotechnology team to the activity of herding cats. To avoid this outcome, the prudent project manager establishes, at the first team meeting, decision-making rules. The process itself should be identified and recorded so that it is followed in the future. Criteria for reaching a decision (e.g., by majority rule, by consensus, etc.) and time limits are established. The project manager needs the support of every team member, even those who argued against the chosen course of action, once a decision has been reached. Meeting minutes reflect each decision taken by the team. Finally, the team should consider how decisions will be communicated to stakeholders and how failure to reach a decision will be resolved.

Project Risk Assessment and Management

Risk identification and management, important assets to a biotechnology project, are often project team functions, at least for projects under the team's purview, and led by the project manager. Processes inherent to project management lend themselves to identification and control of risks. The project manager is assisted in risk management efforts by others on the team, notably individuals representing quality assurance and regulatory affairs.

Each biotechnology operation faces a number of risks and team meetings provide a forum to discuss each risk, perceived or actual. Yet it may be difficult for the project team to hold open and frank discussions of risks, even those with high probability to confound, delay or impede a project. Perhaps this stems from the nature of biotechnology itself, an entrepreneurial endeavor operating in an environment with many inherent financial, technical and operational risks. However, boards of directors and a biotechnology firm's investors expect that significant risks be revealed, at the time of investment and throughout the operational phases of a project. Regulatory agencies increasingly demand that risk analysis be part of any product development program because such efforts result in safer and more efficacious biopharmaceutical products. Hence, project teams in biotechnology development are increasingly becoming the clearinghouse for risk assessment and mitigation and with this the project manager becomes a leader in risk identification, assessment and mitigation activities.

The practical aspects of project team risk management require considerable time and effort. A project team is aware of their role in risk management and ensures defined processes are in place so each response is appropriate and all responses are consistent. The process allows for the initial assessment of potential risk elements and for prioritization of the risks as well as for implementation of mitigating actions and periodic reassessment of risks. A project team performs an initial risk assessment at the outset of a project and renews and revises this assessment at predetermined milestones or whenever significant changes are made on a project. For example, product and process risks are reexamined prior to clinical trials, in the case of medical products, or before field studies, for agricultural products. Biotechnology operational projects establish milestones and risk assessment is considered a milestone-related task. The project manager ensures that risk assessment programs are established with milestones and scheduled for completion. Whenever a project team is involved, risk management is part of the project management plan. At an early meeting, the team agrees to the most likely possible risks, these proposed by individual team members. Next, each risk is subjected to one of three types of risk assessment—fault-tree analysis, informal assessment or effects analysis and failure modes—by the appropriate professional or committee. Informal risk analysis is used whenever the likelihood of a risk is low, if it is a highly technical problem or if the risk has minor consequences. One individual or a few people may review the issue and report back to the project team.

Fault-tree analysis is a complex process often used in engineering systems, but can be quite helpful with some biotechnology operational endeavors, such as biomanufacturing. Failure mode and effects analysis, or FMEA, is most commonly applied to projects in the biotechnology, drug and medical device industries. A team usually performs this analysis. Every step of a process or feature of a product is listed and possible failure modes identified. Then each possible failure is assigned a score and the team makes plans to measure and address risks that exceed a certain score. FMEA can be time consuming upfront but, under the guidance of a project team, it ultimately saves considerable time and resources. The project manager must drive these processes and ensure they are brought to completion. Also, it is often the project manager, speaking on behalf of the team, who communicates risk information to executive management.

Metrics and Tracking Progress

Once the project is underway, the project manager is responsible for tracking progress of each task against the plan and schedule. While some tracking is done informally, using memory and communication with colleagues, most is performed using written schedules and lists of interrelated tasks and milestones, these prepared with computer programs, such as Microsoft® Project. A scheduled list of tasks forms a "track" or roadmap for the project. Such tools are introduced later in this chapter. The purposes of tracking are both to monitor progress against objectives (e.g., milestones) and to control the process. Tracking allows the project manager to predict if any piece of the project (e.g., a task) seems to be at risk of failure or is heading "off track."

Specific activities in tracking include collecting actual work and cost performance information and estimates to completion of milestones. The project manager then compares actual performance with the plan and, if necessary, revises tasks, working with project team members in an effort to bring the project back "on track." Indeed, a diligent project manager uses tracking to identify problems before they delay or limit progress. Tracking methods also are used to communicate issues to the project team. Reports to senior management and to investors are based on data obtained by timely and accurate project tracking.

Additional guidelines for effective project tracking by the project manager are to:

- Review all aspects of a project at regularly scheduled meetings.
- Make changes between meetings and notify team members.
- Use various tracking tools, timelines and charts, to communicate with the project team.

Keep management and team members informed through meeting summaries and minutes. Consider that changes in project/schedules necessitate both reassessment by the team and changes in resources. Because biotechnology development projects frequently encounter issues that must be addressed by the team and because potential delays are not uncommon to product development operations, tracking is an intensive but important aspect of project management.

Resources: Planning and Usage

Resources are the people, facilities, equipment, raw materials and money applied to a biotechnology development project. The process of allocating or reallocating resources is referred to as *budgeting*. Budgets themselves are negotiated during the project's planning stage and executive management provides a team with a specified amount of resources. Each team member, having outlined the pathway and schedule to the objective, has, for his/her functional area, identified the resources his/her department needs to achieve objectives. This process requires great time and effort, some negotiation and full justification. The project manager is responsible for preparing an overall budget, by task, year (or quarter) and functional area manager. The project manager tracks progress of the project against consumption of resources and is responsible for identifying budgetary risks and overages. This is done periodically and reports are provided to team members and executive management. Project management programs, such as Microsoft Project, are very helpful, as they provide a means of entering and tracking resources as usage and as compared to completion of tasks or achievement of milestones. The resulting charts and graphs allow the project manager and team members to visualize this information.

Finances are important to a project, but people are the primary resource and so a project manager considers the many facets of human resource planning and use. The project manager is aware of several human resource factors. First, most staff assigned to a project actually report to functional area supervisors. Ultimately, it is the responsibility of functional area managers or supervisors, not of the project manager, to manage staff and performance issues on a day-to-day basis. Indeed, many members of a project team are responsible to more than one project and each of their staff has additional responsibilities as well. Furthermore, because human resources are typically the greatest expense at any company, salaries have a tremendous impact on the overall project budget. The project manager pays particular attention, through tracking, to utilization of staff and supervisors assigned to his/her project. Also, if not enough staff are assigned to a particular technical task, then the project is at risk of failure. Even if a staff member is assigned to a team full time, his/her time is defined and limited. Planning and tracking of human resources is ultimately and routinely the responsibility of functional area supervisors, but, for the overall team effort, they are always major concerns to the team's project manager.

Budgeting monetary requirements is a process referred to as "costing" by project managers. As noted above, there are people costs (internal employees), but there are also external (e.g., consultant), capital, revenue operating, raw material, energy and other project costs. Accurate costing may be beyond the training or experience of some project managers. Hence, a prudent project manager will ask for assistance from the finance department to plan cost budgets and to serve on or advise the project management team. Because the financial department is ultimately responsible for the annual corporate budget and because financial staff have experience in estimating costs, they can be immensely helpful, even indispensible, to a project manager and the team. Indeed, a financial officer is a great asset to any project team, no matter how technical the objective. Even with the assistance of financial staff, the project manager, representing the team, has continuing responsibilities for planning and tracking a project's resources. Project planning tools provide the project manager with a means of accomplishing these objectives throughout the life cycle of a project.

Human Factors in Project Management

A newly appointed project manager either adopts a preexisting team, builds or, in some cases, rebuilds a project team. We have already discussed administrative aspects of building teams and human resource functions, but there is another side to the issue. Even though the project manager does not directly supervise team members, he or she still takes on numerous responsibilities related to coordinating activities of team members and he/she does this in both the planning and performance stages of a project. A short list of a project manager's human resource duties might include the following:

- Integrate human resource planning and incorporate strategic planning into the project; structure teams; and foster working relationships between team members
- Interact with supervisors and human resource professionals

- Respect dual roles or career paths (e.g., discovery scientist and development leader) of team members
- Lead creative and innovative people
- Remain open to new ideas from team members
- Manage unique personalities
- Anticipate and then manage change in teams and team members
- Manage conflict within the team
- Trust team members with scientific and technical expertise

Interestingly, most of these responsibilities match skills that were listed for effective project managers (Box 2.2 and Box 2.3). While managing a team can be a daunting task for anyone, it may quickly overwhelm a new project manager, the individual with little experience or training in interpersonal relationships or in human resource management. For those with no supervisory authority, human issues can be the leading cause of anxiety, frustration and stress. This is particularly true in smaller biotechnology firms, where the project manager may have limited experience and little authority, yet have the responsibility of managing team members, some of whom might be brilliant, opinionated, scientifically experienced, extremely busy or hold lofty titles. Certainly this is a challenge for any project manager.

The project manager also ensures a balanced team, one composed of members suitable for achieving the objective. Balance must be established at the project's initiation and maintained through the project's lifetime, even as the need for certain skills varies from phase to phase. The project manager spells out roles and expectations for each team member and, because executive management may not agree on every proposed position, filling out a team may involve explanation and negotiation.

The most effective teams have a project champion, sometimes called a project leader, in addition to the appointed project manager. The project manager must ensure that the roles of project champion, a proponent of the technology and often a stakeholder or influential scientist, do not conflict with his/hers. It is important for executive management and everyone on the team to recognize that the role of project manager differs from that of project champion or from those of each functional area director serving on the team. These roles are based, to a great degree, on each organization's philosophy, organization and policies. The project manager must understand this and adapt his/her team leadership style accordingly.

Thus, there is a need for the project manager to manage egos, the team and the project. Not long after project teams are formed, the project manager and functional area managers or directors may come to view each other as different and even "difficult" individuals in each other's minds. Yet, in a successful biotechnology operation, each person will realize the importance and the contributions of others on the team.

BOX 2.7 COMMON HUMAN SOURCES OF CONFLICT ON PROJECT TEAMS

Team size or composition not suited to project
Lag in schedule
Technical failures
Inherent tensions (especially true with matrix organizations)
Disagreements, often longstanding, on results or decisions
Individual background or developed styles of team members (e.g., communication)
Difficult individual behaviors (e.g., divisive, passive-aggressive)
Senior management intervention or micromanagement

However, there are instances in which true emotional hostilities break out between team members or toward the project manager. Dealing with "difficult people" or "difficult situations" is a critical issue in project management. People skills are essential for effective project management and leadership. Some argue that conflict is inevitable on any project team, even a well-managed one. However, the project manager is well advised to never ignore conflict, but to recognize it, identify the sources, and work with the team to manage disputes, overt or hidden and simmering. There are many sources of conflict, a few of which are listed in Box 2.7. Conflict is recognized by the project manager in many ways and some, such as body language or facial expressions, are subtle. Conflict must be differentiated from disagreements, which can actually be a positive for team performance as it engages team members in healthy debates and discussions. However, real conflict is never to be ignored because it can get out of control and disrupt progress. More often than not, executive management usually expects the project manager to resolve conflicts that involve the team or the project and to lead the team down a pathway bordered by productive behaviors.

Project Completion

Yes, projects are actually completed, perhaps not always on time or within budget, but, like a movie, each project does have an ending and some are happy and others sad. Wrap-up is an important part of any project and should be considered by the project manager as a separate task. What better way to indicate to the team that an end is in sight than to reveal, early on and in the project schedule, a date for the final project meeting. The last meeting includes a "project wrap," simply a review of what has happened—the good, the bad and the ugly aspects of the project. The final meeting is a learning experience as well as cause for celebration. Paperwork, such as reports, and other outstanding responsibilities are assigned and scheduled.

From the meeting, a "lessons learned" document is produced for management. It need not be long or detailed, but it must be honest and reflect actual team performance and project outcomes. Whether technical aspects of the project succeeded or failed, the team is recognized by the project manager and executive management for a job well done.

Lessons learned reports and meetings need not wait for total completion of a project. They can be used as well whenever major milestones are reached or when a significant risk or issue has been resolved. Such look-back exercises allow team members to share experiences and better prepare for future challenges.

Project Management with Contracts and Collaborations

Outsourcing of technical efforts, such as manufacturing, quality testing and nonclinical animal or clinical studies, is a common, indeed an important, practice in the biotechnology industry. Managers from functional areas are responsible for their contractors and they assign one individual the responsibility of outsourcing a piece of work and managing the agreement. This person has specific technical, project management and contractual experience and can best ensure a successful outcome to partnerships and agreements. The project manager seldom has direct responsibility for a contractor or consultant, but he/she does consider all contractual efforts as integral to the overall project. Hence, the budgets and schedule of tasks and milestones for a contractual, collaborative or consultant's efforts are, in all respects, part of the project and considered in the PDP, WBS and schedules and reports.

Numerous outsourcing models are available to the project manager. Some examples include:

- Competing several vendors with similar and acceptable capabilities
- Selecting vendors from a list of prequalified contractors
- Partnering with a particular vendor or sole-source contracting with an established vendor

Virtually any service or material may be outsourced by a firm. Vendors provide functional and technical services, such as manufacture and regulatory affairs. Indeed, the virtual biotechnology firm may even use a consultant or contractor to provide project management services and to manage their other contractors, consultants and vendors.

In addition to managing contractors or vendors, biotechnology firms often collaborate and co-develop products. These business arrangements are usually between two biotechnology companies or between a biotechnology firm and a large pharmaceutical firm. Another business model is partnership between a biotechnology firm and a contract research organization (CRO). Some CROs provide services in part for equity in the product and sponsoring firm, but in most instances the agreement is fee for services rendered. Whatever the business arrangement, the project team at a biotechnology firm must follow progress of each and every aspect of co-development.

Biotechnology firms often enter into partnerships with larger biotechnology or pharmaceutical companies. Here, interfaces may be quite broad and also have depth, extending well into highly technical endeavors. In a partnership example, one party manufactures product while both parties provide quality control testing services. Each partner must understand the nature of the product and the full scope of manufacturing and control. In addition, they both must have a clear understanding of all technical details in these areas. Teams can become large, with many functional specialists, up to 50 residing at several locations, comprising a single team. This in itself presents a challenge to the project manager. When developing partnerships in operational areas, individuals in business may not appreciate future needs for professional project management to guide the relationship. To ensure success in a highly technical partnership, there is often a need to meld two different technical and business cultures (e.g., the culture of a large pharmaceutical firm with the culture of the biotechnology firm) and this requires a great deal of coordination between many individuals at both organizations. It also involves contract or legal specialists for both parties. Project managers are often responsible for forming integrated project teams and ensuring effective project leadership. Business development must work with project managers well before a partnership deal is consummated and project managers are well advised to include business developers on their teams. Alternatively, arrangements with partners may have collaborating parties working quite independently from each other. In such cases, interface between the parties is more commonly one-on-one, that is, between project managers and functional area managers from each organization.

No matter what the business or management relationship, strong project management experience, leadership and negotiating skills are absolute musts if these arrangements are to succeed. Hence, the project team and manager are fully aware of all contracts and collaborations as well as the statements of work, and roles and responsibilities.

Tools for Effective Project Management

How is it possible to put together a project and then communicate and track it over several years especially given the complexity, size, and inevitable changes to many biotechnology efforts? Today, project managers have at their disposal and at reasonable cost comprehensive and powerful project management tools to help in these efforts. Microprocessors and project management software provide four areas of project support. They are used to define plans and schedules, identify resources, track tasks and milestones and produce reports. Software will assist the project manager in establishing a WBS, listing each task and placing under it any number of subtasks. This organizes the project so that the reader discerns project structure and definition.

The process itself is rather simple. The user first prepares a list of what must be accomplished, breaking the list down in outline format. The list, a breakdown of the project by tasks and milestones and estimates of the schedules for each, becomes the input for entering each task and subtask into a project management computer software program. Tasks and milestones are also linked to each other as a means of identifying and demonstrating dependencies and interrelationships. The software then presents this information in both graphical and written format to the user; this is the draft or initial output of a WBS. It is then shared by the project manager with other members of the team for review and comment. With little training, each team member may now visualize both the overall project plan and WBS and they then appreciate their designated role and responsibilities for the project. Task and milestone relationships and integration into the overall schedule are also clear. Potential risks and problems of the intended project become apparent to each team member when witnessed in graphical format. Weak points can be identified and corrected. The next step in using this tool is for the reviewer to identify specific steps that might be taken to resolve issues over the course of the project. The draft WBS, referred to as output, may now be revised based on recommendations of project team members. This review-to-revision process is repeated several times before project tasks, milestones, integration and schedules are finally established to the team's satisfaction.

The computer software or program most widely used by project managers today is Microsoft Project, but other good programs are on the market. A Gantt chart, shown in Figure 2.1, depicts the output from MS Project and provides an example of how this chart demonstrates a work breakdown structure. For this example, the project was divided into functional areas and each was entered as a line into Gantt format as research, quality control, manufacturing, quality assurance, clinical or regulatory affairs. Then tasks, each a technical or administrative step in a project, was entered under a respective functional area. A task is a piece of work, clearly definable in terms of technical requirements and schedule. In the example, the tasks Write Protocol, Perform Study and Write Report were listed under the functional area Nonclinical. Tasks were further broken down into subtasks and this is shown in the example (Figure 2.1) under the area of Nonclinical, and task Perform Study. Here three subtasks were entered as Order Animals, Treat Animals and Observe Animals. As each task or subtask was entered, it was assigned a start and finish date or duration and predecessor tasks were assigned. These are seen in the columns to the right of the task name and identification number (ID). For the example of the Nonclinical task ID #21, Write Report, it began on October 18 and took six weeks, until November 28. It had one predecessor task, ID #19, and several successor tasks. A milestone, completion of the report, was added to indicate the date on which a series of tasks were completed, a major event in the overall project schedule.

The software produces a visual in Gantt chart format, to the right of these entries. Horizontal clear bars identify each task against the schedule or calendar while the overall schedule for each functional area is shown in a dark bar. Milestones are visualized in the Gantt chart as dark diamond shapes. Vertical lines in the chart outline the dependencies of tasks, shown as arrows leading from a predecessor to a successor task. Other information, such as notes and resources, may be inserted by adding columns. If resources are included, budgets are then calculated by task or subtask and by specific time period. The integrated nature of each task and milestone is readily apparent from the chart.

Once a project is underway, tracking functions of the program allow the project manager to compare actual to planned progress. It is also a simple matter to enter proposed changes and determine how any given change or set of changes will impact the overall timeline or any other task. Once a change is made to one task, the hierarchy and schedule are automatically recalculated and the outcome is made immediately obvious on the revised Gantt chart. Let us suppose the animals do not arrive on July 15, as scheduled in task ID #18, but instead arrive two weeks later on July 29. By simply changing this date from July 15 to July 29, it is possible to learn if this will delay any other task or even delay the project completion date The project manager is able to demonstrate this to the project team and this allows the team to consider alternative arrangements or risk mitigation strategies. Indeed, project managers often project Gantt charts at team meetings to demonstrate to the whole team exactly how a change or delay might impact the overall sequence of events or schedule. This tool is a very powerful means of communication.

Project management software also assists the project manager with preparation of reports, outlining various elements and adding to the report specific examples and illustrations. In addition to the Gantt chart, these programs are capable of presenting the project or any stage in the project in other illustrative formats, such as scheduling charts. As is the case for much of today's software, project management software has thousands of other helpful functions, too numerous to mention here.

Project managers use other methods to visualize a project or to present this information to various individuals or groups, such as the project team, investors or senior management. Although the Gantt chart may be familiar to members of a project team it can be foreign to stakeholders or individuals not familiar with project management. How then does the project manager present the visual representation of a complex biotechnology project to an audience that is unfamiliar with the Gantt format and the project itself? To speak to such an audience the project manager uses simpler formats. These are found in commonly available and easy to use software programs. Examples of two formats, a timeline and a PERT chart, are shown in Figure 2.2 and Figure 2.3, respectively. Notice how these simple formats of the information presented in the Gantt chart (Figure 2.1) allow one to tailor a format to the audience.

Other direct communication tools are available to the project manager and many were mentioned earlier in this chapter. Some are better for a particular purpose or situation than are others, but they all serve the purpose of serving both project manager and team by communicating the project to those outside the team. Telephones remain a frequent means of communicating, one-on-one, and teleconference meetings are a common audio tool used for communication by the team. Face-to-face meetings are co-located at one site, but, for many teams, meetings in one room are infrequent, occurring quarterly or annually. Today, video conferencing is frequently used and this technology has become more dependable and user friendly in the past decade. Electronic mail, e-mail, is an excellent means of communication because it is nearly instantaneous (if the recipients read their messages) and can accommodate a group of any size. It allows individual or group responses. However, it is rather impersonal and is not good when cross-interactions must be rapid and fruitful and completed rapidly. Also, it is more difficult to reach decisions or clarify complex technical or scheduling issues by e-mail, and we are all too well aware of misunderstandings that arise using this method of communication. Project managers employ a variety of communication methods, enhancing communication with a mix. And they use with each the tools for graphically demonstrating a project with tasks and schedule.

Summary of Project Management in Biotechnology Development

This chapter has focused on project management, the endeavor that pulls together the biotechnology operation, integrating functional areas into an organized whole, this aimed at achieving a common goal. Project management is performed as team work and this team is managed and led by a project manager, the individual responsible for organizing, orchestrating and monitoring the various processes, tasks or work activities. To do this, the team and project manager rely on sociotechnical skills and tools to plan, put the project into motion, measure progress and communicate the project. In each of these endeavors, there is reliance on credibility, trust, knowledge, experience and, most importantly, on project objectives and plans.

3

Regulatory Affairs

The U.S. Food and Drug Administration: Law and Regulations for Biopharmaceuticals

Historical Basis for FDA Regulation

Food and drug regulations evolved in the twentieth century, a reflection of major changes in the way in which foods and drugs were processed and sold. In the nineteenth century, these products were processed on a small scale. Grain was milled locally, community butchers slaughtered animals and sold meat to neighbors, and local pharmacists and physicians formulated and dispensed medications. This changed late in the nineteenth century, notably in the food industry, as large mills and slaughterhouses became a part of the industrial revolution. But abuses, like the sale of adulterated foods in some instances, led to social revolt and the desire for governmental regulatory controls. Upton Sinclair's book, *The Jungle* (Doubleday, Page & Co., 1906), a revealing look at practices in the meat industry, is thought to have stimulated the U.S. Congress to pass the Pure Food and Drugs Act in 1906. For drugs, this Act focused on the need to inform the public about foods and drugs through the use of honest labeling. A label reveals the contents of a container and cannot provide false or misleading information in a fraudulent manner. The 1906 Act did not, however, establish the need for review of each product by a federal agency before it could be marketed and sold to the public.

This initiative was to come later, in 1938, when the U.S. Congress passed the Food, Drug and Cosmetic (FD&C Act or the Act) Act of 1938. This Act, passed in the Great Depression, changed the Food and Drugs Act of 1906 in several key ways. First, it no longer required the government to prove fraud was committed if a drug claimed a curative or therapeutic effect. It also required premarket drug review of a New Drug Application (NDA) in which the sponsor of a product, which is the company distributing and selling a drug, provides written evidence that its product was safe and effective. It authorized other government actions as well: the Federal Trade Commission (FTC) was to review drug advertising, promotional claims and material; the Food and Drug Administration (FDA) would inspect drug manufacturing facilities and enforce the law and levy fines and punishments, and it prohibited false therapeutic claims; and it defined classes of regulated products—biologics, medical devices and cosmetics—as existing under these rules. The FD&C Act of 1938 is the foundation for today's regulation of drugs, biologics and medical devices and it was the basis for establishment of the U.S. FDA.

The FD&C Act and other food and drug laws are often responses by the public and government to tragic and avoidable situations. The Biologics Control Act of 1902 was the result of 10 children contracting tetanus after taking a poorly made antitoxin. The "Cutter Incident" of 1955, in which children were exposed to live polio virus from a poorly manufactured vaccine, was the basis for expansion of biologics regulation. Other amendments often followed problems, abuses, or deficiencies, perceived and real, in drug, biologics and medical device manufacture, control, evaluation and marketing. This trend has continued unabated for over 50 years and, unfortunately, it may be horror stories that lead to additional food and drug laws or amendments in the future. A few of the many acts and amendments in the past 100 years are shown in Box 3.1.

BOX 3.1 EXAMPLES OF FOOD AND DRUG LAWS: 1906 TO 2007

Year and Name	Purpose
1906. Pure Food and Drugs Act	Prohibits interstate commerce of adulterated or mislabeled food or drugs
1938. Food Drug and Cosmetic Act	Provides for safety testing prior to marketing, adequate labeling, appoints FDA responsibility
1944. Public Health Service Act	Regulation of biological products
1962. Kefauver–Harris Amendment	Requires drugs have proven efficacy
1966. Fair Packaging and Labeling Act	Honest and informative labeling on consumer products with FDA responsibilities
1983. Orphan Drug Act	Encourages development of products to treat rare diseases
1984. Waxman–Hatch Act	Drug price and completion and patent restoration; generic drugs
1987. Prescription Drug Marketing Act	Requires licensing of drug wholesalers, bans diversion of drugs
1992. Prescription Drug User Fee Act	User fees established for FDA review of applications
1996. FDA Export Reform and Enhancement Act	Controls for imported and exported products
2007. Food and Drugs Administration Amendments Act	Broaden and upgrade drug safety programs

Many products resulting from biotechnology are considered biologicals or biopharmaceuticals. Two Acts of Congress, the Biologics Control Act of 1902 and the Public Health Service Act, PHS Act, established special rules for biologicals or biopharmaceuticals. Today, many biologicals result from biotechnology endeavors. Through these Acts and amendments, Congress has delegated to FDA the responsibility of ensuring compliance of biopharmaceuticals. This, in turn, has profoundly impacted the means by which many biotechnology products are developed, manufactured, tested, distributed and sold in the United States.

Regulatory Organization of FDA

The FDA is responsible for protecting the public health by assuring the safety, efficacy and security of human and veterinary drugs, biological products, medical devices, our nation's food supply, cosmetics and products that emit radiation. The FDA is also responsible for advancing the public health by helping to speed innovations that make medicines and foods more effective, safer, and more affordable, and helping the public get the accurate, science-based information they need to use medicines and foods to improve health.

As such, the FDA is a regulatory agency responsible for many consumer products used in the United States today; a huge task, especially considering the impact these products have on our health and livelihood. Clearly, it would be impossible for the FDA to oversee or individually inspect each product item that is sent to consumers. The FDA also communicates information to various interest groups, pharmaceutical, biological and medical device industries, and to those who distribute or prescribe the products, such as pharmacies and physicians, and to consumers. For industry oversight, the FDA inspections represent but a small and selected fraction of the material that is distributed to the user.

The FDA is an agency within the Department of Health and Human Services (DHHS) and is composed of various organizational units or offices and seven centers. These are shown in the current organizational chart in Figure 3.1. The responsibilities of centers or offices are listed in Box 3.2. Specific responsibilities for biologicals and biotechnology products are outlined in Box 3.3. Amendments to the FD&C Act typically apply to biologics as well as drugs. Recently, the FDA transferred some of the therapeutic biological products that had been reviewed and regulated by Center for Biologics Evaluation and Review (CBER) to the FDA's Center for Drug Evaluation and Review, CDER. Finally, biotechnology product development is regulated by agencies, local, state, national and international, other than the FDA. These regulatory agencies and responsibilities are reviewed in Chapter 4.

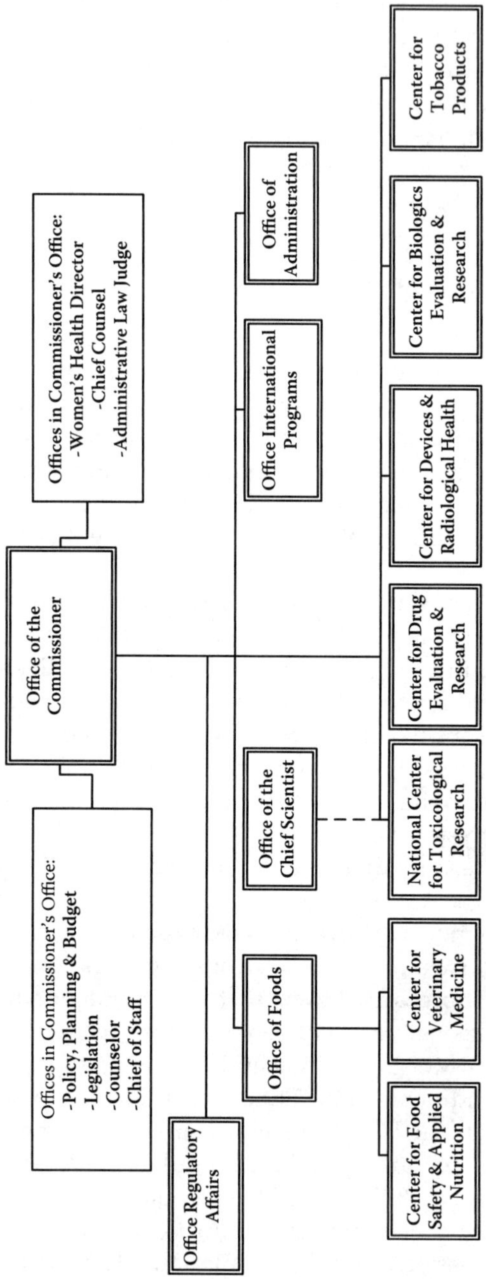

FIGURE 3.1
Organization of the U.S. Food and Drug Administration.

BOX 3.2 RESPONSIBILITIES OF SELECTED CENTERS AND OFFICES AT THE FDA

Office of the Commissioner	
Office of Orphan Products Development, OPD	Review applications for orphan drug designation
Office of Combination Products, OCP	Identify primary review office for combination products
Office of Regulatory Affairs, ORA	Ensure regulatory infrastructure and enforcement
Center for Biologics Evaluation and Research (CBER)	
Office of Biostatistics and Epidemiology, OBE	Statistical review and support
Office of Blood Research and Review, OBRR	Blood products and device review and research
Office of Vaccines Research and Review, OVRR	Vaccine products review and research
Office of Compliance and Biologics Quality, OCBQ	Inspections of biologics facilities
Office of Cellular, Tissue and Gene Therapies, OCTGT	Review of genetic therapy and cell and tissue products
Center for Drug Evaluation and Research (CDER)	
Office of Compliance	Surveillance, monitor, inspections
Office of Surveillance and Epidemiology	Epidemiological review and support of new and approved drugs
Office of Clinical Pharmacology and Biopharmaceutics	Pharmacology review and support
Office of New Drugs	Review of IND, NDA, ANDA
Office of Nonprescription Drugs	Review of OTC drugs
Office of Oncology Drug Products	Review of drugs to treat or prevent cancers
Office of Pharmaceutical Science	Drug development and testing
Office of Generic Drugs	Review of ANDA applications
Office of Biotechnology Products	Review biotechnology products, therapeutic proteins and monoclonal antibodies
Office of Testing and Research	Analyze drugs, ensure product quality
Office of New Drug Quality Assessment	Ensure critical pharmacological attributes, tests
Office of Translational Sciences	Statistics, clinical pharmacology
Division of Drug Information	Public and professional information
Office of Medical Policy	Review advertising and promotions

Center for Devices and Radiological Health (CDRH)	
Office of Compliance	Inspections of device manufacturers
Office of Device Evaluation	Review marketing applications for medical devices
Office of *In Vitro* Diagnostic Device Evaluation and Safety	Review marketing applications for *in vitro* diagnostic devices
Center for Veterinary Medicine (CVM)	
Office of New Animal Drug Evaluation	Review marketing applications for veterinary drugs

BOX 3.3 REVIEW OF BIOTECHNOLOGY PRODUCTS AT THE FDA: RESPONSIBILITIES BY CENTER

Center for Biologics Research and Evaluation (CBER)	Center for Drug Research and Evaluation (CDER)	Center for Devices and Radiological Health (CDRH)
• Vaccines • Plasma or serum products • Blood products • *In vitro* diagnostics for blood • Genetic therapies • Somatic human or animal cells or tissues • Plueripotent cell-derived products • Allergenic materials • Antitoxins, toxoids and toxins • Antivenoms • Combination products in which the biologic is primary mode of action	• Monoclonal antibodies • Therapeutic immune-therapies • Cytokines • Therapeutic proteins derived by biotechnology • Enzymes • Interferons • Growth factors • Peptides • Small molecule drugs • Combination product in which the drug is primary mode of action	• Medical devices of biotechnology origin • Combination products in which the device is dominant mode of action • Radiation-emitting devices

Food and Drug Law, Regulation and Guidance

Laws enacted by Congress and signed by the president are the basis for FDA functions. Regulations are established requirements, developed by an authorized federal agency. Regulations interpret the law, considering the intent of Congress when the law was established, and they apply technical, scientific and administrative best practices to fulfill the law. Regulations cannot be simply mandated by the FDA, but must go through a rulemaking process, this established by the Administrative Procedures Act of 1946. The rulemaking process mandates that a regulatory agency propose to the public every regulation and seriously consider the comments received in response. Hence, rulemaking is a very transparent process with significant influence by citizens and organizations. Once the discussion period is completed, the regulation is then published and goes into effect. Regulations are placed into the Code of Federal Regulations (CFR), which are bound in numbered volumes by functional area. Food, drug, biologic and medical device regulations are published in Part 21 of the CFR; hence, reference to 21 CFR. Regulations have the impact of law and, if violated, are enforceable by authorized law enforcement agencies (see Chapter 4).

Regulations alone may fall short in their ability to fully interpret the law and provide scientific or technical guidance in highly specialized areas. Thus, regulatory agencies interpret regulations through the use of highly technical publications referred to as *guidelines*. A guideline, unlike a regulation, does not carry the weight of law but instead suggests to an interested party the best technical, scientific or administrative practices that may be considered as a means of achieving the intent of a regulation. The FDA has dozens of guidelines, written and updated by scientists working on cutting-edge technologies; they are then made available to the public. For individuals working in biotechnology, guidelines are extremely important resources because they facilitate targeted planning of product development by identifying scientific, technical and regulatory processes. Consideration of guidelines can greatly increase regulatory compliance and prevent loss of time in development.

FDA-Regulated Products

FDA regulates a vast array of products and these will be described as individual classes of products along with information about the FDA center that regulates them. An important aspect of the regulatory planning process is to understand exactly the nature of a product and to focus regulatory efforts to that area. Jurisdiction for product review and responsibility at the FDA would seem obvious, but with the many biotechnology products, this may not be the case. In developing a regulatory plan, it is extremely important to determine jurisdiction and this is often based on both the biological class and/or intended use (indication) of the product, information that should be available in the targeted product profile (TPP) (see Chapter 1).

Biologics

The PHS Act defines a biologic as "any virus, therapeutic serum, toxin, antitoxin, vaccine, blood, blood component or derivative, allergenic product or analogous product … that is intended for use in the diagnosis, cure, mitigation, treatment or prevention of disease." For many years, this was construed to mean virtually all biotechnology products. However, over time and with the advent of biotechnology, biological products became very diverse. As noted earlier, some biotechnology-derived products, such as monoclonal antibodies, enzymes, cytokines and simple protein therapeutics, are considered "well characterized" in that their molecular nature is well known. These well-characterized products are reviewed by CDER. In this book, the term biopharmaceutical refers to products under the former definition of a diverse universe composed of biologicals or biologically derived molecules, cells, tissues or organisms intended for use in the diagnosis, cure, mitigation, treatment or prevention of disease. Because a biopharmaceutical may be reviewed either by CBER or CDER (Box 3.3), depending in part on nature or indication, it is important to determine for each product the most likely route of regulatory review.

CBER is the FDA lead regulatory office for many biologicals, notably the blood and blood-derived products, cellular, tissue and gene therapies and vaccines. Three offices within CBER (Box 3.2) manage activities and regulate these products.

The Office of Blood Research and Review (OBRR) handles blood products, most of which are derived from human whole blood or plasma. Even in a high biotechnology world, human-derived blood and blood products, such as red blood cells, plasma, platelets and clotting factors, comprise a large industry. In addition, recombinant blood products, such as clotting factors, are approved and others are in development. Human-derived blood products are highly regulated from their sources (blood and plasma donor centers) to the finished product; this is, in part, because of risk that such products could contain adventitious agents. Blood establishments must be registered, and they are routinely inspected by the FDA. Blood products are rigorously tested for a wide variety of infectious agents. Materials such as blood bags and infusion lines used in blood collection, handling and testing, are also regulated by OBRR and CBER. Blood products used for further manufacture, such as platelets or plasma, are subjected to rigorous methods to remove and inactivate viruses, such as hepatitis or human immunodeficiency viruses.

The Office of Cellular, Tissue and Gene Therapies (OCTGT) regulates cellular and gene therapies and human cell and tissue products. Most of these products are derived from biotechnology while others directly apply biotechnology methods. Many are live microbial products, such as viral vectors that have been engineered to carry a therapeutic gene. Others are macromolecules, such as plasmid DNA, intended to have a therapeutic effect on cell entry.

Somatic tissues or cells, such as allogenic skin grown in culture to replace that of a burn patient, or bone marrow-derived cells, selected for a trait and expanded in culture, are reviewed by OCTGT as well.

Human organs, such as kidneys, lungs or livers, intended for whole organ transplantation, are not regulated by the FDA. However, the agency does regulate human cells or tissues, or products made from cells or tissues, intended for infusion, implantation, transfer or transplantation. The agency also regulates animal whole organs, such as pig skin or liver, that might be transplanted into humans. It matters not that these are autologous or allogeneic, as the primary determinant is that they are manipulated and have a systemic effect or distribution, which are primary reasons for their regulation. As with blood products, safety of tissue products is based on registration of donor centers, careful selection of donors and testing of the donated cells or tissues. Current Good Tissue Practices require special handling of cells or tissues and any added material, such as supplements, must be considered a part of that product. Human organs, such as kidneys, livers or lungs, intended for whole organ transplantation, are not regulated by the FDA.

The FDA has recently proffered guidelines for pluripotent (or stem cell-derived) products and these are only now approaching early clinical trials. However, the early development pathways that will encourage use in humans of only safe and effective products do reflect the nature of pluripotent cells and their great potential as therapeutic products. For example, the FDA has made clear that nonclinical animal studies are critical to understanding possible risks. Nonclinical testing must be designed based on the source of the cell, the intended use and route of delivery and data from laboratory studies. The cells themselves must be fully characterized and major issues, some of which are listed below, must be defined in the laboratory and in animal studies.

- Mechanisms of action, physiological parameters
- Distribution in the body or migration to tissues other than the target
- Ectopic growth potential
- Tumor, benign or malignant, formation

Interestingly, even though these issues may seem unique to stem cells, they are, in general, the same questions posed by regulatory agencies to sponsors of a wide variety of biotechnology products.

The Office of Vaccine Research and Review (OVRR) reviews biological products that are intended to protect from or cure disease via an immunological affect. Hence, virtually all vaccines and allergenics, with the possible exception of cancer vaccines, are under the purview of OVRR. Vaccines, like genetic therapeutics, represent a wide variety of biotechnologies, too numerous to mention here. All products reviewed by OVRR have in common an intended mode of action—to elicit an immune response. The indication may be therapeutic or it might be preventative. Product types come to OVRR in a vast array of technologies and today most are derived from biotechnology;

only a very few candidate vaccines, such as influenza, remain as natural products. Live bacterial or viral vectors, like *Escherichia coli* or adenovirus, are used both to stimulate an immune response to that organism or as carriers, intended to stimulated immune responses to other proteins genetically engineered into the host. Vaccines may be given by oral, intranasal, intramuscular, epidermal or other routes. Vaccines today are often partnered with a delivery device, making a combination product. Most biotechnology products engineered and intended to stimulate an immune response go to OVRR for review. Vaccine adjuvants, molecules intended to improve immune response when given with an antigen, are also reviewed. Allergenic products, often natural substances purified and then used to treat allergies or in hypersensitivity testing, are products reviewed by OVRR as well.

CBER also reviews unique types of biological products. Exact regulatory pathways have not been developed for every type of product that might be conceived by the biotechnology industry. It is worth noting one class of biologics, those derived from plants, that are also reviewed by CBER. Drugs and biologicals may be derived from bioengineered and selectively bred plants and these may be reviewed by CBER, by CDER, or both. An antimalarial drug, currently in short supply worldwide, is one such drug. In this situation, the product receives review with special attention paid to the recombinant nature of the plant from which the product, not itself a recombinant molecule, is derived. The basis for regulation and review may be based largely on the indication and manner in which the product will be used and less on the nature, drug or biological, of the product.

All centers and offices within the FDA and the drug regulatory agencies of other developed countries play a significant role in providing support for therapeutic products and vaccines used in global health, even if these products will not be licensed in the United States. This is due in part to the FDA's leading international reputation for scientific product review and also because many target countries cannot afford to have a food and drug authority in their governments. When a biopharmaceutical firm applies to test a product indicated for global health in the United States, it will be reviewed by the FDA and comment provided. Indeed, most global health products are first tested in the United States or another developed country prior to being fielded in developing countries, where it is necessary to conduct field trials and it is desirable to manufacture commercial product. Hence, the FDA, along with institutes and academic centers in the United States and other Western countries, provide early product development support and advice that can be transferred to target countries.

Drugs

Drugs are broadly defined as products used to mitigate, treat or prevent disease in man and that affect physiology or anatomy of the human body. The industry standard for a drug is a small or large molecule that has a well-defined chemical structure. Following market approval, drugs are recognized as such in reference texts, such as the *United States Pharmacopoeia (USP)* or the *National Formulary (NF)* (see Chapter 7). The *USP* and *NF* are published by a government charted laboratory, the United States Pharmacopeia. Pharmaceutical is synonymous with the word *drug*. The historical definition for drug is: *a small molecule, or ethical, pharmaceutical, a compound that is synthesized from nonbiological sources.* Examples of drugs under this definition are acetaminophen or aspirin; antibiotics, such as penicillin; and the anesthetic ether. Drugs of this nature are under the purview of the FDA's CDER (see Box 3.2). Certain molecules of biological derivation, such as monoclonal antibodies and therapeutic proteins, are reviewed under CDER even though they do not meet the historical definition of drug. Prescription drugs or biopharmaceuticals are those products that must be prescribed by a licensed medical professional and dispensed by a licensed pharmacist. Over-the-counter (OTC) drugs are those that do not require a prescription. Most biopharmaceuticals are dispensed by prescription.

Generic drugs are drugs that no longer have marketing exclusivity, but are still regulated under drug regulations. Generics must be tested in adequate clinical studies, albeit abbreviated, and they must receive marketing approval from the FDA under the Abbreviated New Drug Application, or ANDA. The Waxman–Hatch Act of 1984 was instrumental in establishing regulations that guide marketing approval of generic drugs.

CDER, in addition to reviewing small molecule drugs or pharmaceuticals, also takes the lead for review of a number of biopharmaceuticals. Examples are shown in Box 3.3, but the list is growing as new biopharmaceuticals enter clinical trials or receive market approval. Recombinant therapeutic proteins are macromolecules or peptide products intended to treat disease. Responsibility for review or co-review within CDER may be determined also by the indication or intended use. This is because CDER is organized by disease area. Hence, a therapeutic recombinant protein intended to treat gastrointestinal cancer might be reviewed by the oncology group, specifically by the Division of Biologic Oncology Products group. Another example of a biopharmaceutical product falling under CDER is monoclonal antibodies. They represent a class of molecules that are produced *in vitro* as a result of genetic engineering. Monoclonal antibodies are produced to react with a variety of target proteins and to have a therapeutic effect on the patient, for example, antibodies directed against inflammatory cytokines; these are engineered to stop undesirable inflammation due to autoimmune diseases. They are produced in genetically engineered, immortalized and cloned cells in much the same way as other recombinant proteins. Therapeutic monoclonal antibodies are reviewed by CDER with primary responsibility resting in its Office of Biotechnology Products (OBP), but also may be reviewed by experts in the disease. In the case of the monoclonal antibody directed against an inflammatory cytokine, rheumatologists at CDER would work closely with OBP on review of regulatory documents. In conclusion, the division of review responsibility for a given biopharmaceutical may, in CDER, be more diffuse than it is for one at CBER.

Whichever division or office within CDER or CBER has primary responsibility for review of a given product, there might be specialists from another center involved in the review of that product as well. There are specialists within each of the centers for functional areas, such as nonclinical studies and toxicology, clinical trials, manufacturing and quality control; therefore, review of applications by any FDA center represents a team effort.

Medical Devices

This major class of products includes instruments, prostheses, delivery technologies, *in vitro* diagnostic tests, implants, apparatus and a host of other engineered yet nondrug and nonbiologic items. Regulatory responsibility for medical devices rests with the FDA's Center for Devices and Radiological Health (CDRH). The range of products is huge. Examples of medical devices range from tongue depressors to cardiac pacemakers, dental amalgams to CAT scanners, and software to transmit x-rays to HIV test kits. Unlike drugs and biologics, medical devices are classified, *a priori*, based on their risk to the user and the level of regulatory control, review and safety concern. Under an established hierarchy, medical devices with the greatest risk to the user are given the highest classification. A tongue depressor is Class I, General Controls; a test for the common flu virus is Class II, General and Special Controls; and a heart–lung machine is Class III, Special Controls and Premarket Clearance or Premarket Approval. In general, the FD&C Act requires safety testing for all devices, but does not require testing for efficacy studies of Class I and Class II medical devices. Most Class III devices need some assurance of adequate performance. Many Class II devices are eligible for regulatory approval under a regulation referred to as 510(k) approval if they are substantially equivalent to a predicate device. Manufacturers find this an attractive means of seeking regulatory approval from CDRH, but it only applies to certain devices. There are many other differences in the regulatory requirements and review and structure of processes leading to market application and approval for devices as compared to drugs and biologics. While few biotechnology products are considered medical devices, some are a combination of a device and a biologic.

Combination Products

Many biopharmaceutical products are not simply classified as biologics, drugs or medical devices, but instead are mixture of these product classes. For example, a drug aimed at a blood cancer cell, historically a drug, may be attached to a protein, a biologic, that targets the cancerous cell. A recombinant vaccine protein, a biologic, may be delivered to the skin using a novel jet injector, a medical device. The combinations and permutations seem limitless, with new biotechnology combination products appearing daily.

The FDA accommodates these so-called combination products through the Office of Combination Products, responsible for defining jurisdiction and ensuring a coordinated review of applications. Jurisdiction is based on the primary, or most important, mode of action for the product. It is critical to identify and then to evaluate every potential combination product early in the planning process so as to establish the most likely path forward. The first step is to determine if a product is, in fact, a combination product in the eyes of the FDA. Referring to examples cited earlier, the drug would provide the primary mode of action because it would kill the cancer cells and CDER is likely the lead for review. The vaccine protein would be primary for that example because it elicits, and, thus, provides the desired outcome: protective immunity. While a paradigm, such as the one shown in Figure 3.2, helps a sponsor to identify the primary review office, consultation with the FDA, through a Request for Designation, is always recommended.

Other Classes of Biotechnology Products and Their Review at the FDA

Products for Veterinary Use

Biologicals, including biotechnology products used to treat or prevent disease in domestic animals, are largely regulated by the U.S. Department of Agriculture (USDA) and not by the FDA. However, the Center for Veterinary Medicine (CVM) of the FDA is designated for animal drugs and medicated feeds. The CVM reviews marketing applications, ensuring such products are safe and effective for their intended use. The process is similar to that applied to human drugs, and includes the need to file an Investigational New Animal Drug Exemption to support investigational use or a New Animal Drug Application (NADA) to receive product marketing approval. Animal feeds, both those that contain medications, such as antibiotics, and those without supplements, are under the purview of CVM because they should not contain harmful substances and must be properly labeled. Biotechnology-derived products could fall under both the medicated and nonmedicated classifications of animal feeds and these products must receive marketing approval from CVM.

Cosmetics, Food, Dietary Supplements, Homeopathic or Nutritional Products

Regulations do exist for cosmetics, products that are applied externally, but have no claim of therapeutic value. Cosmetic regulations, as compared to drug rules, are relatively simple yet they ensure these products are safe and not adulterated. A biotechnology-derived product could, by intended use, be considered a cosmetic and, if so, cosmetics regulations would apply.

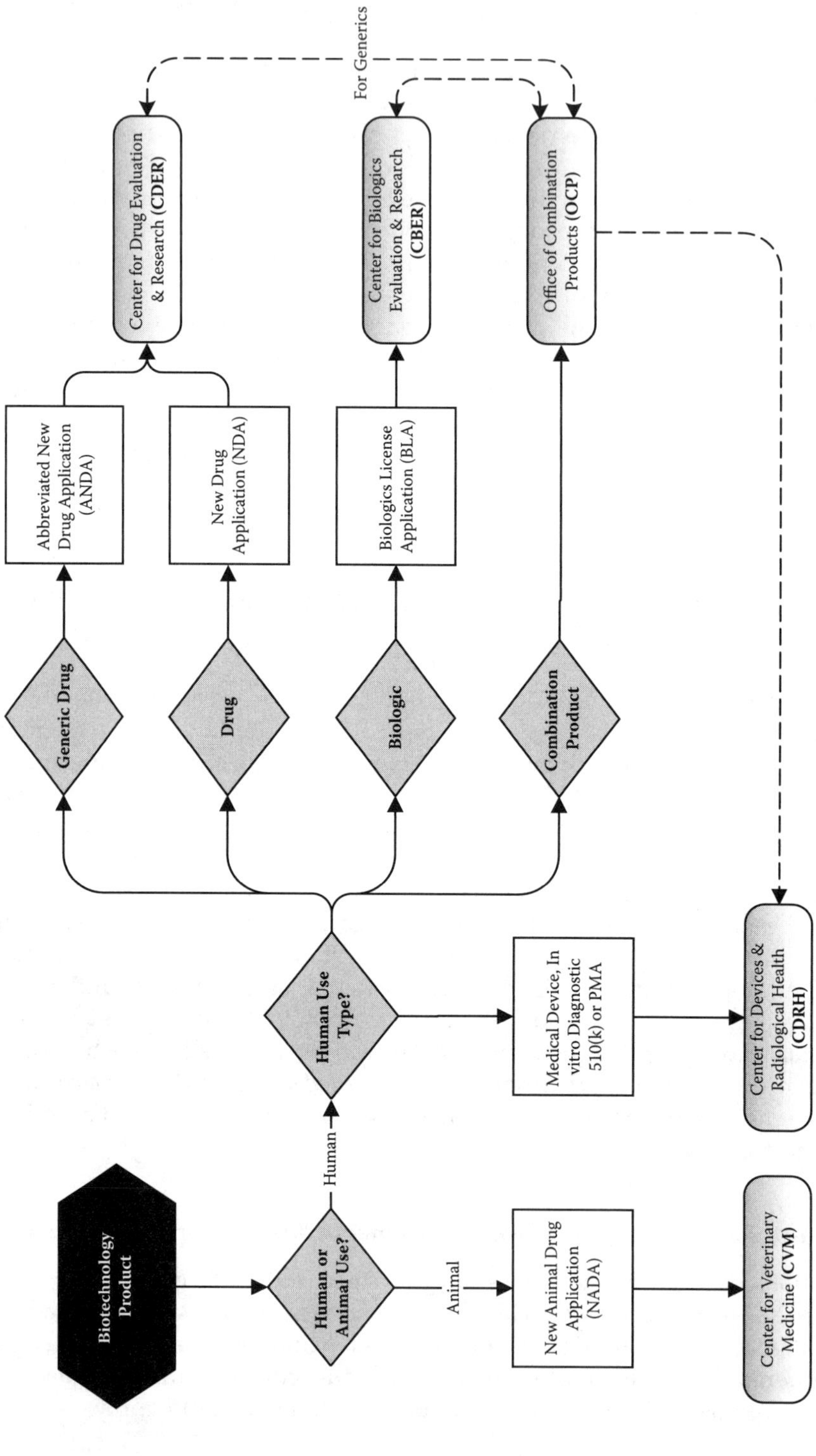

FIGURE 3.2
Paradigm to establish the class or type of biopharmaceutical for regulatory review.

Products that are nutritional in nature and for which no claims are made of a therapeutic or medicinal affect are regarded by the FDA as a food. FDA regulates most food products, with the exception of meat and a few other animal-derived items that fall largely under the U.S. Department of Agriculture's purview. Additives for food and food colors do fall under the purview of FDA. Foods and additives are not considered here.

Dietary ingredients and dietary supplements are not subject to premarket approval by the FDA, but the agency must be notified of marketing intention nonetheless. These products are regulated by the FDA based on their nature, but largely upon the indication and the therapeutic or preventative claims made by the sponsor. Classification of such products requires experience as the complex guidelines are based on nutritional or medicinal claims made by the sponsor, which themselves are often difficult to parse or comprehend.

Dietary supplements and nutritional products are now generated by biotechnology, although the vast majority are still naturally derived or chemically synthesized. Dietary supplements, regulated largely as food and not as drugs, are products taken by mouth and "intended to supplement the diet" and "not intended to treat, diagnose, cure, or alleviate the effects of disease." Again, the claim made on the product label is instrumental in determining the level of product regulation and the designated regulations and review process at the FDA. Certain "homeopathic" compounds may fall under similar rules and have their own set of definitions, depending on claims made on therapeutic value. Certain homeopathic compounds are formulated in pharmacies for individual patients, a process that is legal in the United States under specific circumstances. As with nutritional supplements, the indication determines the level of regulation because pharmacies typically fall under the laws of individual states. With homeopathic medicines, there is a point at which the distribution or volume of sales can conflict with this definition. Then FDA regulations may be brought to bear and the product declared a drug, biological or device. Few biotechnology products produced for therapeutic purposes would be considered homeopathic medicines or nutritional products.

U.S. FDA Regulatory Information and Resources: Regulatory Intelligence

This section focuses on the function of the FDA regulatory intelligence and a biotechnology firm's need to continually obtain information that allows an understanding of the regulatory environment as it applies to each product. Also, a prudent biotechnology firm will have in place processes to abstract and communicate FDA regulatory information to operational staff and upper management. Sources of information on non-FDA regulatory bodies are described in Chapter 4.

Regulatory intelligence is the process of finding and analyzing publically available regulatory information. It is not necessary, nor is it possible, to know all regulations that apply to every biotechnology product; it is more important to understand the overall environment and possible sources of regulatory information and to be able to locate information on particular types of products.

Gaining regulatory intelligence is always a first step in preparing a regulatory plan. To find pertinent regulatory information on a particular product, it is necessary to first understand the product and indication, as provided in a TPP, then a structured search is initiated. The search begins as a broad investigation to understand the regulatory environment using general and specific sources. This provides a general background on the subject. It then focuses on areas of interest, moving from one regulatory source to another until it seems all regulatory databases have been exhausted. Scientific databases list additional product information and these are examined. The result is a bibliography that provides a regulatory history and scientific background of the product and product class, an exact idea of the current regulatory environment and insight into future regulatory initiatives. Taken together, this background information serves as a foundation for preparing the regulatory plan and a basis for later updated information on the product and its regulation.

Today, public regulatory information, with the exception of regulatory textbooks and some international guidelines, is available on the Internet at no cost to the user. Hence, the key to finding the information is to use a library of Web sites for searches and to apply search engines. Some searchers use regulatory information blogs to network with colleagues, an approach that is particularly helpful for locating very specific bits of information. Also, before beginning a search it is a good idea to develop a method to catalog the information so that it is easily retrieved for review and reference. Most regulatory libraries are electronic databases. Simple systems, such as using office support software with desktop search capabilities, are fine for smaller databases while more complex and dedicated software is available for regulatory libraries consisting of thousands of references.

National regulatory agency Web sites, such as www.FDA.gov, and international or harmonization sites, e.g., www.ich.org, are a wealth of information on FDA regulations and international standards, sometimes providing far too many resources outside the intended scope. Scientific literature, through PubMed or a government institution or university library search engine, provide journal articles of a regulatory or technical nature. Professional and industry or trade journals post articles on various regulatory topics and recent articles offer insight into recent or hot topics.

Regulatory intelligence does not end with an initial search, but continues through the product life cycle. Information is obtained through periodic searches of Web sites; by using commercial regulatory intelligence software; through e-mail alerts, many provided by various government agencies including the FDA; by using blogs and through diligent personal networking; by e-mail from professional meetings; and by networking with other professionals. Some long-standing and helpful U.S. sources of intelligence on biopharmaceuticals are given in Box 3.4 while other sources, notably International Conference of Harmonization (ICH) guidelines, are described in Chapter 4.

BOX 3.4 WEB-BASED SOURCES OF U.S. REGULATORY INFORMATION

U.S. Government	
Food & Drug Administration	www.fda.gov
National Institutes of Health	www.nih.gov
U.S. Pharmacopeia	www.usp.org
Clinical Trials Registry	www.clinicaltrials.gov
Pub Med (National Library of Medicine)	www.ncbi.nlm.nih.gov/pubmed
Regulatory Trade Organizations	
Regulatory Affairs Professional Society (RAPS)	www.raps.org
Drug Information Association (DIA)	www.diahome.org
Food & Drug Law Institute (FDLI)	www.fdli.org
American Association Pharmaceutical Scientists (AAPS)	www.aapspharmaceutica.org
Pharmaceutical Education and Research Organization (PERI)	www.peri.org
Parenteral Drug Association (PDA)	www.pda.org
Biotechnology Organization (BIO)	www.bio.org
Pharmaceuticals for Practitioners	www.pharmaportal.com
Association of Clinical Research Professionals (ACRP)	www.acrpnet.org
The Organization for Professionals in Regulatory Affairs	www.topra.org
Regulatory Newsletters	
www.fdcreports.com	www.foi.com
www.foi.com	www.bioworld.com
www.fdanews.com	www.fdaadvisorycommittee.com

Regulatory Operations for FDA Applications

Regulatory Planning and the Regulatory Environment

Biotechnology endeavors are influenced in many ways by regulations and the regulatory environment. To succeed, a biotechnology operation must understand the regulatory landscape lying ahead before embarking on product development. This is achieved by preparing a regulatory plan, an early and important part of any product development strategy (PDS). Benefits are obvious. It builds a framework for the overall operational plan and serves as a foundation upon which many aspects of nonclinical, clinical, manufacturing and control planning may be built. A regulatory plan identifies potential regulatory hurdles through inspection of product-specific regulations and regulatory agencies. It allows a biotechnology firm to communicate a message to outside parties, investors or potential partners that they will be successful, in part, because they have identified regulatory issues specific to their product and intend and planned to address those matters. A well-considered regulatory plan goes beyond reacting to the environment and also identifies regulatory opportunities, such as unmet needs or accelerated pathways, that enhance product value. Indeed, a well researched and reasoned regulatory plan seeds new ideas into a biotechnology firm's business plan.

Novel biotechnology products often sail into uncharted regulatory waters. There are new diseases or unique indications for novel diseases, product lines never before marketed and radically new technology approaches to unsolved problems. As noted earlier, the breadth of biotechnology products is enormous, meaning that regulatory requirements cut a very wide and deep swath. The regulatory plan for a given product may need to cover a number of diverse areas and also delve into each of these areas, perhaps setting precedent in one or more areas. Elements of a regulatory plan are given in Box 3.5 and discussed further in Chapter 1. In biotechnology, the regulatory planner is challenged from many directions. Even if regulatory pathways are clear and simple with much precedent, then a well-conceived regulatory plan adds product value by demonstrating how the product may be a regulatory "slam dunk." The time and effort involved in preparing a regulatory plan is justified for any biotechnology product.

Risk versus Benefit

The FDA is chartered to protect the public health. Hence, at the heart of regulatory judgment for any given product is the need to weigh "risk versus benefit," as it relates to individual and public health. This can be challenging for a biotechnology firm, especially the small or virtual operation, to appreciate. Nonetheless, it is critical at some point to think like FDA staff and consider the risk-to-benefit balance that a product must bring to the

BOX 3.5 ELEMENTS OF A REGULATORY PLAN

The Product

- Characteristics
- Class
- Competitor or look-alike
- Indications: primary, secondary
- Limitations in safety or effectiveness
- Special safety considerations
- Product "blemishes"

The Regulatory Environment

- Scope of applicable regulations: global, national, state, local
- Purposes for regulations: medical, agricultural, environmental, safety
- Regulatory agencies or guidelines: FDA, USDA, ICH, European, Japan, rest of world
- Laws, regulations, guidelines
- Precedence for product class or competitor products

Global Strategy

- Global market targets
- Sequence or timing of applications
- Exclusivity by market
- Application and approval methods by market or country, ex-U.S.

Influences on Regulatory Environment

- Political: U.S. and ex-U.S.
- Social: U.S. and ex-U.S.
- Advisory committees
- Local or state authorities

Regulatory Communication

- Formal applications: IND, NDA, BLA
- Written communication: letters, amendments
- Meetings or teleconferences
- Public conferences
- Timing of each major communication

Options for Special Designations and Pathways

- Orphan product
- Accelerated
- Emergency use
- Fast track

Regulatory Risks

market. Any sponsor has surely considered the benefit that its product might provide to mankind. In fact, the benefits of a biotechnology product, most probably, have been touted to the world in an effort to garner financial and public support. But, it is not typical, certainly not in the early stages of biotechnology development, to admit that a product also carries risks; it is counterintuitive to the entrepreneurial environment to delve deeply into the possibility of product-associated risk. Also, it is not good business practice to tout this potential risk to the public. However, the FDA must, by law, consider the possibility that any product can cause harm: the risk outweighing the benefits. Regulatory agencies look at any product as possessing possible risk as well as benefit and they have an obligation, as part of their mission to public health, to evaluate and, sometimes, to publicize product risks. This is a reason some biotechnology firms are asked by the FDA to place a "black box warning" onto the labels of a marketed product. These different ways of thinking create a tension between biotechnology firms and regulatory agencies.

In developing a regulatory plan and throughout the development cycle, it falls on regulatory and quality professionals to carefully weigh product benefits against risks and pass the results of this analysis along to members of the product development team. This is done so that risks can be addressed before they become an issue with regulatory authorities or, worse, cause harm to public health. Regulatory risk identification and management begins early in the development cycle, as a formal and broader process of product and project risk analysis and management, as discussed in Chapter 1 and Chapter 5. Here, we point out the simple fact that regulatory agencies perceive potential risks early in a product's life cycle. If risks are not recognized as such by the sponsor then the FDA may demand that they be both recognized and mitigated. This process may go on through the entire cycle of product development. Failure to heed warnings by the FDA on risk can lead to regulatory action by the agency. The regulatory professional is often the biotechnology team member to elicit support of product development team members to identify and manage technical risks so that they do not later become product or regulatory risks.

Applications Seeking FDA Investigational Use or Marketing Approval

The development life cycle for a biopharmaceutical is heavily influenced by regulatory requirements: hurdles, some might say. These requirements are considered met only after a biotechnology firm has communicated scientific plans or results to the FDA. Indeed, the regulatory process is a dialogue between the agency and the sponsoring firm. The sponsor, a legal entity responsible for the product, may be an individual, but, in most cases, it is a corporation or an institute. A sponsor, such as a biotechnology firm, has one individual responsible for signing regulatory documents and this person is the designated sponsor's representative. During the development life cycle, dialogue between the FDA and the sponsor is carried on through a series of meetings, teleconferences and written documents. These steps in communication with the FDA can and should represent important milestones in meeting objectives in the overall product development cycle. The overall sponsor–FDA communication scheme process, shown in Figure 3.3, is much the same for drugs and biologics even though the nomenclature differs in some respects. Certain aspects of medical device development, including major documents, can differ significantly from drugs or biologics.

The FDA regulatory process begins with a designation by the sponsor that a candidate product will be used in humans for the treatment or prevention of disease. The process of developing a TPP (see Chapter 1) is critically important in part because the sponsor clearly identifies the intended use, indication and nature of the product. Consideration as a medical device, a drug or a biologic locks it into a regulatory pathway that has been established by precedent and regulation or guidelines. Thus, the TPP process is a critical first step in development of a PDS and for regulatory planning.

Investigational Use Applications: Investigational New Drug (IND) Application

The IND application is a request to the FDA to perform human safety and effectiveness studies on a biopharmaceutical or drug. The IND application often represents the first legally binding document that is submitted by a sponsor to the FDA. However, the IND filing at the FDA is often preceded by less formal means of communication, such as a pre-IND meeting or teleconference. A "new drug" is a construct or molecule that has an ingredient or combination of ingredients that have not yet received marketing approval. While under development and an IND, a product is considered "investigational." Indeed, even compounds with slight molecular variations on an established product may necessitate filing an IND with the FDA. And products approved for an indication also are considered a new drug if they are to be used (investigated) outside the approved label claim that is used for another indication. Investigational use of a new drug is represented by a clinical study or studies of that drug performed in humans by or on behalf of the sponsor. The IND process focuses on submission, by the sponsor to the FDA, of a protocol, a written plan to test that drug in the first clinical study ("first-time-in man") and the scientific data and background information related to that drug. The format for an IND can be found at www.fda.gov.

Many other countries have investigational applications, but the format varies across borders. Consider that the sponsor and the FDA are both responsible to the public for ensuring that a new drug will not present unnecessary risks to those receiving it and that any risks are balanced with potential benefits. The contents of an IND, outlined in Box 3.6, are standard for any IND and the sponsor is responsible for writing this document and submitting it

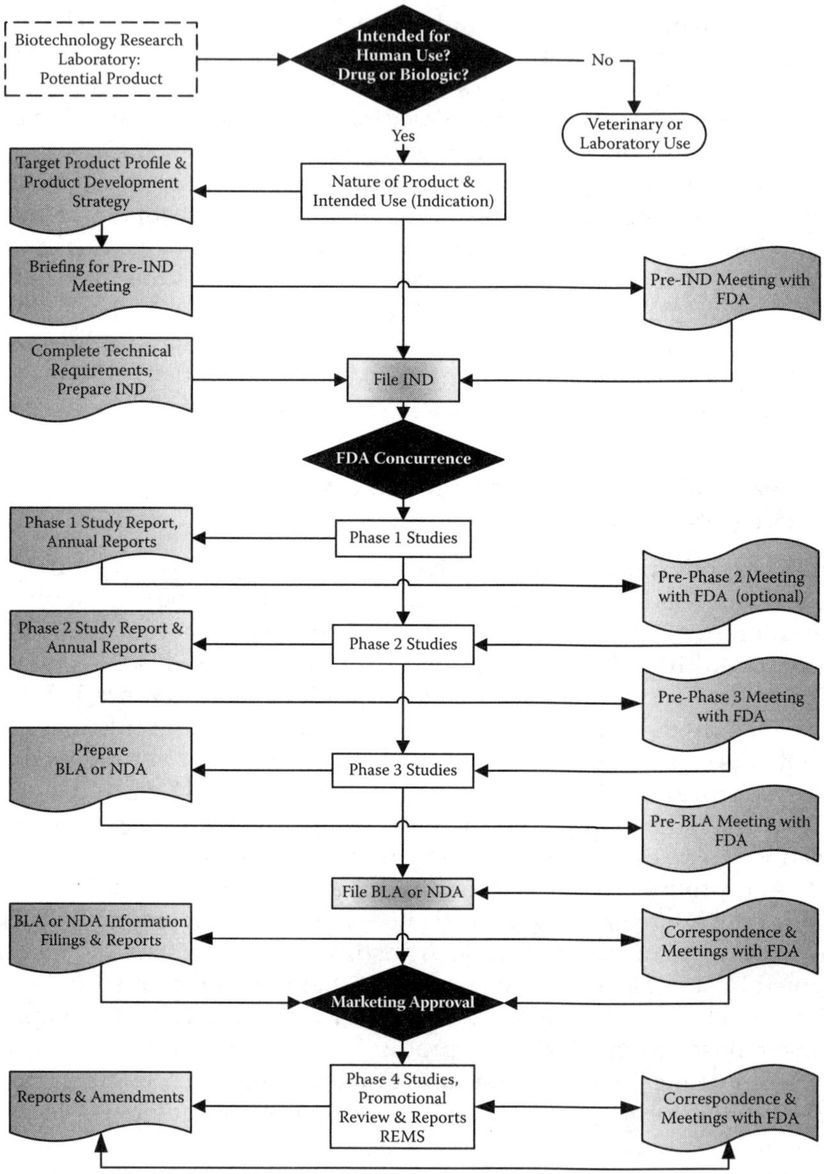

FIGURE 3.3
Regulatory activities and milestones in biopharmaceutical development.

BOX 3.6 CONTENTS OF AN IND

Section A(1)*. Cover page is FDA Form 1571

The sponsor's statement to certain questions about the product and responsible individuals is placed onto a standard form provided by the FDA. Signature sponsor's representative.

Section B(2). Table of Contents

Section C(3). Introductory Statement and General Investigational Plan

A review of the background, perhaps noting the research basis for the product, information on the disease or condition, and an explanation of why the product was chosen to treat or prevent the indication.

Section (4). Reserved for FDA Requests

Section D(5). Investigator's Brochure

The "IB" is a summary of critical information found in the IND, and is written by the sponsor to fully inform the clinical investigators and staff.

Section E(6). Clinical Protocols

The protocol describes in detail the nature of the study and all procedures that will be taken during the clinical trial. Supportive documents include the FDA Form 1572, a signed statement of the principal investigator, résumés of clinical staff, and other documents related to the clinical investigation.

Section F(7). Chemistry, Manufacturing and Control Information

Here the sponsor describes the manufacture of the product and provides detailed information on the raw materials, process and facilities. Quality control assays are identified and Certificates of Analysis presented. An environmental statement is required.

Section G(8). Pharmacology and Toxicology Information

All nonclinical studies are described, with full study reports. All nonclinical or preclinical studies that related to safety or toxicology and key studies that demonstrate potential efficacy, such as pharmacokinetic and pharmacodynamic studies, are included.

Section H(9). Previous Human Experience with the Investigational Drug

* The FDA has used two systems, one numbering (1 through 10) and the other lettering (A through I) to identify the elements of an IND.

Descriptions and references to all previous use of this or closely related products with special attention to safety issues in humans.

Section I (10). Additional Information

References

A list of scientific publications cited in the text, letters granting permission to reference Master Files or other INDs.

Appendices

Large documents, such as nonclinical toxicology reports or pertinent scientific publications may be included if they have been abstracted or cited in an earlier section.

to the FDA. In doing so, the sponsor must attest to the validity of data and background information provided to the FDA by completing and signing a form (FDA Form 1571). It is the information provided in an IND that is critically important to the FDA, as this represents the basis for their review and it is key to their making a risk versus benefit assessment. As shown in the IND format, Box 3.6, the information comes from a variety of sources, this described in other chapters in this book. Of considerable importance is Section 6 of the IND, providing the clinical protocol and other clinical investigation documents (these are described in Chapter 9). The nonclinical information (described in Chapter 8) and any laboratory studies related to laboratory studies, such as biodistribution, are provided in Section 8 of the IND. Previous human use of this or closely related products is communicated in Section 9 of the IND. Section 7 of the IND is for chemistry, manufacturing and control of the product (described in Chapter 6 and Chapter 7). The IND format is being replaced by an international format, the Common Technical Document (CTD) investigational application format. The contents of the IND and CTD will be the same, but the reporting format sectional structure will differ.

The IND is a legal document and it is important for the sponsor to provide only information that is true and accurate and to disclose any and all information relevant to the product, notably safety data. Upon receipt, the FDA has 30 calendar days to review the IND and decide whether or not to allow the sponsor to proceed with the proposed clinical study. If the FDA decides against allowing the sponsor to proceed, and such decisions are most often based on findings related to product safety or potential undue risk to human volunteers, then the application is placed on "clinical hold," meaning the clinical study may not be conducted unless the FDA says otherwise in a future letter. A sponsor may appeal a clinical hold, but, in doing so, must provide significant data to refute or better explain the safety concerns brought forth by the FDA. This process of scientific negotiation takes some time and, thus, a clinical hold is an undesirable event to any biopharmaceutical development program.

To better ensure the chances of a successful IND, a sponsor often communicates and discuses with the FDA its intentions to submit an IND well before the IND is written. This is a formal process with several steps. The FDA provides guidelines for meetings with sponsors, classifying each meeting type as either A, B or C, and offering guidance on how to plan a successful meeting at various stages of development and for different purposes. Once a sponsor has some scientific information about a product and can define specific questions, it submits a request to the FDA for a pre-IND meeting. A brief written document, a "premeeting package" outlining information and data relevant to a product, is submitted to the FDA well prior to the meeting date. This is accompanied with a request that the FDA meet in person or by teleconference with the sponsor to discuss issues. The sponsor also queries the FDA staff's opinion on specific items of concern. For example, the sponsor may ask whether a particular nonclinical study was necessary or if a certain manufacturing method or the concept clinical protocol design was acceptable. Timing of the request for a pre-IND meeting is crucial because the sponsor must have enough information acquired about a product to allow the FDA to make an informed response to questions. Yet the sponsor does not approach the FDA shortly before filing an IND. The FDA responds to requests with staff comments in a letter and either a meeting or a teleconference is held for clarification purposes. The result of a well-considered meeting with the FDA is thoughtful recommendations from the agency to the sponsor, these based on questions raised in the letter to the FDA. Pre-IND communication greatly reduces the chances of a clinical hold on the future IND if, in fact, the sponsor addresses the issues raised by the FDA in its pre-IND responses.

Responsibilities for the sponsor and the FDA to communicate do not end with filing an IND and receiving clearance to begin a clinical study. Once the IND has been accepted by FDA and the product is investigational, the process of submissions, responses, meetings and teleconferences has only just begun. Meetings between sponsor and the FDA continue throughout the life of an IND and the development cycle it supports. These are shown in Figure 3.3. A few commonly used meeting venues, in addition to pre-IND, are noted here. A meeting is sometimes held prior to initiating Phase 2 clinical studies (pre-Phase 2) and one is always held prior to performing a pivotal clinical study, Phase 3 (pre-Phase 3 meeting), and prior to submitting the marketing application, a New Drug Application (NDA) or Biologics License Application (BLA) (pre-NDA or pre-BLA meeting). These meetings allow the FDA and the sponsor to agree to clinical study designs that, if successful, support marketing applications. Sponsors also request, during the course of development, a meeting with the FDA to discuss a special issue, such as the need to do an additional nonclinical study or the design of a new manufacturing plant. While these subject matter-specific agendas may be combined with a clinical or milestone meeting, they typically require input of experts outside the clinical arena and, thus, result in individual meetings or teleconferences.

Throughout the life of the IND, and this may represent over 10 years, the sponsor must file written reports and data with the FDA. Annual reports, outlining progress on the program and changes in technology, and summarizing clinical and nonclinical studies, are provided on the anniversary date of the IND. In addition and at anytime during the year, significant changes made or findings related to safety of the investigational product, notably toxicology or adverse reactions seen in clinical or nonclinical studies, must be immediately communicated to the FDA in writing. There is also a need to keep the FDA abreast, by written correspondence, of matters arising in development. In any given year, a sponsor is likely to submit between 10 and 50 letters, many containing protocols or data, to the IND. Each letter is considered an amendment under that IND and, therefore, represents a legally binding statement. For example, the tenth letter or amendment to an IND would be numbered by the FDA as BB-IND32401-010, where BB stands for a biological, IND32401 is a sequential number given to that original IND upon submission, and 010 is the tenth filing past the original IND. Ensuring the accuracy and completeness of each submission, managing meetings and premeeting information and maintaining an IND are important tasks and require professional regulatory support.

If undue risk is noted by the FDA, the agency can, at anytime, impose a clinical hold and halt ongoing or prevent planned clinical studies. Clearly, there is a need for the sponsor to be diligent and to have a means, typically through a formal regulatory process and professional staff, to monitor all aspects of product development and to report findings and changes to the FDA in a timely manner.

Marketing Applications: Biologics License Application (BLA) and New Drug Application (NDA)

A marketing application, NDA or BLA, from a sponsor is a request from the sponsor to the FDA for approval to enter a product into interstate commerce and to make claims of safety and efficacy about that product in the labeling. Hence, a marketing application is composed of all the information that is relevant to that product, all clinical and nonclinical study designs and results, and everything else known or discovered about the product, its research, manufacture and control. NDAs and BLAs, submitted to CDER and CBER, respectively, are massive documents, containing narrative summaries, tabulated information, raw data and explanations of how the data were generated, analyzed and submitted. The sponsor also proposes the product labeling, critical documents stating the indication and making claims for the product and the product label, the written information affixed to the container of product. The proposed label must provide warnings, contraindications, directions for use and other information. An outline of the most important (and required) product labeling, the package insert, is shown in Box 3.7. The *Physician's Desk Reference* and other publications provide a collection of

BOX 3.7 ELEMENTS OF PRODUCT LABELING (PACKAGE INSERT)

Trade Name and Chemical Name

Description. Describes the drug's nature and classification. Identifies how it is supplied in the final container and lists any excipients.

Clinical Pharmacology. This gives the mechanism of action, pharmacodynamics, pharmacokinetics and known drug interactions.

Clinical Studies. All pivotal clinical studies and important smaller studies are described, to include endpoints and outcomes. Data are often summarized in graphic or tabular format and all important results, efficacy and safety are identified.

Indications and Usage. The exact indication is given exactly in a very few brief sentences.

Contraindications. Here are listed any situations in which the product should not be used.

Warnings. Messages to the prescriber and user are noted, sometimes in bold and capitalized text to stress special safety issues. Warnings may be listed by organ system (e.g., cardiopulmonary events) or by disease (e.g., malignancies). These may be in bold and surrounded by a black box.

Precautions. These are items that the user or physician should watch for, issues less common or important than those given in warnings. They are general, written for physician and user, or they are information for patients, specifically written for the user. Instructions for special situations also are placed here and specific items are highlighted as paragraph headings. Recommendations may be given to stop using a product, for example, if a disease progresses or if certain symptoms are noted. Drug interactions, use in nursing mothers or in pregnancy, pediatric and geriatric use or use in other special populations are generally included in this section.

Adverse Reactions. General adverse reactions are first described, these coming from reactions seen in the clinical studies. Here again key results of clinical trials are often presented in graphic or tabular format. A table presenting the most common adverse events is usually provided here. The most common adverse events are then described in a narrative paragraph. Finally, adverse events are listed by body system, cardiovascular through urogenital.

Overdosage. This describes what is known if a patient should take more than the prescribed amount.

Dosage and Administration. Information on how product is provided to a patient in final or dosage format is provided. This repeats and provides details on what has been given under Description.

How Supplied. If multiple formats exist (e.g., liquid or a tablet), then each is described. The NDC number, a product-specific identifier, is given for each.

References. A few key scientific publications are cited.

Administrative. Manufacturer's name, address and license number with dates are given.

Patient Information. Instructions that are for the patient, especially concerning proper handling, storage or use, may be given on a separate but attached pamphlet.

product labeling for FDA-approved drugs and biologicals, bound into one volume. Product labeling also may be found at the Web site of the FDA and those of product sponsors or manufacturers. Biotechnology firms seldom make draft product labeling available to the public until product approval.

The proposed labeling drives the application review process at the FDA. Each of the product claims in the proposed labeling must be supported by the data submitted to the FDA in the NDA or BLA. For example, if the sponsor claims in the product labeling that a monoclonal antibody would stop the growth of prostate tumors for a period of two years (an endpoint) and prolong the life expectancy of the patient for four years (a second endpoint), then the design of the pivotal clinical trials must be focused on acquiring data to confirm these endpoints. In another example, if product labeling submitted to the FDA makes the claim that a monoclonal antibody is 98.7% pure, then the manufacturing and quality control data provided in the BLA must show, with data from testing multiple lots of product, that it did, in fact, reach that purity level. The BLA must further demonstrate that the monoclonal antibody will be manufactured in such a way that there is high probability that future lots of product will achieve 98.7% purity. And, if the product labeling claims this monoclonal antibody did not result in autoimmune disease in nonhuman primates when it had been given over four years, then that claim must be evident from the nonclinical data submitted under the BLA.

Further, for each NDA or BLA, it is typical for the FDA to request additional data from the sponsor, sometimes asking that an additional study be performed. And inspections of clinical, nonclinical and manufacturing facilities are routine during the examination product. These inspections are referred to as preapproval (PAI), biomonitoring (BIMO) or post-approval. Exact wording on labeling claims may be open to negotiation between the sponsor and the FDA, but the claims must always be supported by data. Inevitably, the FDA asks a sponsor to make changes to its proposed labeling; discussions can be contentious, but negotiations inevitably result in fair and balanced wording to labeling and the FDA always makes the final decision. Note the cyclic nature of the product labeling in a well-conceived biopharmaceutical development project. In Chapter 1, we described the need for the TPP, draft labeling, to drive the development process and now, at the NDA or BLA stage, we finalize that labeling. All the effort of the biopharmaceutical development team went into generating data that would support the claims made in the draft labeling, presumably the same claims seen in the proposed and final labeling. With product labeling, the development process is truly a cycle, beginning with a visualization of what will appear in labeling and ending with proposed and approved labeling, this supported by many years of effort and investment in between.

Upon submission of the NDA or BLA, the sponsor must pay a significant sum of money, referred to as an FDA user fee. This fee supplements the FDA resources for product review and is not refundable if the application is denied by the FDA. Additional fees must be paid for each manufacturing facility at application and periodically over the market life of the product. Review by the FDA takes months because each claim must be carefully examined in light of the data presented in the application. No wonder that a prudent biotechnology operation carefully prepares an NDA or BLA and submits it to the FDA only once all claims are fully supported by data. Failure is expensive for a sponsor in many respects.

The format for an NDA and BLA has changed to harmonize with the international community. Most countries have adopted the CTD format. The eCTD is encouraged over the hard copy CTD. A general outline of the CTD is provided in Box 3.8; the full outline is much more detailed and specific and can be found at www.fda.gov.

Medical Device Applications: 510(k) and Premarket Approval (PMA)

Sponsors of new medical devices, regulated through the FDA's Center for Devices and Radiological Health, are faced with a variety of pathways to market approval. For any given device, the pathway is determined largely based on the risk posed to the user by that class of device and, to a lesser extent, on the nature of the device and previous experience with that device. As noted earlier, medical devices are in one of three classes based on the risk posed by the device. Class I, generally low risk devices, must be developed under "general controls," processes that include the quality systems regulation for manufacturing and record keeping. Very few Class I devices require premarket notification, or 510(k) process. Class II devices, of moderate risk, are subject to both "general controls" and "special controls" and most are subject to premarket review and clearance under 510(k) premarket notification process, a process described in a previous section. Class III devices

BOX 3.8 COMMON TECHNICAL DOCUMENT (CTD) FORMAT

Module I. Regional and Administrative Information
Information and documents, such as forms and certificates that are specific for the region, e.g., United States or Europe, are placed here.

Module 2. Summary Documents
These are summary documents of the other modules and are provided so all reviewers, no matter their specialty, may gain an overview of all information related to the product.

Module 3. Quality
All chemistry, manufacturing and control information is presented in this module.

Module 4. Nonclinical study reports
Nonclinical study reports focused upon safety, toxicology, pharmacokinetics and pharmacodynamics, and references cited are placed in this module.

Module 5. Clinical study reports.
Clinical protocols, full clinical study reports, integrated safety or efficacy reports, case report forms, previous human use reports, FDA meeting minutes and references cited are included here.

carry potentially more risk and include many life-sustaining or life-supporting implantable devices. These are subjected to the most rigorous controls. In addition to general and special controls, they are subject to premarket approval. Some Class III devices may be marketed under the 510(k) rules, but most must undergo the premarket approval (PMA) process. The PMA process demands proof that the device is both safe and effective, which usually means clinical studies are required. For human studies of medical devices, an investigational device exemption (IDE) is required of devices presenting the highest risk to the user. No matter which route is taken to market, medical device labeling is required in a device marketing application and it cannot be false or misleading.

Few biotechnology products are considered medical devices alone, but some are combination products and so the sponsor of a combination device must consider medical device regulations in its development scheme. However, the inclusion of a medical device into a combination product does not lessen the regulatory hurdles, rather it may increase them. As noted earlier, the final determination of a biopharmaceutical combination product is based on which component, device or biological, primarily produces the major effect. More often than not, it is the biological that is the lead actor and, in such cases, the product will follow the IND to BLA (or IND to NDA for a drug–device combination) route to approval with concurrent review by CDRH.

A device submitted to the FDA for clearance under 510(k) must demonstrate substantial equivalence to a legally marketed predicate device. The device must have performance standards, it must have an indication for use, the proposed labels must be accurate, and it must be fully described to CDRH in the 510(k) application. There must be evidence that it will be legally marketed and information on the device's safety and efficacy profile must be provided. The PMA and its contents are equivalent to those of the BLA or NDA because they demand a significant amount of nonclinical and clinical study data to support label claims and because the manufacturing controls are quite stringent. Clearly, most device manufacturers would prefer to register their devices under 510(k).

In vitro diagnostics (IVDs) are a class of medical devices that have a few special rules. Because many biotechnology products are used as key reagents with IVDs, their regulation is of great interest to biotechnology firms. IVDs encompass a host of products, from complex instruments to test kits used to diagnose life-threatening diseases to simple diagnostic laboratory reagents. Many IVD biotechnology products are marketed following 510(k) premarket notification and a few must enter the market through the PMA route; either pathway for IVD is under review by CDRH. Also under the review of CDRH are biotechnology products, reagents and instruments that are designated for "research use only," "investigational use only," "analyate-specific reagents" and "laboratory developed tests." Biotechnology firms may find a ready market for their products, sometimes originally developed as laboratory reagents, with IVD manufacturers who purchase reagents to include in their test kits. Before making a decision to enter the IVD field, firms should understand the possible impact of entering a regulated arena. Considerable diligence, research and regulatory planning are advised.

Special Documents, Pathways or Exemptions

- **Master Files**. The FDA provides a means to file confidential information for any type of product; the submission is called a master file (MF) and the contents are maintained in strictest confidence, as are other regulatory filings. The advantage of filing an MF is to allow the FDA to review important information without divulging confidential information on manufacturing, testing, etc. The MF may be referenced by the FDA in support of an IND, NDA or BLA, with a letter of permission from the sponsor of the MF to the FDA. The FDA reviews the MF only as informational sources and in no case does an MF constitute an application for investigational use or market approval. For example, a biotechnology firm, A, produces a synthetic molecule that is sold to another biotechnology firm, B, for use in firm B's *in vitro* diagnostics. Firm A prepares an MF that describes in detail specific proprietary information, for example, the manufacture and control methods used to make this molecule, and firm A then submits or "files" it with the FDA. Firm A also prepares a letter of reference for firm B and in it they give the FDA permission to read firm A's MF as part of firm B's IND 510(k) filing. Firm B provides firm A's letter of reference in its application. In effect, firm B does not have access to the technical details of firm A's proprietary information, yet the FDA can review this information in great detail to ensure that it is pure and potent and suitable for use in firm B's IVD. MFs are commonly used in a number of ways to support both the regulatory and business interests of companies.

- **Animal Rule**. This regulation applies to biopharmaceuticals for which there is no possible or ethical way to test efficacy in humans. Examples are products to prevent or treat diseases caused by weapons of terrorism or mass destruction. It allows the sponsor to test the efficacy of the product, a countermeasure, in a well-developed and surrogate (to man) animal model. The product's safety is then studied in human clinical studies under IND. The IND or PMA process is used throughout development with the unique features being that efficacy studies are conducted in at least two animal species.
- **Accelerated Approval**. Drugs or biologicals indicated for the prevention or cure of serious or life-threatening diseases are eligible for a program known as accelerated approval. Under this regulation, the FDA may approve such products on the basis of a surrogate or clinical endpoints if these are likely to predict clinical benefit. Clinical studies are still required, but they may not require the stringent and long clinical processes for other drugs. There are restrictions on the approval, however, and sponsors carefully choose this route in special projects.
- **Emergency Use and Treatment IND**. An FDA regulation allows investigational drugs to be used outside the standard clinical protocol in serious or life-threatening situations and with FDA concurrence. This regulation may speed a drug to a patient that might otherwise be ineligible to enroll in the study, such as someone living far from the study site. There are caveats to this as the sponsor must supply product at no cost and there is risk that the drug may be misused or result in harm to the patient. Biotechnology firms developing life-saving products agree to emergency use with caution, but consider it important as a good faith effort to speed a product to patients. Treatment IND is a slightly different approach for new biopharmaceuticals. There must be preliminary evidence of efficacy and an indication for a serious or life-threatening disease or it may be used if there is no alternative drug available and if death is expected from the disease within months. Examples are advanced cases of AIDS and cancers. Typically, the product is in Phase 3 clinical studies and the information on its use must be reported to the FDA.

- **Orphan Drug Designation.** This leverage option is commonly used by biotechnology firms because many biopharmaceuticals are developed for the diagnosis or treatment of rare diseases or conditions. By definition, a rare disease or condition affects fewer than 200,000 people in the United States per year. It provides incentives for firms to develop biopharmaceuticals that are used less frequently and, therefore, have a smaller market value. This program, which has been quite successful for almost 30 years, is coordinated by the FDA's Office of Orphan Products Development. A common European and FDA application is available. The sponsor benefits by receiving both seven years of exclusive market rights for the indication and a tax credit. Also, FDA awards development grants to sponsors of orphan products.
- **New Drug Product Exclusivity.** This is protection from competition, for three or five years, for the holder of an NDA or BLA when the drug is a new chemical entity, i.e., a product unique both in nature and the marketplace. This exclusivity, in addition to that provided by a patent, encourages biotechnology firms to develop novel products.
- **Fast Track.** Another program to expedite products to patients with serious or life-threatening diseases, and where there is an unmet medical need, is called Fast Track designation. The sponsor must request Fast Track designation from the FDA and certain criteria must be met. However, a sponsor with Fast Track designation receives special consideration from the FDA, such as additional meetings in which guidance may be provided, priority review of market applications, and even a program to review certain sections of the NDA or BLA incrementally, thus saving time in the review process.

Generic Drugs and Biosimiliar or Follow-On Biologics

Generic drugs have the same active ingredient as brand name, marketed drugs and represent look-alikes that enter the market following expiration of patent protection. They are often made by manufacturers other than the company that originally made the brand-name drug and they must be tested in small, head-to-head studies—laboratory and clinical pharmacokinetics and pharmacodynamics—to demonstrate chemical identity and pharmacokinetic and pharmacodynamic similarity. The approval process for a generic drug at the FDA follows the sponsor filing an Abbreviated New Drug Application (ANDA) and indeed it is just that.

It is unlikely that generic drugs will be derived from biotechnology. However, their equivalent, biosimilars, also referred to as follow-on biologics, biogenerics or generic biologics, are biologic look- and perform-alikes and will certainly be marketed someday. While not yet approved by the U.S.

Congress, it is only a matter of time before they enter the market and compete with brand name biologicals that no longer hold patents. They have met stiff resistance on two fronts: first, from biotechnology firms that currently hold the market in absence of patent protection, and, secondly, because there are serious questions regarding whether or not biological molecules can, in fact, be reproduced to mimic, exactly, the purity, potency, efficacy and safety of the predecessor molecule. Clearly, any biosimilar will need to be thoroughly tested in adequate and well-controlled clinical studies. The extent of testing is much debated, but the biogenerics will, at some time, reach the market.

Other Regulatory Activities

Many additional regulatory activities must be completed during the product life cycle, some before and others after product registration. Several examples include:

- **Establishment Registration.** Establishments manufacturing any drug or biopharmaceutical or medical device, whether U.S. or foreign, must be registered with the U.S. FDA.
- **Licensing Issues.** Divided, shared or contract manufacturing. To accommodate the complex and sometimes specialized manufacturing schemes required for biopharmaceuticals, the FDA allows manufacturing of one product at two or more sites. For example, a biopharmaceutical might be produced by fermentation at one site, then shipped to a second site for purification, formulated at a third site and filled and labeled at a fourth site. Such divided or shared manufacturing, generally done largely by contractors for the sponsor, is allowed if it is carefully controlled and defined and if each site is a registered establishment.
- **Proprietary Name.** A sponsor wishes to have a unique name, apart from the often long and confusing chemical name, for their biopharmaceutical. To avoid duplication for confusion in labeling, the FDA is responsible for approving the proprietary name for each biopharmaceutical.
- **National Drug Code.** In addition to the unique name, the FDA issues with marketing approval a unique drug number, a National Drug Code or NDC, and this is clearly marked on all labeling.

Public Meetings and Advisory Committees

A cornerstone of good government is the right to speak in public for or against an issue, especially if that issue arises from a government or government-regulated activity. Biotechnology products and the marketing approval of biopharmaceuticals and medical devices are no strangers to the public arena, and their use and release into the environment have been a matter of debate ever since recombinant *Pseudomonas syringae* was sprayed on strawberry fields of California in 1983. We noted earlier that a key part of the process to make or to change a regulation is the public rulemaking process in which the public has an opportunity to review and comment on proposed regulations prior to rule publication and codification. In addition, regulatory agency processes may be influenced by public petition, requests made to produce, remove or change a regulation. Other rules allow the public to demand economic or environmental impacts for a regulation and these processes also mandate the public be informed and allowed to influence the government's decision. All of this applies to the FDA and other regulatory agencies. States have similar rules that result in public hearings or meetings.

The FDA also uses advisory committees to its and the public's advantage by asking expert panels to review data on safety and efficacy for products near completion of review and recommend approval. The FDA has established these committees for every class of drugs, biologicals and medical devices under their purview. Members meet at established intervals to make recommendations on a variety of subjects. At the top of their list are product-specific recommendations, notably whether or not a product should receive market approval. This is typically done after the FDA has completed the review. For this, the committee is asked to answer a series of questions, such as: "Does the committee view this product to be safe for its intended use?" Panel members vote, but the recommendation is not binding to the FDA and the agency will sometimes decide in a manner not consistent with the panel's majority recommendation. Clearly, such meetings and the committee's voting record and recommendation are extremely important to the biotechnology firm sponsoring the product put before a committee. Another function of advisory committees is to make recommendations on groups or classes of products. For example, if a class of monoclonal antibody, represented by several similar products, appears to cause an unexpectedly high number of allergic reactions, the FDA might ask a committee to meet and discuss the situation and perhaps make a recommendation, such as posting a warning on each label. Advisory committees also perform more mundane tasks, such as reviewing research laboratories at the FDA. It is very important that members of these committees be experts yet have no conflict of interest, such as working in a commercial environment with the products on which they make recommendations or receive money from the sponsor. Indeed, a member should have no strong personal bias for or against a technology. It is a challenge for the FDA to find the right experts to serve on advisory committees. Committee meetings are open and announced to the public, comments are solicited and minutes and votes are a matter of the public record.

Having read this chapter, one might ask, "What is there about a biopharmaceutical operation that is not made public and that can be kept confidential and proprietary?" The answer is quite a lot; much, perhaps most, of the technical information that the firm considers proprietary and all of the financial facts are kept from public view. The FDA does not delve into a firm's finances or the public or private nature of a company; such is territory for the Securities and Exchange Commission or the Internal Revenue Service. The FDA does not consider marketing, other than whether or not promotion is in line with approved labeling and in a few other areas related to market approval. There is information that is considered proprietary to the biotechnology firm, but it must be disclosed to the FDA in applications and correspondence; however, the FDA has a wonderful record of keeping information hidden from public view. Regulatory agencies, unlike many organizations in political capitals, do not "leak" confidential information to the public or to the media.

Postmarketing Requirements and Activities

Interaction with regulatory authorities does not stop once market approval is given. The FDA can and has withdrawn approval after it has been granted; this whenever the agency can prove cause. This happens when a sponsor (holder of the market approval letter) fails to meet reporting requirements or if a product proves unsafe. What are postmarketing requirements of the sponsor? Some of them include:

- **Postapproval Maintenance of the Approved NDA or BLA.** Sponsors must file annual reports with the FDA as long as they are marketing an approved drug. Elements of an annual report include, but are not limited to, labeling, chemistry, manufacturing and control, nonclinical testing and clinical data. Reporting of changes from the original market approval are especially important to regulatory agencies and it may be necessary for the sponsor to report significant changes immediately and not wait for the annual report. At the time of market approval, it is normal for the FDA and sponsor to agree to certain postmarketing clinical studies, such as Phase IV or monitoring of special patient populations. These commitments by the sponsor also include advertising and labeling changes, product complaint reporting schemes or events that trigger product recalls. The FDA and the public take these commitments quite seriously.
- **Reports of Adverse Drug Events (Experiences).** ADEs are explained in Chapter 9. Briefly, they involve reactions in patients using a biopharmaceutical and they must be reviewed by the sponsor if they fall under certain guidelines for severity or frequency and might be related to the product. Physicians and users may report these experiences to the sponsor and sometimes directly to the FDA. Direct communication is referred to as MedWatch for many medications, while specific products, such as vaccines, have a unique reporting

system. The sponsor, in turn, must report serious experience situations to the FDA in an alert report within 15 days. The rules are not complex, but they are considered extremely important to maintaining public health and a safe source of biopharmaceuticals as well as keeping a positive image for the firm and the biopharmaceutical industry as a whole.

- **Risk Evaluation and Mitigation Strategy.** The FDA has instituted this program, referred to as REMS, to improve postmarket approval safety of medical products. REMS includes guides for medications, patient-friendly labeling and improved communication from the FDA or the sponsor with healthcare providers to better ensure proper use of products. The document is a plan that is submitted by the sponsor to the FDA for approval as part of the marketing application.
- **Dear Doctor Letters.** The FDA believes that an effective means of communicating new information, especially risks, for marketed prescription biopharmaceuticals is by ensuring well-informed prescribing physicians. Letters to doctors often fill that objective, along with announcements in medical journals and through the public media.
- **FDA Letters to Manufacturers.** Letters to manufacturers are another matter because these are targeted directly to the sponsor and are often compliance issues. Communication between sponsors and the FDA are discussed in Chapter 4.

Advertising and Promotion

Drugs, biologics and medical devices are heavily marketed to various target populations—physicians, nurses, pharmacists and users—and we accept this in our society. As compared to many other countries, biopharmaceutical marketing and promotion is lightly regulated in the United States. As discussed above, marketing and advertising activities are regulated through claims made on the label. A biopharmaceutical label is defined as: "A display of written, printed or graphic material on the immediate container of a drug." Labeling, as defined by the FDA and used as a noun, is: "Any written, printed or graphic material on the drug, on any of its containers or wrappers, or on any material accompanying it." Hence, the package insert, that lengthy document that contains prescribing, safety and dosage information, and is stuffed into boxes of over-the-counter or prescription medications, is labeling. Promotional labeling is any labeling used in advertising or marketing activities. Promotional labeling is at the heart of biopharmaceutical sales and so it becomes a point of contention whenever it does not reflect the approved labeling. For example, one of the most egregious violations of the FDA marketing rules is to promote the use of a biopharmaceutical for an indication or use that is not given in the approved label. This is known as promotion for "off label" or unapproved use. While the FDA does not restrict licensed medical practitioners from prescribing medications for or advising patients to take medications outside the labeled information, the FDA does not allow sponsors to, in anyway, promote this practice. Hence, it is only legal to promote biopharmaceuticals, or other drugs and biologicals, in accordance with the approved label. Advertisements must be balanced and complete, again as driven by information included in the label.

A product's label will almost certainly change during the postmarketing period and this results in refined definitions for what can and cannot be included in promotional materials. To ensure that biopharmaceutical promotional information is in line with current labeling, the FDA insists that it be provided for review by the agency at certain times. Investigational products may not ever be promoted. Most or all promotional materials for approved products must be submitted to the FDA as they will be used immediately after market approval. This first advertising campaign is referred to as a "launch." Also, most other promotional materials, generated postlaunch, also must be submitted for review. Dissemination of scientific and medical information is closely monitored as well by the FDA, and direct-to-consumer advertisements, such as television and newspaper ads, are controlled. An important subject is whether or not the consumer or the health practitioner is given adequate information about safety issues that are known to be or could even possibly be related to a product.

Summary on Regulatory Affairs Activities in Biotechnology Operations

This discussion completes the chapter on regulatory affairs operational activities and it reiterates themes that were introduced in Chapter 1. Regulatory planning starts early and involves successful product development through the very end. Planning must consider all regulatory aspects of the product and its development—each step in the cycle. The product labeling is central to the development life cycle, beginning with a draft labeling or TPP and ending with approved labeling for the marketed product.

4

Regulatory Compliance

Regulatory Compliance

Information in this chapter builds on an understanding of regulatory operations (see Chapter 3) by examining the broad world of regulatory compliance, by discussing the U.S. Food and Drug Administration (FDA) requirement to integrate quality into all aspects of biopharmaceutical development programs and by reviewing many regulations outside of the FDA that impact most biotechnology operations.

Quality Systems to Meet Regulatory Compliance

Compliance and Quality Systems

The Oxford Dictionary defines compliance as "... the act or instance of complying; obedience to a request or command." Further defined for biotechnology product development, compliance is the act of meeting a plethora of rules, regulations and directives. Compliance impacts each biotechnology development function every day. It involves constant vigilance to identify and understand each applicable regulation and, most importantly, it drives the biotechnology firm to institute and integrate programs that ensure obedience to these requests and commands. Compliance is achieved largely by ensuring quality in all aspects of development and at every step in the development cycle. This is best done by instituting quality systems (described in Chapter 5). Indeed, results of FDA inspections repeatedly demonstrate that firms with mature and effective quality systems consistently have, in the eyes of regulatory agencies, fewer deficiencies than do operations with deficient quality systems. With this in mind, it is easy to consider the need to integrate (into a biotechnology operation) scientific and technical skills, regulatory guidance and quality systems.

Quality systems are composed of quality hallmarks, features of a well-established, compliant and smooth operation (see Chapter 5). In this chapter, we discuss the intersection of compliance with quality systems. Three examples of quality systems, current Good Manufacturing Practices (cGMPs), current Good Laboratory Practices (cGLPs) and current Good Clinical Practices (cGCPs), are outlined below and each will be further described in Chapter 6 and Chapter 7 (Manufacture; Quality Control), in Chapter 8 (Nonclinical) and in Chapter 9 (Clinical), respectively. These are excellent examples of systems applied by most nations to protect public health by ensuring safety and efficacy of biopharmaceuticals. They are presented here as U.S. FDA regulations, but similar good practices of many nations are currently being harmonized into international compliance guidelines and so these elements now reach worldwide.

cGMPs: Current Good Manufacturing Practices for Manufacture and Quality Control

cGMPs were established to prevent drug and medical device manufacturers from producing and selling adulterated product to the public. This, in turn, was desired by the public because adulteration had occurred in medical product manufacture, and the practice was not tolerated. cGMP is an established quality system that has been shown to have a positive effect on the quality of biopharmaceuticals and, hence, cGMPs have been adopted as the manufacturing standard worldwide.

In the United States, biopharmaceutical manufacturing compliance is based on regulatory requirements for manufacturing processes and utilities, codified for human and animal drugs and in 21 CFR (Code of Federal Regulations) 210 and 211 as well as in other sections of FDA regulations. Important elements of cGMP are listed in Box 4.1. cGMPs are now quite well harmonized worldwide and any differences in cGMPs largely reflect the nature of the product or differences in its usage. It is repeatedly stated that quality cannot be tested into a product but that the sum total of what constitutes that product must be of the highest quality. Biopharmaceutical manufacture and control are, under cGMP, based on the idea that a product, the process to make it and the laboratory control tests must be designed in a manner that meets the intended use. cGMPs strive to meet that standard. These quality systems concepts are discussed further in Chapter 5 and examples or application of cGMPs to actual manufacturing and control processes are demonstrated in other chapters.

cGLPs: Current Good Laboratory Practices for Nonclinical Lab Studies

cGLP regulations were established because certain individuals were performing nonclinical laboratory studies in an unscientific or careless manner and the results of these studies could not be trusted. Important toxicology

BOX 4.1 HIGHLIGHTS OF CURRENT GOOD MANUFACTURING PRACTICES (CGMPs)

21 CFR 210 Current Good Manufacturing Practice in Manufacturing, Processing, Packaging or Holding of Drugs
Status, applicability and definitions

21 CFR 211 Current Good Manufacturing Practice for Finished Pharmaceuticals

A. General Provisions
B. Organization and Personnel
Responsibilities of quality control (assurance) unit, personnel qualifications and responsibilities
C. Buildings and Facilities
Design, construction, lighting, ventilation, plumbing, sewage, washing and toilet, and sanitation
D. Equipment
Equipment design size, location, construction, cleaning, maintenance; automatic, mechanical and electronic equipment, filters
E. Control of Components and Drug Product Containers and Closures
Receipt and storage, testing and use of components, containers and closures; retesting; rejection; drug product containers and closures
F. Production and Process Controls
Written procedures, deviations; yield; equipment identification; sampling and testing of in-process materials and drug products; time limitations; control of microbial contamination; reprocessing.
G. Packaging and Labeling Control
Materials examination; issuance of labels; tamper-evident packaging; inspection; expiration
H. Holding and Distribution. Warehousing and distribution
I. Laboratory Controls
Testing and release for distribution; stability testing; special testing; reserve samples; laboratory animals
J. Records and Reports
Cleaning and use logs, component, container, closure and labeling records; master production and control, laboratory, distribution and complaint records and review
K. Returned and Salvaged Drug Product

21 CFR 600 Biological Products

A. General Provisions

B. Establishment Standards
Personnel; establishment, equipment, animals; records, retention samples, product deviations; temperatures during shipment
C. Establishment Inspection
D. Reporting Adverse Experiences

21 CFR 610 General Biological Products Standards
A. Release Requirements
B. General Provisions
Methods and processes: General safety, inactivation, sterility, purity, identity, constituent materials, combinations, cultures.
C. Dating Period Limitations
D. Labeling Standards
Container and package labels, name of product, manufacturer and distributor, export

21 CFR 630 and 640 Standards for Human Blood and Blood Products

21 CFR 660 Standards for Diagnostic Substances for Laboratory Tests (Blood products)

21 CFR 680 Additional Standards for Miscellaneous (Biological) Products

21 CFR 820 Quality System Regulation (for design and manufacture of medical devices)

21 CFR 1270 Good Tissue Practices

21 CFR 11 Electronic Records; Electronic Signatures

Controls for closed and open systems, signature manifestations and record-linking, electronic signature components and controls, identification and passwords

data were found to be questionable. In response, a regulation, 21 CFR 58, was established in 1979 with the purpose of ensuring quality of nonclinical safety studies for medical substances. Key elements of FDA cGLP regulations are outlined in Box 4.2. Taken together, cGLPs ensure that testing is done in a sound scientific manner and with an established quality system. Notable are requirements for study protocols, accurate reports, internal quality audits and acceptance of results by both scientists and a quality assurance professional.

cGCPs: Current Good Clinical Practices for Clinical Studies

cGCPs represent a quality system that ensures the highest quality science and ethical treatment of human subjects for clinical studies of all types and at all phases of development. With cGCPs, the burden for quality is shared

BOX 4.2 ELEMENTS OF CURRENT GOOD LABORATORY PRACTICES (CGLPs)

21 CFR 58 Current Good Laboratory Practices for Nonclinical Laboratory Studies

A. General Provisions
   Definitions, applicability and inspections
B. Organization and Personnel
   Personnel, management, study director, quality assurance unit
C. Facilities
   Animal care and supply; handling test and control articles, laboratory areas, specimen and data storage
D. Equipment
   Equipment design, maintenance and calibration
E. Testing Facilities Operations
   Standard operating procedures, reagents and solutions, animal care
F. Test and Control Articles
   Test and control article characterization, handling and mixtures
G. Protocol for and Conduct of a Nonclinical Laboratory Study
H. Records and Reports
   Reporting study results, storage of records and data, retention of records
I. Disqualification of Testing Facilities
   Grounds for disqualification, notices, final orders, actions, public disclosure and suspension

21 CFR 11 Electronic Records; Electronic Signatures

Controls or closed and open systems, signature manifestations and record-linking, electronic signature components and controls, identification and passwords

between the principal parties conducting a clinical trial: sponsor, investigator and, if one is used, contract research organization (CRO). Unlike cGLPs and cGMPs, which are to be found in one or a few sections of 21 CFR, cGCPs are codified in a number of chapters and sections of the regulations. This is due in large part to the broad scope of clinical trials overall, the fact that they involve FDA functions, and generally recognized and codified rules for the conduct of research that involves human subjects no matter what the reason for their enrollment. The key components of cGCPs are outlined in Box 4.3. In the biopharmaceutical industry, an important foundation of cGCP is that the regulation both protects users of biopharmaceutical products and

BOX 4.3 REGULATIONS FOR CURRENT GOOD CLINICAL PRACTICE (GCP) AND CLINICAL TRIALS

21 CFR 11 Electronic Records; Electronic Signatures

Controls or closed and open systems, signature manifestations and record-linking, electronic signature components and controls, identification and passwords

21 CFR 50 Protection of Human Subjects. Informed Consent

General requirements, elements and exception for informed consent, additional safeguards for children

21 CFR 54 Financial Disclosure by Clinical Investigators

21 CFR 56 Institutional Review Boards

Organization, personnel, functions, operations records and reports, administrative action for noncompliance

21 CFR 312 Investigational New Drug Application and Foreign Clinical Trials

Responsibilities of sponsors and Investigators: responsibilities of sponsors, transfer of obligations to a contract research organization, selection of investigators and monitors, informing investigators, review of investigations, recordkeeping and retention, inspection of records and reports, disposition of investigational drug, assurance of IRB review, disqualification of clinical investigator

Drugs intended to treat life-threatening and severely debilitating illnesses, emergency use

Foreign clinical studies not conducted under an IND, public disclosure of data and information

21 CFR 314 Applications for FDA Approval to Market a New Drug

21 CFR 320 Bioavailability and Bioequivalence Requirements

also safeguards the well-being of human subjects, those individuals taking personal risk by volunteering to test new products. Indeed, the protection of human subjects is paramount in cGCPs, as it should be.

Compliance for Biopharmaceuticals: Other Regulations of Importance

Compliance for Import of Biopharmaceuticals into the United States

Importation of biopharmaceuticals is regulated by a number of agencies in every country of the world. For the United States, the Center for Biologics Evaluation and Review (CBER) or the Center for Drug Evaluation and Review (CDER) oversee importation and exportation of biologics or drugs, respectively, to ensure they comply with all U.S. laws and regulations. The FDA works closely with Customs and Border Protection (CBP). Inbound shipments in violation are detained by CBP on behalf of the FDA or the United States Department of Agriculture (USDA). The FDA must be advised if a final biological or drug product is manufactured overseas. A foreign manufacturer must have a U.S. FDA license to manufacture and distribute that product. This means the product must have an approved marketing application and, before this is granted, the foreign manufacturer must usually pass FDA inspection. A product approved and manufactured in the United States may, however, be exported from this country to another country without additional FDA authorization to export. In such cases, the FDA provides, on behalf of a biopharmaceutical sponsor and to a foreign regulatory agency, a Certificate to Foreign Government to substantiate marketing approval in the United States. Investigational biopharmaceuticals are another matter. The manufacturer need not have a U.S. FDA license, but it must declare a valid and active Investigational New Drug (IND) by number and name. CBP screens such shipments carefully, notifying the FDA and USDA if paperwork is in any way out of order or incomplete.

Compliance for Medical Devices

There are aspects of medical device compliance that differ from other FDA products, and the biotechnology firm developing a combination product with device components is well advised to understand these nuances. As noted earlier, registration and listing of U.S. establishments developing or manufacturing devices is critical to understanding medical device compliance. Also, devices are classified according to level of risk to the user. Quality systems regulations (QSR) and guidelines demand a strict quality system for development and production of medical devices. While some aspects of device QSR, identified in Chapter 5, now apply to drugs and biologics, they are quite detailed for devices. Certain reporting requirements also are unique to medical devices, as identified under the FDA's Medical Device Reporting (MDR) regulations. Additionally, there are stringent rules on tracking of certain medical devices. Medical device import and export compliance has many similarities to drugs and biologicals, but some processes do differ and are important to firms engaged in international transport, manufacture or marketing of medical devices or combination products.

Inspection and Enforcement

We are certainly all aware that, in any society, it is necessary to enforce laws and regulations. Yet, skirting or blatantly disobeying regulations just seems to come naturally to certain individuals and so societies have established means of ensuring, or trying to ensure, compliance by everyone. These are (1) enforcements, a means of imposing on individuals the observance of law; and (2) inspection, the official and careful examination of an item or an activity. Biological, drug and medical device activities have, in the past, been found to be deficient and, in some cases, there has been proved a serious intent to produce adulterated product, to falsify nonclinical or clinical study data or to avoid providing human subjects with their legal rights. Such behavior does, unfortunately, exist. In an effort to ensure that all biopharmaceutical products are both safe and effective and to increase public confidence in the biotechnology industry, the FDA inspects virtually all aspects of regulated development and enforces regulations intending to keep biopharmaceutical products safe and effective.

Inspections

Inspections provide one means of ensuring compliance and most countries have enacted laws to allow regulatory inspectors to review facilities, records and operations that produce or distribute investigational or approved products. U.S. FDA inspections are typically conducted for the following reasons:

- Periodic review of an operation to ensure continuing compliance
- Supplier of products to the government
- Directed review due to issues related to a product
- Revisit, following finding of deficiencies on an earlier inspection
- Following a recall
- Preapproval visit based on a market application or amendment or, more rarely (in the case of a new technology), an IND application

The Food, Drug and Cosmetic Act of 1938 gives the FDA broad authority in what may be inspected as long as the items—facilities, records, even vehicles—bear on whether a product (e.g., a biopharmaceutical or active ingredient) or service (e.g., a nonclinical or clinical study) is in compliance with the Act. Personal, financial or business information is not a target of inspections and technical information is kept confidential for inspectional reports. Individuals are not required to sign affidavits, but information they disclose may be used in a case against the firm.

Let us examine the FDA inspection process as it might happen at a biopharmaceutical firm. An inspection typically begins in the morning of a weekday as FDA inspectors present their credentials and state the reason for their visit. If a firm refuses entry, then the FDA will seek an administrative inspection warrant or, if serious breaches of the law are suspected, a criminal search warrant. The inspection itself involves a review of the plant, facilities and records. FDA inspectors are highly trained, and inspection teams, varying in size from one to a dozen FDA employees, include individuals with various expertise. For example, the team sent to a biomanufacturing plant might include individuals experienced in record review, others with expertise in technologies used at the plant, and specialists with a deep understanding of general manufacturing processes and regulations. Inspections may be brief, lasting less than a day and conducted by one individual, or they may take weeks and involve teams of inspectors, visiting continuously or sporadically. Inspectors carefully research the history of a product and the facility before they visit, and they are guided by FDA's *Inspection Operations Manual*. Further, inspectors now use a systems approach when visiting an operation. There are compliance "trends" that lead to investigational emphasis, and these issues should be evident to the biopharmaceutical community through meetings and from press releases by the FDA. Today, for example, this is corrective and preventive action (CAPA), sources of Active Pharmaceutical Ingredients (API), and production and process or facilities and equipment controls, but in the future it may be other topics that the FDA and the public believe require immediate attention to ensure a supply of safe and efficacious products. A thorough investigation begins at the top, looking at management responsibility and involvement, moving to design control and always touching on the hot topics. Upon completion, the inspector conducts an exit interview with management and provides a list of observations.

Documents may be prepared by the FDA as the result of an inspection.

- FDA Form 483, Inspectional Results, lists observations made by the inspectors. It is issued to the firm before the inspector leaves on the final day.
- Upon returning to their FDA offices, inspectors file details of their findings and present evidence or exhibits of deficiencies, uncovered in the Establishment Inspection Report (EIR).

Inspections result in one of three courses of action, as recommended by the FDA. For the biopharmaceutical firm, the preferred outcome is NAI (or no action indicated), a clean bill of compliance health, if you will. Another possible outcome is VAI (or voluntary action indicated) and the third is OAI (or official action indicated). A prudent biotechnology firm will take any inspectional findings of VAI or OAI very seriously. The report on their firm becomes a matter of public record and investors, competitors, or customers may file a Freedom of Information Act (FOI) request to obtain FDA Form 483 or the EIR, redacted of confidential information. Management and quality staff of the firm are involved in all reviews of and responses to inspectional findings. In the ideal situation, a team of supervisors carefully examines the inspectional findings and compares them to regulations cited and to company records or procedures identified and described. Indeed, a systems approach is applied and the firm typically generates a voluntary plan to correct each deficiency cited by inspectors; this plan is submitted by the sponsor to the FDA for review. Negotiations between the agency and management of the firm may follow and there is usually a final resolution and agreement, satisfactory to regulatory authorities. Such is the outcome for most VAI situations. Enforcement action is, however, indicated for OAI determinations. Time and again it has been shown that the FDA has the upper hand in these matters and rarely does a firm avoid the need to admit to and correct OAI deficiencies found on an FDA inspection.

Enforcement Actions

The Food, Drug and Cosmetic Act of 1938 went beyond inspectional authority and action and delegated certain authority for enforcement to the FDA. This gave the agency authority for seizure, injunction, civil penalties, criminal prosecution or import and export restrictions for certain products. FDA enforcement actions may, however, only be applied to certain acts, with the most common being production or delivery of adulterated or misbranded product into interstate commerce or of adulterating or misbranding the product once it is in commerce, refusing to permit an inspection, failing to register a manufacturing facility and adulterating or removing labeling. Adulteration and misbranding require further definition as these actions apply to biopharmaceuticals, but not every nuance can be listed here. Adjectives used in specific definitions of adulterated include: putrid, filthy or decomposed, lacking indicated strength, quality or purity, out of cGMP compliance or having a deficient container. Phrases used to define misbranded labeling include: false or misleading, failure to list essential elements such as name of drug or manufacturer and directions for use and directions that result in a dangerous situation when followed. The point is clear and the public agrees that products meeting definitions of adulterated or misbranded should be pulled from the market. Consider now how this discussion on FDA inspection and enforcement directly relates to discussions regarding regulatory operations, quality systems, manufacturing and control and nonclinical or clinical studies.

The FDA may only bring to bear enforcement actions if the product is introduced into interstate commerce. Courts have interpreted the terms adulterated, misbranded and interstate commerce quite broadly and it is virtually impossible for a firm or even a university or institution to avoid compliance with FDA regulations. So, what might the FDA do if the sponsor fails to correct deficiencies uncovered by the FDA? There are many possibilities, but those most commonly used are an enforcement letter to the sponsor, forced recall of the product and judicial enforcement. Debarment is an option when individuals are responsible. It is not unusual for the FDA to take two or more of these actions before resolving a case.

Debarment is imposed when action is sought against individuals. For example, an officer of a biotechnology firm may be debarred from working in the industry for a period of time. The investigator of a clinical study or the director of a nonclinical may be prohibited from conducting further studies after proven egregious behavior.

Letters. For OAIs or when VAIs are not resolved to the satisfaction of the FDA, the sponsor is sent a strongly worded "warning letter" in which the FDA states the case against the biopharmaceutical firm. The letter is addressed to an individual, usually an executive, at the firm. The agency then posts this letter at www.fda.gov for all the world to see. Letters also list additional enforcement action, including possible criminal action that could be taken against the firm or high-level individuals at that firm unless the matter is resolved to the satisfaction of the FDA. Not surprisingly, many issues are resolved to the FDA's satisfaction shortly after a warning letter is issued, the ultimate step in ensuring compliance through administrative means.

The FDA always has at its disposal judicial actions. The FDA, like many other regulatory agencies, is not alone authorized to bring enforcement action, but must use judicial tools in conjunction with the Department of Justice. Judicial enforcement is reserved for situations that cannot be resolved by other means or those in which public health or safety of individuals is at risk. The FDA may apply the following judicial actions:

Seizures or Recalls. The FDA may send federal marshals to a plant with instructions to seize all product and the FDA may then order the company to announce and to pay for a complete recall of all product: sold, on the shelf or in distribution.

Injunctions. Injunctions—temporary, preliminary or permanent—are legal tools used to keep a party from doing something or to proactively make them do something. Unlike an administrative action, injunctions carry the force of criminal penalties. Consent decrees of permanent injunctions may result and they can remain in effect forever or they can expire on a particular date.

Criminal Prosecution. It really happens. Firms and individuals will go so far as to face criminal prosecution over a disagreement with the FDA. Others flee the country before they can be prosecuted. The Department of Justice is always involved and there is coordination with the FDA's Office of Criminal Investigations. The capstone to this process is that criminal prosecution involves strict liability, which means that a corporate officer need not commit the act or even know that a specific act was committed. Prosecution can rest on failure of a responsible individual to "seek out and remedy" when situations occur or have occurred. In most cases of criminal prosecution, the FDA targets officers of a firm and may or may not involve technical operators or supervisors. Food for thought.

Product Liability

The biopharmaceutical firm also must be concerned about another legal issue involved in possible adulteration or misbranding issues. Product liability, or other civil actions related to poorly designed products, incorrect manufacture or control, or inadequate or misleading clinical or nonclinical studies that result in harm to a private party, such as the user of a biopharmaceutical, can result in civil actions. Civil suits are commonly pursued in the United States and it does not require FDA action for a biotechnology firm or for the officers of that firm to end up in a court of law accused of selling bad product or of putting a human subject at risk.

Compliance with Non-FDA Regulations: International, National, State and Local

When we think of compliance in biotechnology operations, most people have a mental image of meeting regulations of the Food and Drug Administration. While U.S. FDA compliance is important to most biotechnology firms, some will never need to consider 21 CFR. Yet every company will face non-FDA compliance issues. These issues can arrive with little warning and they can have a tremendous and, unfortunately, negative impact on operations. For example, virtually every biotechnology firm ships biologicals and chemicals across state lines and international borders. Shipping such materials is highly regulated by several agencies at the national level and, perhaps, also at the state level as well. Another example is disposal of waste generated in laboratories and during nonclinical and clinical studies.

This chapter provides an overview of non-FDA regulatory compliance situations that are frequently encountered by biotechnology firms in the United States. The subject matter has been organized under headings related to a particular activity, but the reader will find that a single activity may be regulated by two or more agencies at the local, state and federal levels.

International and Foreign National Regulatory Authorities for Medical Biotechnology Products

National interests and international political differences can be major hurdles to multinational regulatory approval of biotechnology products. Attempts are underway by both regulatory agencies and biopharmaceutical firms to eliminate these differences through transnational harmonization and, for much of the world, by strengthening national regulatory authorities (NRA) in some countries. While these efforts might not bring every national agency into agreement, they are making a difference in many international biotechnology markets, notably for countries consuming the greatest amounts of biopharmaceuticals. Despite the lack of movement by some nations, NRAs are generally moving in the direction of harmonization. Today it is prudent to assume the fastest route to multinational approval for biotechnology products in a multinational marketplace is though the early application of harmonized documents.

Perhaps the organization most involved in harmonization is the International Federation of Pharmaceutical Manufacturers and Associations (IFPMA) and the tool used by them and NRAs to harmonize international regulations is the International Committee for Harmonization (ICH). Notably, the ICH promulgates harmonized guidelines on a variety of subjects critical to biopharmaceutical development (http://www.ifpma.org and www.ich.org). ICH topics are divided into four major categories and ICH topic codes are assigned according to these categories: (1) Q, or quality topics, are those relating to chemical and pharmaceutical quality assurance (e.g., quality control test validation and stability testing); (2) S, or safety topics, are those relating to *in vitro* and *in vivo* preclinical studies (e.g., carcinogenicity testing); (3) E, efficacy topics, are those relating to clinical studies in human subject (e.g., dose response studies, good clinical practices); and (4) M, multidisciplinary topics, are cross-cutting topics, which do not fit uniquely into one of the above categories (e.g., medical terminology, or MedDRA, and the common technical document, or CTD). A partial list of ICH documents relevant to biotechnology product development is in Box 4.4. Most ICH guidelines are accepted by the FDA and so they are a particularly helpful guidance for ensuring safe and effective biopharmaceuticals enter the U.S. marketplace. The ICH guidelines recognized by the FDA and hundreds of guidance documents for virtually every aspect of biotechnology product development are available at the FDA Web site http://www.fda.gov/cder/guidance/index.htm.

The FDA also provides dozens of guidelines specific for the United States and these are listed in Box 4.5.

The World Health Organization (WHO) also certifies producers of products in international commerce and provides international standards for nonclinical and clinical testing of drugs and biologicals (http://www.who.int). As a general rule, WHO standards are neither as detailed nor stringent as the national standards of developed countries, but WHO certification does, in the case of a manufacturing facility, provide a level of assurance that product from an operation is not adulterated and will be accepted in other countries. The WHO document, "Good Practices in the Manufacture and Quality Control of Drugs," embodies GMP as recommended by WHO

BOX 4.4 EXAMPLES OF INTERNATIONAL CONFERENCE ON HARMONIZATION (ICH) GUIDELINES USED IN BIOTECHNOLOGY DEVELOPMENT

Quality

Q1A: Stability Testing of New Drug Substances and Products
Q2: Validation of Analytical Procedures
Q3: Impurities in New Drug Substances
Q5A: Viral Safety of Biotechnological Products
Q5B: Quality of Biotechnological Products: Analysis of Expression Construct, Cells for Production of r-Proteins
Q5C: Quality of Biotechnological Products: Stability Testing
Q6B: Specifications: Test Procedures and Acceptance Criteria for Biotechnological Products
Q7: Good Manufacturing Practice Guide for Active Pharmaceutical Ingredients
Q8: Pharmaceutical Development
Q9: Quality Risk Management
Q10: Pharmaceutical Quality System

Safety

S1B: Testing for Carcinogenicity of Pharmaceuticals
S2: Guidance on Genotoxicity Testing
S4: Single Dose Toxicity Tests
S6: Preclinical Safety Evaluation of Biotechnology-Derived Pharmaceuticals
S9: Nonclinical Evaluation for Anticancer Pharmaceuticals

Efficacy

E2: Clinical Safety Data Management
E3: Structure and Content of Clinical Study Reports
E6: Good Clinical Practice
E10: Statistical Principles for Clinical Trials

Multidisciplinary

M1: Medical Terminology
M2: Electronic Standards for the Transfer of Regulatory Information
M4: The Common Technical Document

and is found at http://www.who.int/medicines/areas/quality_safety/quality_assurance/en/index.html. WHO also provides guidelines for clinical trials, including an international registry of clinical research and templates for study documents. WHO provides guidelines in other aspects of drug and biological development, including those for nonclinical safety testing, formulation, distribution and purchase.

BOX 4.5 EXAMPLES OF US FDA GUIDELINES FOR DEVELOPMENT OF BIOPHARMACEUTICAL PRODUCTS

- Quality Systems Approach to Pharmaceutical CGMP Regulations. FDA. September 2006.
- Points to Consider in the Manufacture and Testing of Monoclonal Antibody Products for Human Use. February 1997.
- Process Validation. General Principles and Practices. Guidance to Industry. November 2008.
- CGMP for Phase 1 Investigational Drugs. July 2008.
- INDs for Phase 3 and Phase 3 Studies—Chemistry, Manufacturing and Controls Information. FDA. May 2003.
- Assay Development for Immunogenicity Testing of Therapeutic Proteins. December 2009.
- Q10. Pharmaceutical Quality System. April 2009.*
- Medical Device Quality Systems Manual. December 1996.
- Protocol Development Guideline for Clinical Effectiveness and Target Safety Trials. July 2001.
- Structure and Content of Clinical Study Reports. July 1996.*
- Labeling for Human Prescription Drug and Biological Products—Implementing the New Content and Format Requirements. January 2006.

* FDA guidelines based upon ICH guidelines.

European and Japanese regulations continue to have, in much of the world, a direct impact on development of biopharmaceuticals. While each member of the European Economic Community (EEC) has a national regulatory authority and national regulations, harmonization within the EEC is being led by the European Agency for the Evaluation of Medical Products (EMEA) with the Committee for Proprietary Medical Products (CPMP), with a Web site at http://www.emea.europa.eu . The Japanese Ministry of Health, Labor and Welfare, Web site at http://www.mhlw.go.jp/english/, provides regulations and guidelines for biopharmaceutical products marketed in Japan and this regulatory guidance also influences regulatory agencies of countries in East Asia and the Pacific region. Australia and Canada have strong regulatory infrastructure for biomedical products as well. In addition, many trade organizations, such as the Biotechnology Industry Organization (BIO) and the Pharmaceutical Research and Manufacturers of America (PhRMA), actively pursue harmonization because it is good business and good regulatory practice for member firms.

Finally, marketing approval is important if biotechnology products are to be used in these developing countries, many burdened with much of the world's infectious diseases. This is especially important for drugs and biologicals developed to treat diseases, such as AIDS and malaria. As noted earlier, the FDA and the drug regulatory agencies of other developed countries are influential in the development and regulation of products for global health. Also, the WHO is actively involved in searching for means to gain regulatory approval in countries that do not currently have national regulatory authorities or lack a science-based review system for biopharmaceuticals.

Transporting Infectious or Otherwise Hazardous Materials

Successful transportation, national and international, of materials and products is important to any biotechnology firm. The transportation community, such as national and international shippers, and the transportation regulatory agencies like the U.S. Department of Transportation (DOT) are cautious about the materials they allow or accept for shipment (www.dot.gov). Proper shipping procedures for hazardous and infectious materials are enforced because they protect employees of the shipping firms, ensure public health and allow all compliant firms to transport a variety of materials, some considered hazardous. While most biologicals and chemicals can be shipped, many require special precautions in packaging, labeling and handling.

The shipper bears virtually all responsibility for ensuring safe shipping of infectious or otherwise hazardous materials. Therefore, the shipper must be aware of the various and sometimes complex regulations and then properly classify, identify, package, mark, label and document the substance being shipped. References to common shipping regulations or guidelines for the United States are listed in Box 4.6. The International Air Transportation Association (IATA) is a clearinghouse for international air transport and foreign, national regulations (www.iata.org). By regulation, the shipper of dangerous goods must be a trained person and he/she must comply with regulations and also certify that materials will arrive at their destination in good condition and not present any hazards to humans and animals during shipment. Commercial carriers refuse to accept any package that fails to comply with international, national and the shipper's regulations or guidelines. Failure to comply with shipping regulations often means that a material does not reach its destination and, for spills or human exposure to infectious, chemical or radiological substances, can result in substantial fines for noncompliance.

Biological materials are especially important to the biotechnology industry. They are considered as: (1) infectious (etiologic) agents, (2) diagnostic (clinical) specimens or (3) biological products. Infectious substances are those known or reasonably expected to contain pathogens. Pathogens are microorganisms (including bacteria, viruses, rickettsia, parasites and fungi) or recombinant microorganisms (hybrid or mutant) that cause infectious disease in humans or animals. This includes (1) all cultures containing or suspected of containing an agent, which may cause infection; (2) human or

BOX 4.6 SOURCES OF INFORMATION ON TRANSPORTATION OF BIOTECHNOLOGY MATERIALS OR PRODUCTS

Transportation within the United States

- *Hazardous Materials Regulations,* 49 CFR Parts 171-178, U.S. Department of Transportation
- *Interstate Shipment of Etiologic Agents,* 42 CFR Part 72, U.S. Public Health Service, Centers for Disease Control and Prevention
- *Occupational Exposure to Blood-Borne Pathogens,* 29 CFR Part 1910.1030, The Department of Labor, Occupational Safety and Health Administration

International Transportation

- Recommendations of Infectious Substances and Diagnostic Specimens, United Nations. http://www.who.int/csr/emc97_3m.pdf
- Technical Instructions for the Safe Transport of Dangerous Goods by Air, International Civil Aviation Organization (ICAO)

animal samples that contain such an agent in quantities sufficient to cause infection should an exposure to them occur due to a transport mishap; (3) sample from a patient with a serious disease of unknown cause; and (4) other specimens not included above, but designated as infectious by a qualified person, e.g., a physician.

Diagnostic specimens are any human or animal material including, but not limited to, excreta, secreta, blood and its components, tissue or tissue fluid, being transported for diagnostic or investigational purposes, but excluding live infected animals. Diagnostic specimens resulting from medical practice and research are not considered a threat to public health. An example is a serum sample not suspected of containing an infectious agent that is shipped to a laboratory for routine testing.

Biological products may have special licensing requirements. These specimens are further defined as those products derived from living organisms that are manufactured and distributed in accordance with the requirements of national governmental authorities. They are used either for prevention, treatment or diagnosis of disease in humans or animals, or for development, experimental or investigational purposes and include, but are not limited to, finished or unfinished products, such as vaccines and diagnostic products. This general definition would include many biotechnology products.

Hence, diligence and thorough research of any and all guidelines is required by any biologics manufacturer prior to shipment in an effort to ensure that special regulations do not apply to any product. As noted earlier, import of infectious agents is highly controlled. For infectious materials and vectors imported from foreign countries, there are requirements for importation permits and shipping labels issued by the U.S. Public Health Service (42 CFR Part 71.54).

Biotechnology firms also ship a variety of chemical substances and dangerous goods, defined as a substance capable of posing an unreasonable risk to health, safety or property when transported by commercial carrier or by air, and identified as explosive, corrosive, flammable liquid, oxidizer or compressed gas. Indeed, one of the most often overlooked chemical dangerous substances transported by biotechnology firms is solid carbon dioxide, or dry ice. Several steps are taken before a chemical substance is transported. First, the shipper must determine if the good can be shipped by commercial carrier. If so, then written guidelines are followed for packaging, marking and declaring the good. A Dangerous Goods Declaration is an essential part of every shipment and it appears outside on the carton, along with clear labeling as to contents, risk and response in case of spill or carton damage.

Importing, Possessing or Transferring Controlled Biotechnology Materials

The Animal and Plant Health Inspection Service (APHIS), U. S. Department of Agriculture (USDA), regulates the importation, possession and transfer of many controlled plant and animal materials (http://www.aphis.usda.gov). Importation is controlled for virtually all animal- and plant-origin materials and animal products and byproducts as well as biological materials that contain or have been in contact with materials of animal origin. The USDA also controls selected microbial agents that could pose a risk to animals or plants in this country. This includes *in vitro* materials, such as cell lines. Requests for permits authorizing the importation of such controlled materials must be submitted to APHIS and importation cannot commence until the application is approved.

APHIS regulates genetically engineered plants by administering the Federal Plant Protection Act. This legislation authorizes APHIS to control interstate movement, imports to the United States, and release (for field testing) of "organisms and products altered or produced through genetic engineering, which are plant pests or for which there is reason to believe are plant pests." A plant pest is a risk to other plants and ecosystems. The term is generally applied to weeds, insects, diseases or untested genetically modified organisms (GMOs). Applying the term *plant pest* to a genetically engineered plant means only that the nonpest nature of the plant has yet to be demonstrated. APHIS requires a permit and concurrence of individual state Departments of Agriculture for movements across state lines. For field testing of a new plant, referred to as environmental release, a permit also may be required from APHIS. For selected plants, one of two other processes, the notification process or the petition process, may be used in place of a permit.

Today, many firms are developing genetically engineered plants that produce drugs or biological compounds intended for medical or veterinary treatments. The FDA has responsibility for regulating the active ingredients produced by these plants. APHIS ensures engineered plants do not pose a significant plant pest risk, a risk to threatened and endangered species, or a risk to people working with them. To reduce the risk of harm to other organisms and to prevent such plants from entering the food supply, an APHIS permit is required to take such plants to the field.

Under the regulation 7 CFR 340, the Biotechnology Regulatory Services (BRS) division of APHIS is responsible for importation, interstate movement and field release of genetically engineered plants. The BRS Web site, http://www.aphis.usda.gov/biotechnology/about.shtml, is the most complete and up-to-date source of information for those engaged in agricultural biotechnology.

The National Center for Import and Export (NCIE), another center under APHIS, is responsible for protecting animals important to agriculture (http://www.aphis.usda.gov/import_export/index.shtml). This agency facilitates international trade, monitors health of animals before they enter the United States and regulates the import and export of animals, animal products and biologics. It is in the import and export area that NCIE has the greatest interaction with the biotechnology industry. Generally, a USDA veterinary permit is needed for import of nonhuman materials derived from animals or exposed to animal-source materials. A wide range of materials, for example, animal tissues, RNA/DNA extracts, hormones, antisera and monoclonal antibodies for *in vivo* use, are regulated.

The FDA also regulates the care and use of laboratory animals (www.usda.gov). Any biotechnology firm that does business with a research animal breeder or vendor or itself houses or uses animals is familiar with this extensive set of regulations.

The U.S. Fish and Wildlife Service, part of the U.S. Department of Interior, enforces possession or transfer of certain species (e.g., endangered primates) or any part of those species (e.g., feathers, blood or tissue) under the Endangered Species Program for the Endangered Species Act (http://endangered.fws.gov). Permits are required to transport or hold specimens and the law is enforced at U.S. borders by Customs and Border Protection, part of the Department of Homeland Security.

The Bureau of Industry and Security (BIS), of the U.S. Department of Commerce, formerly the Bureau of Export Administration, oversees U.S. exports of dual-use commodities, technology and software (http://www.bis.doc.gov). The bureau has the lead role in both the export licensing process and in enforcement operations. Its mission, based on national security, is to control exports of sensitive products to entities that could misuse U.S. technologies and products. BIS licenses exporters of certain products, including biotechnology-related products, such as fermentation equipment. In doing so, it requires exporters to notify other parties of the sale and the conditions of sale and to obtain written acknowledgment from the end user of the intended use. Licensing conditions are sometimes necessary to make certain that approved items are in the correct location and being used in an appropriate manner. This bureau does not regulate all goods, services and technologies, but it does control the export of certain microorganisms, toxins and equipment used to make these items. The items are provided in the Commerce Controlled List. To further complicate matters, other U.S. government agencies regulate more specialized exports. For example, the U.S. State Department has authority over defense articles and defense services. A list of other agencies involved in export controls can be found on the Web site for BIS.

Public Health Security and Bioterrorism Preparedness and Response Act of 2002

This law has certainly complicated movement or use of many biotechnology products within the United States and across its borders, but also it has made the public feel more secure. It includes sections on "Enhancing Controls on Dangerous Biological Agents and Toxins" and providing for regulation of these biologicals by the Department of Health and Human Services and the Department of Agriculture. It recommends interagency coordination between the two departments regarding overlap agents and toxins and provides for criminal penalties regarding certain biological agents and toxins. For the Department of Health and Human Services, the Centers for Disease Control and Prevention (CDC) has primary responsibility for implementing the provisions of this Act (http://www.cdc.gov). APHIS is the agency fulfilling the roles assigned to USDA.

The USDA regulations are within 7 CFR and 9 CFR, while the CDC regulations are 42 CFR 73. In general, the regulations are aimed at animal and plant agricultural and human health threats, respectively, but there is some overlap. The regulations establish and enforce safety procedures for listed agents and toxins, including:

- Measures to ensure proper training and appropriate skills to handle agents and toxins, and proper laboratory facilities to contain and dispose of agents and toxins.
- Safeguards and security measures to prevent access to listed agents and toxins for use in domestic or international terrorism or for any other criminal purpose.

- Procedures to protect animal and plant health, and animal and plant products, in the event of a transfer or potential transfer of a listed agent or toxin in violation of the safety procedures as well as safeguards and security measures.
- Appropriate availability of biological agents and toxins for research, education and other legitimate purposes.

The regulations themselves cover requirements for registration, security safety and emergency response plans, training, transfer, record keeping, inspections and notifications. They regulate molecular parts of the organisms because particular genes or proteins from these organisms also might constitute a risk to public health. Even a small amount of nucleic acid from a select agent may be regulated. A permit system allows a research investigator or biopharmaceutical product developer to import, keep, transfer or test (e.g., field test a genetically engineered plant) an agent. Because the penalties for improper or illegal possession, use or transfer of the agents are severe, those biotechnology firms using even seemingly safe and innocuous agents or molecules should become familiar with the select agent list and the regulations well before they consider transferring the material to their laboratory.

Importation or Exportation of Biotechnology Products for the Purpose of Treatment of Diseases in Man

Earlier in this chapter, FDA regulation for import or export of any "virus, therapeutic serum, toxin, antitoxin or analogous product" for the "prevention, treatment or cure of diseases or injuries of man" was discussed. It is important to note the important roles that customs and other federal agencies play in the importation and exportation of these products because the scope of regulations encompasses many biotechnology products. Labeling requirements are absolute because they inform customs and other government officials. If a product is intended for human use, then it must be labeled and may be inspected and samples taken by Customs and Border Protection (http://www.cbp.gov). The FDA may be contacted and may even inspect the shipment. If there is no evidence that the product is licensed by the FDA, then it is held by CBP. If the product is of animal origin (e.g., a horse antiserum against snake venom), it may require a USDA permit as well. If it is from an endangered species, it also will need a permit from Fish and Wildlife Service under the Endangered Species Program. In effect, CBP serves as a gatekeeper at the U.S. borders, acting on behalf of several federal agencies. In summary, importation and exportation of all biotechnology materials and products must be carefully researched by the shipper with the expectation that the regulations of multiple federal agencies could complicate, delay or stop the movement of these goods.

Occupational Health and Safety

A biotechnology firm must have effective health and safety policies and practices for one simple reason: it protects employees, their most valuable asset. We all know that work can affect our health and, when queried, people state that good health is a leading factor in quality of life. If a workplace is safe, people enjoy their jobs and are more interested and involved in their employment.

The biotechnology laboratory work environment includes hazards, but it need not be unsafe. We work with harsh chemicals, acids, corrosives, radiochemicals and biological agents, such as viruses and toxins. In research laboratories, individuals are often in close contact with these materials and in biopharmaceutical manufacturing there may be large volumes of potentially hazardous materials in the workplace. The work environment also may have carcinogens, flammable gases or liquids, steam and hot fluids.

As with just about everything in biotechnology, the key to providing a safe and comfortable work environment is good planning. Every biotechnology firm should have a health and safety policy and plan and procedural documents, all receiving the full support of upper management and line supervisors. Good policies emphasize prevention rather than reaction to incidents or accidents. Standards for health and safety are based on risk assessments and regulatory requirements. Health and safety plans state objectives or goals and standards or specifications. The results, based on measurable outcomes, are compared at regular intervals against health and safety objectives.

Biotechnology firms should have a visible organization or structure to support a health and safety plan. An environmental health and safety specialist is the individual responsible for developing, implementing and monitoring industrial safety programs within the biotechnology company. While a smaller firm may not require a full-time health and environmental safety officer, a consultant, such as an occupational safety specialist, is an important member of the corporate team. These professionals inspect laboratories and product development, manufacturing and testing areas to ensure compliance with federal Occupational Safety and Health Administration (OSHA), state, and local regulations and corporate policies. They evaluate new equipment and raw materials for safety, and monitor employee exposure to chemicals and other toxic substances. A safety specialist also conducts training programs in hazardous waste collection, disposal and radiation safety.

Finally, management must make every effort to encourage a safe and healthy culture within the biotechnology firm. Communication is an important part of the process, with periodic seminars and, most importantly, an effective means for employees to express their health and safety concerns to management. A safety training program gives employees an opportunity to learn more about safety as it relates to their particular job assignments.

Local, state and federal agencies, notably OSHA of the Department of Labor, regulate health and safety in the workplace (www.osha.gov). As early as 1985, OSHA began to examine the health and safety issues related to biotechnology. OSHA originally felt that no additional regulations were needed for such workplaces since other standards, such as those for general laboratory safety, provided an adequate basis for protection and safety. These OSHA regulations and standards are at 29 CFR 17. In addition, blood-borne pathogen guidelines, which apply to all occupational exposure to blood or other potentially infectious materials, and exposure to other infectious organisms, are responsibilities of both OSHA (http://www.osha.gov/SLTC/biologicalagents/index.html) and the National Institute for Occupational Safety and Health (NIOSH), Centers for Disease Control (http://www.cdc.gov/niosh/homepage.html).

In summary, one of the most important aspects of safety, from the point of view of the employee in a biotechnology firm, is to have a clear understanding of hazards in his/her workplace. Simple acts, such as participation in training, wearing safety glasses, proper disposal of waste and review of material data safety information, are operational keys to health and safety in the biotechnology work environment. A health and safety plan and effective training can go a long way in reaching these objectives.

Environmental Regulations in Biotechnology

There are many environmental hazards in biotechnology operations and there also are numerous federal, state and local regulations as well as agencies to enforce them. Failure to heed environmental guidelines by a firm puts both the company and community at possible risk. These issues, or even perceived problems, are often highly publicized within a state or community and, when a violation becomes known, it creates a negative image of the biotechnology industry as a whole. To make matters worse, there already are concerns, worldwide, about the release of genetically modified molecules or organisms into the environment. There are also the less publicized, but still very real issues related to environmental release of materials from biotechnology laboratories or operations.

A relevant example is the receipt, handling and disposal of radioisotopes, functions regulated by the Nuclear Regulatory Commission (NRC), Department of Energy (http://www.nrc.gov/) and by state and, sometimes, local agencies. The NRC manages radioactive materials by controlling the production, shipment, use and disposal of these materials. It does so through licensing responsible entities, such as universities or biotechnology firms. The NRC also allows individual states to regulate certain activities through the Agreement State Program. A biotechnology firm wishing to purchase, receive, use or dispose of a radioisotope must apply for and receive a license and agree to keep careful records, train employees, follow detailed rules, accept unannounced inspections and pay fines for noncompliance.

Sponsor's of INDs, Biologics License Applications (BLAs) or New Drug Applications (NDAs) are required to file an environmental impact statement or seek categorical exclusion for each product under each of these regulatory filings with the FDA.

Biotechnology firms also face complex regulations, many local, dealing with the disposal of chemicals and biological substances. With a few exceptions, it is not difficult to dispose of small amounts of nonhazardous chemicals or biological materials. However, in biotechnology operations, notably manufacturing, larger amounts of biological and chemical materials may need to be released into the environment. Examples include disposal of a large mass of recombinant bacteria following fermentation or of large volumes of organic solvents following molecule purification. In most cases, these cannot simply be sent to the local landfill or flushed down the community sewer. Regulations for waste disposal often fall under the purview of the Environmental Protection Agency (EPA) (http://www.epa.gov), but one can expect to find complex and extensive state and local regulations as well. Indeed, local officials often understand all regulations—federal, state and local—that apply to their community.

Major regulatory applications, such as IND, NDA or BLA, must consider environmental issues related to the manufacture and use of each product. National environmental law and policy, notably the National Environmental Policy Act of 1969, drive this requirement. Each of these documents must contain either an environmental assessment or the sponsor must state and show that the actions described in the document are categorically excluded from an assessment. Environmental assessments can be large and complex documents and they are not required for most biopharmaceuticals, at least not in early development. However, with some products in the IND stage and with many products in large-scale manufacture, the sponsor must complete an assessment prior to submitting a regulatory application to the FDA. The agency provides guidelines for those who make a choice between claiming exclusion and an assessment and for preparing either document.

Genetically Modified Organisms or Molecules

Several federal agencies are responsible for regulating the release of genetically modified organisms or chemicals into the environment. The role of USDA has been mentioned. The EPA is also involved because of the Microbial Products of Biotechnology section of the Toxic Substances Control Act (TSCA). TSCA authorizes the EPA to, among other things, review new chemicals before they are introduced into commerce (http://www.epa.gov/opptintr/biotech). Intergeneric microorganisms, i.e., microorganisms created to contain genetic material from organisms in more than one taxonomic genus, are considered new chemicals under TSCA and, therefore, the EPA reviews and regulates the use of intergeneric microorganisms in commerce or for commercial research.

The Office of Biotechnology Activities (OBA), Office of the Director, National Institutes of Health (NIH) is involved in activities that affect many biotechnology firms (http://www4.od.nih.gov/oba). In the areas of gene research and biotechnology OBA (1) monitors programs in human genetics research and development; (2) manages the operations of several important review committees; (3) advises other government agencies or departments; (4) develops policies and procedures, and reviews established Institutional Biosafety Committees (IBC); (5) provides information to the public; and (6) develops registries of activities.

Although the OBA is not chartered by Congress as a regulatory agency, as is the FDA, it has significant influence in several areas of biotechnology and can have an immediate and significant impact on fields of biotechnology, notably emerging technologies, and on biotechnology firms themselves. The OBA is influential largely by providing guidelines to the public. A typical action is a decision as to whether an institution, such as a university, may receive NIH funding. For example, if the OBA decided that a type of experiment, e.g., human cloning, was not approved, then any university or biotechnology firm that did such experimentation could lose all NIH funding for any research purpose. Because many biomedical entities, such as institutes, universities and even some biotechnology firms in the Unite d States, receive NIH funding, the OBA guidelines have significant impact. Furthermore, unlike many FDA activities in which the information is held in confidence, OBA activities are largely public and both the press and interest groups often monitor and publicize issues and the biotechnology firms that are involved with the OBA.

The OBA monitors scientific progress in basic and clinical research involving recombinant DNA and human gene transfer. The OBA promulgates the *NIH Guidelines for Research Involving Recombinant DNA Molecules* (NIH guidelines), which is the premier guidance document for DNA research. It specifies all aspects of genetic engineering and production of genetically modified organisms, outlining responsible research practices in basic, animal and clinical DNA research. The OBA sponsors the Recombinant Advisory Committee (RAC), experts appointed to monitor scientific progress in basic and clinical research involving recombinant DNA and human gene transfer. The RAC recommends changes to the NIH guidelines and its members review gene therapy clinical protocols. In effect, the RAC and the director of NIH exert great influence on all recombinant activities in the United States.

The office also manages compliance with the requirements for IBCs. An IBC must be established by any entity that receives NIH funding and performs genetic engineering. Many biotechnology firms that do not receive NIH funding also use an IBC for the purpose of reviewing experiments involving genetic engineering or transfer of genetically modified organisms to the environment. The primary role of an IBC is to ensure that all recombinant DNA research conducted at, or sponsored by, that institution is conducted in compliance with the NIH guidelines, but the roles of IBCs have been expanded at many institutions to include other aspects of laboratory and research with genetically modified materials or organisms.

The OBA maintains a database on human gene transfer (e.g., gene therapy) clinical trials and this is open to the public at their Web site. They also make recommendations concerning informed consent of human subjects enrolled in human gene transfer studies.

The secretary of Health and Human Services (HHS) has advisory committees managed by OBA. For example, an Advisory Committee on Genetics, Health and Safety provides policy advice to HHS on the broad array of complex medical, ethical, legal and social issues raised by the development and use of genetic technologies. The committee is instrumental in providing advice in areas such as genetic testing. Also, there is a Secretary's Advisory Committee on xenotransplantation for policy advice, to HHS, on medical, ethical, legal and social issues raised by xenotransplantation. Another advisory committee focuses on genetics, health and society, studying the medical, ethical, legal and social implications of technology advances, notably in biotechnology.

Somewhat apart from its original charter, but important to biotechnology firms, is the National Science Advisory Board for Biosecurity (NSABB), hosted by OBA. NSABB advises U.S. federal agencies on security issues related to life sciences, notably on ways to minimize the possibility that knowledge and technologies emanating from vitally important biological research will be misused to threaten public health or national security.

Taken together, the advisory committees under direction of OBA are very influential to the biotechnology industry overall and particularly to cutting-edge technologies as they transition to product development. The prudent biotechnology operation, public or privately held, will carefully monitor the activities of each committee and be constantly and fully aware of guidelines produced or changed by the OBA.

U.S. Regulatory Agencies Unified Biotechnology Web Site

A Web site sponsored by the Department of Health and Human Services (HHS), EPA and USDA has been developed as a coordinated, risk-based system to inform biotechnology firms of the many federal regulations and, thus, ensure new biotechnology products are safe for the environment and human and animal health (http://usbiotechreg.nbii.gov/). To date, it has focused largely on agricultural issues. While this site can be quite helpful when one begins a regulatory search, it is also incomplete in many respects.

International Diligence in Biotechnology Operations

Biotechnology is an international endeavor. Most biotechnology firms expect to sell their products or services overseas as well as in the United States. Sometimes biotechnology research can be completed in a national environment, heeding only U.S. requirements, but this is the rare situation and most biotechnology firms or endeavors are or will become transnational or intercontinental businesses. A biotechnology firm must diligently plan to incorporate their operation and, of course, their services and products, into international environments and markets. Therefore, international awareness and compliance is especially important to the success of biotechnology firms.

International regulations and guidelines are far too numerous to cover in this chapter. However, a single example, genetic engineering of plants from which food is derived, is given to emphasize the value of understanding international, as well as country-specific, information prior to embarking on biotechnology product development. Two entities, the Food and Agricultural Organization (FAO) of the United Nations (http://fao.org) and the Biotechnology Organization (BIO) (http://BIO.org) are actively involved in efforts to harmonize international guidelines for genetically modified foods or organisms that provide genetically modified foods. The biotechnology firm proposing to export recombinant organisms, food produced by recombinant plants or animals, or equipment, supplies and raw materials or services for the production of such products, should consider the guidance provided by these organizations.

An example of an international guideline intended to harmonize the movement of genetically engineered foods is the Codex Alimentarius (http://www.codexalimentarius.net/), a collection of internationally adopted food standards presented in a uniform manner. Codex standards are meant to ensure that consumers receive products that meet internationally accepted and minimally acceptable quality levels, are safe, and do not present a health hazard in accordance with FAO guidelines. The Codex is written by an international committee and is approved by the WHO body, the FAO. The priority of the Codex commission is to protect the health of consumers and ensure fair practices in food trade. Therefore, one would assume that by following the Codex, a biotechnology firm could trade its product or service worldwide. Unfortunately, this is not the case, because national laws and regulations for genetically modified foods still differ considerably on almost every point. As with most international guidelines, the Codex guidelines are binding, in a national sense, only when fully ratified by all parties. However, most international guidelines dealing with biotechnology products and services are not yet, and some may never be, accepted by all nations.

Summary of Non-FDA Compliance

Biotechnology activities of all types are highly regulated. FDA regulations apply to all aspects of developing biopharmaceuticals. In any given country, several federal agencies may regulate a single activity or function, state and local regulations must be considered, and there can be considerable overlap in regulatory authority. Due diligence and careful planning regarding all regulatory compliance is essential to the success of any biotechnology firm. Congress continues to change the laws regarding various aspects of biotechnology operations and at the same time executive agencies change or add regulations, while courts interpret application of laws. Adding to the complexity of regulatory compliance, any given government agency may change its mission or focus or become overwhelmed with regulatory submissions. As biotechnology becomes more global, the U.S. biotechnology firm must consider international guidelines and the regulations of countries other than the United States. To be successful, the biotechnology operation must be aware and diligent of all U.S. and global compliance issues as well as FDA regulations and guidelines.

5

Quality Systems

Overview of Quality in Biotechnology

Quality impacts every aspect of a biotechnology operation. While this might seem like a bold statement, those involved in biotechnology would certainly agree it is true. The requirement for quality in biopharmaceutical development is backed by a host of regulations (see Chapter 3 and Chapter 4). As applied to biotechnology operations, a state of quality is necessary in all endeavors and quality increases the value of services and products. We refer to quality systems, quality by design, quality control and the roles of quality in compliance to name but a few quality terms. Indeed, the word *quality* has various meanings to different individuals and for each situation in which the word is used, so it is perhaps best defined in the context of each usage. This book tries to do just that. Yet, we need to begin this chapter on quality systems by establishing a basic definition for quality. The reader might take his/her choice from any one of five definitions given below.

1. *Quality. The degree of excellence of a thing; general excellence* (Oxford English Reference Dictionary, 1997).
2. *Quality is the totality of features and characteristics of a product or service that bear on its ability to satisfy stated or implied needs* (British Standards Institute, 1987, 1991).
3. … *specified requirements for a product can be stated in terms of an established design … and [where] confidence in product conformance can be attained by demonstration of …. capabilities in production.* (ISO, 9002). The definition, taken from the International Standardization Organization, or ISO, is a bit longer because it is more specific and focused on making products.
4. *High quality is freedom from defects* (source unknown).
5. *Continuous improvement and waste reduction* (Henry Ford).

Practical man that Henry Ford; right to the point of saving money while producing a quality automobile at low cost.

Even though 100 people might give 100 definitions for quality, a common theme in each definition would be that a product is *fit* or *fit for use*, that is it performs as intended and, hence, the user is satisfied. But, how do quality and these definitions affect the biotechnology industry? First, everyone in the business of biotechnology is either producing a product, such as a patent, a vial of therapeutic medicine or a recombinant crop plant, or they are producing a service, such as testing products in animals or humans, manufacturing active ingredients or testing those ingredients. Everyone engaged in biotechnology wants their product to be fit and their customers satisfied, it is as simple as that.

In this chapter, the reader is introduced to the concepts and practices of quality, and terminology associated with this field. A brief history of quality is used by way of introduction. This is followed by a review of the Hallmarks of Quality, i.e., criteria that are common to most quality systems used in biotechnology. To finish, we review modern ways of incorporating quality into biotechnology endeavors, identify how quality systems are generally applied in our industry and provide guidelines for establishing and managing a quality system in a biotechnology firm.

History: Evolution of Quality Concepts and Practices

Quality is rooted in ancient history, beginning when individuals made goods for themselves or, through barter, for their neighbors. Because buyer and seller usually lived in the same village, the act of providing a bad product led to a bad reputation for both product and producer, or seller. As it does today, a negative image resulted in loss of business for the producer and it also led to social and economic pressures to make only quality products. With the industrial revolution, manufacturers were at a distance to the buyer and, thus, there was little face-to-face contact between the parties. New forces ensured quality under this rapidly evolving system. Oral, then written, warrantees were developed. For example, manufacturers began to list themselves as the source of a product and, in some cases, such as food, they also provided consumers with the ingredients. If the product was of high quality, this served as a form of advertisement. However, poor quality products led to a negative image and even to conflicts between buyer and seller. This led to various quality initiatives and, over the past 100 years, a number of ideas for ensuring quality of products and services have been adopted to ease those conflicts. Some of the most familiar include:

- Standards for measuring instruments, e.g., kilogram weight standards.
- Marks or seals, e.g., "UL Seal of Approval" on electrical appliances.

- Supervisory responsibility for quality production on the assembly line, e.g., Henry Ford ordering his foremen to be responsible for quality in each of their assembly line areas.
- Regulatory quality requirements were codified, such as the Food and Drug Act, to mandate quality in drugs and medical device development.
- Worker responsibility for quality production on the assembly line, e.g., as touted by automobile manufacturers in the late 1980s.
- Statistical trend analysis, e.g., medical chart reviews by health insurance firms to ensure that patients received the best healthcare for the price.
- "Fresh Ideas," such as ISO 9001, Total Quality Management and Six-Sigma, as introduced over the past three decades.

So, when and how did the concepts and practices of quality come to be applied to biotechnology? Quality was applied to virtually all aspects of biotechnology operations very early, as soon as biopharmaceutical development began in the 1980s. For what reasons was quality applied to biotechnology? For business reasons—good quality results in an excellent product and that leads to healthy sales. For compliance reasons—various regulatory agencies said quality assurance was necessary for most biotechnology products, certainly for those with a potential impact on the health or safety of the user or on public health, in general.

Long before biotechnology endeavors moved from laboratory benches and into development, the pharmaceutical industry had, for these reasons, applied quality concepts to all aspects of drug development. The medical device field was not far behind, adopting slightly different, but no less stringent, quality systems. Because many biotechnology firms must compete within these industry segments and because they are often regulated by the likes of the Food and Drug Administration (FDA) or the Department of Agriculture, the evolving biotechnology industry had no choice but to be compliant with modern quality standards. Does this mean that most, if not all, biotechnology product development or services must consider quality? Yes, today quality is a must. How then can a biotechnology firm begin to develop a product or service without first instituting a quality system and how could they continue to develop products or services with a weak quality system? They cannot. Pretty blunt language, but it is a fact, you cannot provide services or develop products in biotechnology unless you follow quality systems that are accepted by the consumer and regulatory agencies for that product or service. Most biopharmaceutical firms must follow several quality systems. In summary, compliance with regulations is a major reason for using quality systems in biotechnology operations.

Another compelling reason is that quality constitutes good business practice. When a biotechnology firm provides a quality product or a service, customer satisfaction increases and with it rise sales of the product or service. Under a quality system, the number of complaints is significantly reduced. Within a firm, employee pride and satisfaction are enhanced in a quality environment and significant cost savings are realized from increased productivity and reduced waste of materials. Quality also means speeding products to the market and keeping them on the market with an excellent profit as the result.

Quality Systems Approach to Product Development

Biotechnology firms develop and market products. To reach the marketplace, they adapt or invent various development systems—manufacturing, quality control or nonclinical studies, to name a few—for the purpose of product development. A system that cannot be neglected is a quality system, an organized body of immaterial things, if you will, aimed at ensuring the utmost quality of the product or service. A quality system then takes into consideration the many facets or hallmarks of quality that have been adopted by our industry. Hallmarks of quality are listed in Box 5.1 and discussed further in a later section of this chapter. Hallmarks are tools, really, that are applied to ensure a quality product and, hence, user satisfaction.

BOX 5.1 HALLMARKS OF QUALITY SYSTEMS

Management Responsibility
Defined Quality System(s)
Quality by Design and Design Control
Contractor, Vendor and Consultant Control
Product Identification and Traceability
Process Control
Environmental Control
Inspection and Testing (Quality Control)
Control and Release of Material, Services or Product
Change Control and Corrective or Preventive Actions
Packaging and Labeling
Preservation, Storage and Handling
Servicing
Customer Concerns and Adverse Event Reports
Documentation
Training
Auditing

Several quality systems have been defined, for example ISO 9001 (International Standards Organization) and current Good Manufacturing Practices (cGMPs). Each quality system is defined by an authority (e.g., the FDA), each has a specific objective, and each is applied to a particular functional area. For example, cGMP is a quality system applied to the manufacture of biopharmaceutical products while current Good Laboratory Practices (cGLPs) is applied to nonclinical safety testing of products. Specific quality systems are described throughout this book. Quality systems are sometimes defined by regulatory bodies while others represent consensus within an industry. Today, international committees often define or redefine quality systems.

Five quality systems are often incorporated into biotechnology operations and these are briefly described below. They will be mentioned in greater detail in the referenced chapters of this book and the elements of three, per FDA regulations, are listed in Chapter 4.

- **Current Good Manufacturing Practices**. cGMPs are regulations used worldwide to ensure the quality manufacture and control of drugs, biopharmaceuticals and medical devices, worldwide (see Chapter 4, Chapter 6, and Chapter 7). Despite attempts to harmonize cGMPs between nations, there are differences in national cGMPs and an international version as well. In addition, cGMPs also have unique guidelines pertaining to manufacture of special classes or even of unique types of products. For example, the cGMPs for medical devices considers engineered products and, in some cases, the software to operate those products. Each class of biotechnology product also may have unique aspects of cGMPs, e.g., vaccines have several unique manufacturing guidelines and monoclonal antibodies also have special quality features.
- **Current Good Laboratory Practices.** cGLPs are applied worldwide for evaluating the safety of medical products in nonclinical (*in vitro* or animal) studies. As is the case with cGMPs, cGLPs are not fully harmonized across countries and there are differences in how they are applied to various product classes. Additional information on cGLPs is given in Chapter 4 and Chapter 8.
- **Current Good Clinical Practices.** cGCPs are regulations used worldwide for evaluating the safety of medical products in clinical (human) studies at all phases of development. As compared to cGMPs and cGLPs, cGCP regulations are harmonized across countries, although there are differences in some aspects of this quality system due to various types of products and cultural or political features of national regulations. (Additional information on cGCPs is given in Chapter 4 and Chapter 9.)
- **ISO 9001.** This is an internationally recognized standard for quality of virtually any manufactured product and many classes of

services. While it is not often applied for the development of biotechnology products, it is a foundation quality system for medical device development, manufacture and control in much of the world. Additionally, many service providers to the biotechnology industry are ISO 9001 certified. ISO 9001 emphasizes quality processes, making the case that a desired result is achieved more efficiently if activities and resources are managed as a process rather than as isolated events. ISO 9001, like other quality systems, is based on application of the Hallmarks of Quality. Indeed, the Hallmarks of Quality were originally promulgated by the ISO organization. A firm seeking ISO 9001 certification institutes a quality system under ISO 9001 guidelines and then requests a precertification inspection. If it passes, certification is used to demonstrate the quality nature of its products or services.

- **ISO 17025.** This represents one of many specialty certification programs under the ISO umbrella. It is a certification for testing and metrology laboratories worldwide. A laboratory seeking ISO 17025 certification must meet the general ISO 9001 requirements and also implement additional quality criteria specific to the laboratory environment and mission. Laboratories that do not require cGLP or cGMP as a quality system, but wish to be certified as a quality operation, often seek certification under ISO 17025.

Each of these distinct quality systems serves a purpose and has a unique scope. It should not be surprising that institution of any quality system by a biotechnology firm should be preceded by a conscious decision and careful planning to ensure that it is done correctly.

Planning a Quality System

Defining Objectives and Ensuring Management Support

Like any endeavor in the biotechnology industry, quality is planned before a system is instituted. To initiate quality planning, there is a decision on the exact nature, scope and objectives of the quality system to be instituted. Most firms begin with a single quality system and then grow into additional systems as they are required. If product is to be manufactured in-house, the first quality system would be cGMP. Alternatively, if cGMP manufacture and control were delegated to a qualified contractor, then the firm's first quality system might be cGLP or cGCP. A biotechnology company that does not have FDA-regulated products, such as an agricultural biotechnology firm, might wish to become ISO 9001 certified or it could elect to first establish ISO

17025 for laboratory functions. And a firm producing medical devices would consider the Quality Systems Regulation (QSR).

Once a quality system has been chosen, then formal, written documents, the "Quality Manual" and the "Quality Plan," are considered as early elements of quality development. To begin, management empowers an individual or group of individuals to prepare a quality manual and begin to write the quality plan. These might be drafted by consultants, but more commonly a quality professional is employed by the firm at this juncture. The result is that everyone at the firm understands the purpose and objectives of the proposed quality system and what will be done to incorporate it into daily operations. But, most importantly, everyone is now aware that management currently supports and will continue to support quality initiatives. This ensures that employees, contractors and consultants of the biotechnology firm buy into development of a quality system.

Experience suggests that these first steps are the most critical to success, and yet they are often the most difficult to achieve in a biotechnology firm. This is because managers and executives, especially entrepreneurs, are typically unaware of the importance of quality systems as good business practice and as a means to achieve regulatory compliance. Far too often a firm will hastily develop a quality system without management support and absent either a clear objective or a quality plan. In such cases, there is little planning and the quality system is, more often than not, inappropriate for the firm's business objectives, hastily constructed and wasteful of resources. Getting started on a quality system therefore involves significant thought, management support and planning.

The Quality Manual

Once management has made the decision to establish a quality system at a biotechnology firm, the next step is to create a quality manual. A quality professional is appointed the champion and lead author of this document. The manual is a short, usually under 10 pages, document that provides elements, listed in Box 5.2, of the intended quality system. Each section must be tailored to the particular biotechnology operation. Once written, the quality manual is signed by senior management. It states, for all employees, contractors and consultants of the biotechnology firm, the corporate expectations with regard to the quality system and employee performance under that system. Just as a corporate employee manual spells out human resource policies, the quality manual provides quality policies as guidance to all employees, explaining to each why quality is considered important to them and to their customers.

For example, if the chosen quality system is cGMP, then the manual would discuss the firm's commitment to manufacture quality product, choose competent subcontractors, perform internal audits, apply specifications and follow other hallmarks of a cGMP system. The first chapter of a quality manual, the Statement by Management, is critical and must be

BOX 5.2 ELEMENTS OF A QUALITY MANUAL TO SUPPORT A QUALITY SYSTEM

Chapter 1: Introduction and Statement of Management
Chapter 2: Responsibilities and Organization
Chapter 3: Overview of Studies Performed under the Quality System
Chapter 4: Type of Quality System(s)
Chapter 5: Employee Management
Chapter 6: Safety
Chapter 7: Facilities, Equipment and Reagents
Chapter 8: Special Operations
Chapter 9: Records and Documents
Chapter 10: Other Elements of the Quality System

signed by the highest levels of management. The second chapter spells out responsibilities for everyone in the firm, providing an organization chart and authority and responsibilities for quality assurance and other functional departments (e.g., manufacturing). It clearly defines for everyone the areas involved in the planned quality system. The remainder of the manual spells out quality system functions in a general sense, leaving details for the quality plan.

The Quality Plan

Having stated objectives and elements of the quality system in the quality manual, the firm now identifies a route to best implement the chosen quality system. An initial step is to review the business plan, the Targeted Product Profile (TPP) and Product Development Plan (PDP) (see Chapter 1), and then, under the objectives stated in the quality manual, write a supporting quality plan with objectives. This means the quality system is appropriate for and supports the intended operations and biotechnology products or services that will be developed by the firm. This may sound easy, but far too often previous and inappropriate experiences become the sole basis for preparing the new quality plan. Each quality plan is unique and must focus on the new biotechnology product and development objectives. In other words, the quality shoe must fit the business and operational feet.

To begin, the quality planner studies the objectives and processes of all functional and operational areas, current and intended, for that firm. For example, biotechnology firm A may be focused on plant agricultural products and intend to do all product development in-house. Firm B may develop medical therapeutic products, but, as a virtual company, intend to have most manufacturing, nonclinical and clinical efforts completed by contractors. In contrast, Firm C may plan to produce biological molecules for use in research laboratories and in the manufacture of *in vitro* diagnostic medical devices. Each of these three examples requires a unique quality plan because each firm faces a different set of regulatory requirements, operational processes and business objectives. Operational research is critical to establishing the correct quality plan.

Also, it is very helpful to appoint a quality steering committee and to work with project management committees. Members of this committee instruct the quality planner on the technology and resources in their departments and, at the same time, work closely with quality professionals to establish the corporate quality plan. Typically, there is a bit of "attitude adjustment" required of any product development team embarking on the quality planning process. This is especially true at a firm that has no past experience working with a quality system. A quality committee with regular meetings ensures that cooperation between the functional area managers and the quality planners begins at the outset and continues through the process of quality planning.

Project committee meetings are another means of disseminating quality planning information and for communicating quality objectives and procedures between quality professionals and other functional area supervisors.

Why is there a great need to sell the concept of quality to staff? Managers of other departments may view quality as a "threat" because it represents someone from outside their department reviewing critical functions and even suggesting operational changes. The journey to establishing a quality system achieves a milestone the moment managers and supervisors from throughout the firm visibly and fiscally support the draft quality plan. A positive attitude from managers goes far because it instills a team spirit and has a positive influence on everyone serving on the product development team. Using tact and convincing arguments, quality professionals strive to instill this attitude on fellow team members.

Support of peers and managers cannot stop with the quality plan, and quality support must continue beyond establishment of the quality system. This includes continuous sharing of information, full integration of the quality staff into product development teams, provision for additional quality resources as the scope and complexity of efforts expand, and continual dialogue between quality and functional departments. As will be discussed in later chapters, such processes are not unique to the new biotechnology firm, but apply equally to the established firm with a "broken" or deficient quality system.

Once comments of the team members and upper management have been incorporated, the overall scope and purpose of the quality plan should be finished. Regulations are perhaps the most important consideration because a quality system exists, in part, to maintain compliance with the regulations. In Chapter 4, the elements of cGMP, cGLP and cGCP are outlined and their practical applications are discussed in Chapter 6, Chapter 7, Chapter 8, and

Chapter 9, respectively. Next, the quality sections, each focused on a hallmark of quality (see Box 5.1), are written. Not each hallmark applies to every product or quality system, so certain hallmarks may be omitted from any quality plan. But omission of a hallmark must be a conscious decision, and based on the fact that it is not applicable in a given case. Again, consider that each quality system is unique.

Another important element of a quality plan is the quality organization the biotechnology firm wishes to use in daily quality operations. Guidelines for building a quality assurance unit (referred to by the FDA as the Quality Control Unit) or department within a biotechnology firm will be discussed in a later section of this chapter, but a few points are relevant to quality planning. The quality assurance unit must reflect the size and complexity of the overall quality plan. Minimal considerations include quality management, documentation, training and auditing. In a virtual biotechnology firm, these tasks may be completed by a single individual and that person might even be retained as a consultant or contractor, not a full-time employee. However, use of consultants is not always possible once a functioning quality system is in place and operational. In an established biotechnology company, a general rule of thumb is the need for one dedicated quality assurance individual for each 25 employees. Of course this ratio, 1:25, may change greatly if the majority of technical employees are working in development and not basic research. Once again, it is very important to tailor the quality plan and project quality resource needs to business and development plans. A final note is to consider the quality manual as a living document that can be changed if it is not working effectively to support corporate, quality, regulatory or technical objectives and functions.

Hallmarks of Quality: Fundamental Criteria for Building Effective Quality Systems

The term hallmark is derived from the ancient practice of marking precious metals with a stamp that identified the source and quality of these sought-after materials. Over time, hallmark has come to mean a distinctive feature of an item, especially a feature that makes the item stand out as excellent. In the discussion on quality planning, and in Box 5.1, are noted features of a quality system that make a quality operation stand out—the hallmarks of a quality system. Apparently, the term *quality hallmarks* was first used to describe features of the quality system ISO 9001, but today hallmarks are characteristic of any quality system. Indeed, a novel quality system is easily built around the most common hallmarks to be described in this section.

Management Responsibility

As noted earlier, management, both executive management and functional department managers, must support quality efforts of their biotechnology operations. This requirement is so basic to each quality system that regulatory agencies insist management support be clearly identified, and in writing. We noted earlier that management involvement in quality must be specified in the quality plan and other quality documents. The head of the quality assurance unit reports to a high-level executive and this reporting structure is reflected in corporate organization charts. Operational documents also reflect that the quality management-to-executive management relationship exists on a daily or certainly a weekly basis and is functional or even dynamic. This organization of quality resources is driven by regulatory compliance and good quality management practices. Today, inspectors from regulatory agencies, such as the FDA, typically inspect, for quality compliance, biotechnology firms in a "top-down" manner, reviewing operational management records in the first hours of an inspection in an effort to learn if management is actually involved in quality efforts. If this is not the case, then the firm may fail the inspection right from the outset. Indeed, when a firm fails an inspection of a quality system and receives a warning letter (Chapter 4), this unwelcome correspondence from the FDA is inevitably addressed directly to the president or CEO of that firm because this is *the* person with direct responsibility for all aspects of operations and quality activities.

In a practical sense, how then is management involved in quality? Guidelines for management involvement are established in writing and the results are reflected in corporate documents. Executive management understands the biotechnology, regulatory and quality processes to the extent that they can make intelligent, high-level decisions concerning the overall quality, safety and efficacy of their products. Management has set appropriate goals and objectives, has established standards and is clearly receiving feedback from their functional area managers. They are aware of significant changes in all areas of operations and how these changes are executed. Upper management has the authority to make changes when quality issues are brought to their attention and the record should show this to be the case at their firm. An absentee or otherwise disengaged executive is not appreciated in the biotechnology industry, especially by regulatory agencies, since he/she often does not have his/her finger on the pulse of his/her firm.

Defined Quality System

This hallmark simply means that one or more quality systems have been established at the firm; this was mentioned earlier in the chapter as an early step in quality planning. Further, each hallmark identified as part of that chosen system must be important to operations and to the completion of the product development life cycle at that firm and for that product. From a standpoint of compliance, regulators expect each quality system, such as cGMP, cGLP, cGCP, ISO 9001, to be identified in the quality plan, any quality manual and in other quality documents. Also, the chosen quality system must "fit" the product accordingly. For example, cGLP regulations are, by FDA definition, a quality system used for nonclinical safety studies. Yet cGLP is sometimes applied to systems far removed from this scope. This leads to firms touting their compliance with cGLP when, in fact, they do not even meet FDA's intended scope of the regulation. This inevitably leads to inappropriate use of the system and conflicts between the firm and regulatory agencies or customers. Actual quality practices must meet procedural definitions as well. If you tell customers or regulatory authorities your operation is compliant with a particular quality system, then it must indeed meet both the scope and practices for this system.

Quality by Design and Design Control

Quality by Design (QbD) has, in just a few years, become a key component of biotechnology operations and most quality systems. Regulatory agencies now recommend QbD for most medical products just as they demand management involvement in the quality system. It is no coincidence that design control has come to represent the heart of QbD, an important aspect of development for any biotechnology product and a requirement for biopharmaceuticals.

Quality by Design

QbD applies quality concepts to the design and development of biopharmaceutical products of all types and for every indication. It is based on customer needs for quality products, excellent science, design control and risk analysis and management. In the first step and as described in Chapter 1, QbD identifies product requirements to meet the needs of the user, a patient, and the healthcare provider. The next step, designing a product, such as a molecule or a cell, seems counterintuitive. Isn't biotechnology itself based on discovery, the concept of developing a product from a new finding? In the real world, a scientist makes a discovery in the laboratory and then finds investors or another sponsor to take this new molecule or cell forward to the marketplace.

But QbD does not fly in the face of reality. Instead it suggests taking this discovery, new molecule or cell, whatever, and matching it with a requirement or a need and a patient population. Sometimes, however, the match is not good and the molecule must be rediscovered or at least modified to meet the exact need. For example, a new monoclonal antibody against the malaria parasite may be discovered using a mouse system. The intention is to use this as a therapeutic for human malaria in man. In mice, the antibody works as a therapeutic, clearing an infection with mouse malaria parasites. The discoverer wishes to take this finding forward but, in planning the project, discovers that the mouse monoclonal will not work against human malaria parasites and that the mouse antibody might elicit undesirable reactions if it were given to humans. There is no market for a therapeutic for mouse malaria! This is where QbD might come into play. In this example, the scientific concept is solid, but the initial discovery is not useful to the intended user. By redesigning the molecule, and this can actually be done both by generating new monoclonal antibodies or by genetically engineering the mouse antibody to be a humanized molecule, it is possible to "rediscover" the concept and to design a new molecule that will meet requirements of the human user.

Indeed, QbD is not the traditional or twentieth century concept of drug discovery, in which thousands of molecules were first generated by an organic chemist, then screened for any attribute in laboratory and animal studies. This older concept of pharmaceutical discovery has been compared to a funnel with a fine screen and very narrow opening that often led to nowhere for most molecules. A very few made it into the development pipeline yet many of these were a bad fit for their intended purpose and were considered great molecules, but needing a useful and marketable indication.

So, QbD was introduced and, since 2004, the FDA has strongly recommended it as an operational method for modernizing both drug discovery and drug development. It has quickly spread to the biotechnology industry in part because it works quite well and also because the twentieth-century drug development process simply does not suit discovery and development of large molecule and cellular products. The funnel concept just did not work for biopharmaceuticals. Another element of QbD is a need to consider the TPP, discussed in Chapter 1. QbD goes even farther than early stages of discovery and development and continues throughout the life cycle of the product to ensure continuous improvement and innovation as critical quality attributes are applied to product development. QbD also is used with functional areas of development, notably manufacture, quality control, nonclinical studies and clinical trials. An example is given in Chapter 6, where QbD is applied to design of a manufacturing system for a biopharmaceutical product.

The concept of QbD was really an idea whose time had come in the early twenty-first century and both industry and the International Conference of Harmonization (ICH) deserve credit for encouraging its use. ICH describes, through its quality or "Q" series of documents (e.g., Q8, Q9 and Q10), QbD concepts and practices for biopharmaceutical products (see Chapter 4). These arguably are the leading reference documents on QbD today and should be considered by any biopharmaceutical firm entering development and establishing a quality system.

Design Control

QbD relies on a formal means and format for designing a biopharmaceutical product. Design control, at the heart of design, is best designated as the general arrangement or layout of a product. It demands both an active process and results from that process, notably written design documents. Historically,

BOX 5.3 ELEMENTS OR STEPS IN DESIGN CONTROL

Product is specified. The product and indication are clearly described (e.g., using TPP).

Product design plan is drafted. This plan then drives the process.

Design process is fully documented. Detailed records are kept throughout the process.

The process involves the full product development team.

The design control process consists of several elements or steps and these may be repeated until the team is satisfied with the final design. These steps are referred to as design:

- Input
- Review
- Output
- Verification
- Validation
- Change

the process of design control evolved from engineering projects, notably in the medical device industry, where physical design of a product was seen as good business practice and, more recently, as a regulatory requirement. For example, a new heart pacemaker would only function correctly if it was designed to meet certain specifications. To be effective and safe for the user, and thus marketable, an implantable pacemaker is a maximum size, very reliable, e.g., in the accuracy of beats-per-minute, and useable, e.g., have a battery life that supports many years of use by a patient. If management asked for a new pacemaker and did not insist on a user-friendly and certain design, they could find themselves with an implantable device that was 10 cm in diameter, produced 30 ±20 beats/minute and have a battery life of 30 minutes. Hardly desirable to the customer or a regulatory agency and quite difficult to market. Design control takes user needs into consideration *before* the product is produced. It is good business practice.

Today, design control is applied to biopharmaceutical development with the designer following certain steps, these having been adopted by industry and regulatory authorities. These are listed in Box 5.3 and each is discussed below.

- Design control is product specific. The product and indication for the biopharmaceutical are specified and each design control process is for one product with a unique indication and no other. Further, design is repeated each time a change is proposed to that product or its indication. As an example, consider again the monoclonal antibody against a malaria parasite. Having a firm grasp that the concept is feasible from studies in mice, it is now designed as a humanized monoclonal antibody to bind and kill the human malaria parasite. The concept of QbD and, specifically, the process of design control now impacts future development of this monoclonal antibody. Further, quality criteria now take on much greater quality implications, for planning is the first step of biotechnology development.

- A product design plan is drafted. Design planning is a process much like product development planning, but it is much more specific in scope, focusing on how the design process itself will take place. The design plan identifies the product and indication and then it outlines various elements of design, notably input, output, review and decision, serving as an agenda for product design. A product design plan might have an outline as shown in Box 5.3. But, in addition, it would describe who is responsible for each step of design, when it would occur, how it would be completed and what the expectations might be.
- Design documents and records reflect each step of the design control process. Requirements for written records are established. These include process documents, such as agenda and minutes to meetings, and product documents, e.g., the draft meeting minutes and the product's written draft labeling, user's instructions, technical descriptions, specifications and drawings that have been generated during the process. Plans are made to keep product-specific historical files and this is accomplished by following quality procedures.
- Design control involves professionals, technical and management, serving on a design team with responsibility for the product's development. Research, manufacturing, sales, marketing, quality assurance, quality control, senior executives, regulatory, project management, finance and personnel must somehow be involved in the processes referred to as design input, output, review and change. If this design review group seems similar to a project team described in Chapter 2, then you are correct. Very often a project development team, or part of that team, for a biotechnology company will be tasked with functioning as the design committee as well.
- Design input is the next step. The purpose is to enlist opinions of various individuals regarding how the product should be designed and this, of course, takes into consideration the operational plans we have discussed throughout this book. For QbD to be effective, the quality attributes of the product must be a prime consideration in design input. Quality professionals should be actively involved early in the process. Leadership is key to success of design input and here an effective project manager may lead the team effort and ensure constructive communication. In a practical sense, design input involves sitting around a table as a team, speaking, listening

and learning from each team member. Everyone will, at first, be surprised at differences in individual perceptions for a single product and indication. Returning to the example of the humanized monoclonal antibody to treat malaria, marketing may suggest that it be manufactured in final form for $50 per dose and be used to treat three species of human malaria parasites. Manufacturing may disagree, suggesting that it could only be produced and formulated for over $200 per dose. And clinical affairs might suggest that initially it can be marketed to treat only one species of malaria parasite, as initial studies to treat three parasite species would be prohibitively expensive and time consuming. Consensus must be reached in these meetings, otherwise everyone is following their own design and efforts will be disjointed, at best. In effect, design input involves soliciting everyone's opinion and reaching consensus.

- This all leads to design output, which, as the name suggests, is nothing more than producing, in a written report, consensus of opinions, producing a feasible product development objective, a consensus on the design and an idea of how all these efforts might be applied to achieve a common objective. Design output provides important key product specifications, agreed to by the team members and by upper management. For example, the team might agree the monoclonal antibody against human malaria infection must target one malaria parasite species, have a therapeutic effect in 80% of patients, give an excellent safety profile in infants and children and be manufactured for $100 per dose. Of course, the actual output document is much more detailed, but nonetheless the most important attribute is a clear design for the intended product.
- Design reviews are performed throughout the design and operational phases of biopharmaceutical development. Once an output document is established it should be critically reviewed by a much larger audience. Using partners, contractors and consultants is recommended as is review by upper management and technical staff. Reviews often lead to new ideas and improvements and these then become additional input, starting the design process over again or leading to changes in design. This might precipitate another round of design review and revised outputs. Hence, QbD never involves a single meeting, in which everyone agrees and from which design output is completed days later. Instead it requires many meetings, interspersed by additional laboratory and marketing research, and further input from management. During this period, the evolving design is reviewed by the team. Once a design has been agreed upon, design change is inevitable. Problems are encountered in development and they must be reviewed at frequent intervals by the team. Quality issues continually arise and must be addressed in a revised design.

- Design verification follows. Economics suggests that sooner or later the design cycle must end and the product must be produced, at least on a limited scale, and then tested. In biopharmaceutical development, this involves manufacture, quality control and studies, both nonclinical and clinical. The testing aspects are considered design verification. The product, and hence the design, is tested in laboratory animals and man. It determines where and how the design, which is really a model on paper, will be developed and produced. For the monoclonal antibody against human malaria this might mean that extensive laboratory testing reveals, for example, the molecule is stable at a particular temperature, that it can be produced in a 10 L bioreactor and is not toxic to a small animal. Design verification, analogous to Phase I and early Phase II development, involves laboratory and early clinical experimentation and testing. In the end, design verification leads to confirmation or rejection of individual components in the design and, if things are not perfect, it can trigger another round of design input, output, review and verification.
- Once a product has withstood initial testing and is verified, design validation may be used to further substantiate the adequacy of the product's design. Validation typically has a more stringent definition than verification. It requires additional testing and might even involve changes to variables that had not been previously tested. In biopharmaceutical development, we often consider design validation to be middle or late stage (Phase II or Phase III) development encompassing advanced nonclinical and clinical trials.

Design Change

Change is expected during a product's life cycle. Biotechnology products do not speed through the full course of development activities without several changes in TPP, PDP or product design. While change is expected and even good in many cases, change must not be a random or uncontrolled event. Changes in product development must be controlled throughout the design control process and the development life cycle.

In summary, a hallmark of quality is the concept of QbD incorporating the process of design control. We think of this in much the same way as planning development, but it has much broader and deeper implications, taking into account the ideas of design and great technical depth and detail.

Contractor, Vendor and Consultant Control

Every biotechnology development program depends upon acquisition of goods and services. A quality system takes great care to source only the best raw materials, advice and contract support. It does so by incorporating corporate policies and procedures for obtaining, by purchase, collaboration or contract, materials or services. This, along with good business practice and common sense, prevents the purchase of materials or services that might foul part of the product or development scheme. Unlike large pharmaceutical firms with in-house resources and procedures available to closely manage and control vendors, small- to medium-sized biotechnology firms often do not have this expertise and capacity. In fact, a virtual biopharmaceutical firm may rely on very few employees to manage a significant amount of the operational effort that is actually performed or supplied by contractors and vendors. Also, consultants are often retained by biotechnology firms to provide critical advice or prepare important documents, such as regulatory submissions or quality plans. The quality of these materials and services is important to the success of any product development effort, but the buyers must themselves ensure compliance with established quality criteria. How can this be accomplished in a fast paced biotechnology environment? First, a buyer establishes specifications for every material or service considered for purchase or hire. Second, they consider more than one offeror, whenever possible, and carefully evaluate each one by reviewing the corporate history, experience and references. Vendors or providers passing these initial screens then might be evaluated in greater depth through audit of their facilities.

Further to the example of a monoclonal antibody to treat malaria infections, the manufacturing and quality group considers the need to purchase saline for formulation of the product. Specifications, such as USP (U.S. pharmacopeia) grade normal saline for injection in one liter, sealed glass bottles, are established. A request for proposal is sought from three or more contractors. Once the vendor's proposals have been reviewed, further information, such as copies of Certificates of Analysis, for the last three lots of saline, might be requested. Additionally, staff might determine if a supplier complies with a quality system, such as cGMP or ISO, or they may request inspectional results from previous ISO 9001 or FDA inspections. Given that the quality of this saline is important to the overall quality of the monoclonal antibody product, it would be prudent to schedule an audit of the saline manufacturing facility once a likely supplier has been identified from the list of three candidates. (Audit procedures are explained later in this chapter.) Once the saline arrives, samples from each lot might be retested for critical parameters (e.g., sterility, pH, or concentration of sodium chloride) by the firm's quality control laboratory to determine if it does indeed meet the specifications cited in the vendor's Certificate of Analysis. (Quality control testing of raw materials is described in Chapter 7.) Upon receipt, saline is kept at the recommended conditions and it may be tested for stability prior to reaching the expiration date.

Ensuring the quality of consultants or advisors and service providers also is important to a biotechnology firm. Technical requirements and the intended scope of work are clearly stated, to include the amount of control, review and supervision that will be provided by the sponsor to the consultant. Résumés of candidates are reviewed and references are checked to ensure that each consultant applicant is qualified by experience and training.

Service providers, such as contract research organizations (CROs) performing manufacturing, quality control, clinical and nonclinical studies, are thoroughly examined and references reviewed before contracts are signed. These are high-cost and high-profile contracts involving months and years of effort. Failure or delay by a CRO has great negative impact on a biotechnology development program. The steps involved in selecting a CRO are the same as for those used to select a vendor for raw materials. The purchaser establishes exact specifications and fully describes the intended scope of work. At small biotechnology firms, it may be necessary to hire an experienced consultant to write this critical document and then participate in the selection process. The scope includes a significant amount of detail on technical and quality aspects and a request for proposals is advertised. Proposals include elements of technical performance, cost and quality. Once proposals are received, the candidate list is narrowed to only qualified proposals. Again, it may be necessary for the small biotechnology firm to retain consultants to assist in review and to use the expertise of the project team, to include those with finance and contract responsibilities, during the review and selection process. Audits and a check of references is an absolute must when considering CROs for a major contract. Once a vendor is selected, they sign both technical and quality agreements. The sponsor may assign one or more employees on staff to oversee and manage a major development contract.

Quality agreements are based on the quality expectations and specifications for the product or service. They define the quality system and processes to be used by the CRO or vendor to manage quality aspects of the material or service contract. The scope of the quality agreement covers the full scope of quality efforts. The nature of each quality system that applies to the service or material is clearly stated. The vendor–client relationship, with responsibilities, procedures for changes to the deliverables and procedures to resolve disputes, is described in some detail. A responsibility matrix is quite helpful. Because ongoing monitoring and auditing are likely to be part of the contract, the exact nature of these activities is stated. Terms are defined in an effort to prevent misunderstandings as well. For example, quality terms that may be confused are "substantial deviation" or "minor error" and these are either not used or they are clearly defined under the scope of the quality agreement.

Biotechnology firms rely heavily on outsourcing and success or failure of a service or material provided by a vendor can have a huge impact on an operation. It is critical to ensure that quality criteria and functions are considered in all contract, consultant and vendor agreements and activities.

Product Identification and Traceability

The sponsor must have in place a system to identify all materials and product as it moves through the development life cycle. This control applies in-house and to deliverables from services, such as quality control testing and clinical or nonclinical studies. Product identification and traceability are considered quality responsibilities at a biotechnology firm. For manufacture of a product this may be the application of a unique numbering system to identify lots of manufactured product and a method to trace numbered lots through the distribution system. With such systems, a manufacturer may trace "forward" to learn where and how the product was manufactured or to trace "backward" to identify the origin and handling or possession of each raw material, facility and piece of equipment used in its manufacture. Clearly, product identification and traceability necessitate a mature and infallible documentation system, a quality function that will be discussed later. For a service provider, the identification procedures may be more involved, requiring multiple levels of identification. For example, a contractor's quality control testing laboratory has numbering or labeling systems to track each sample and results as they pass through the testing process. A clinical or nonclinical study is based on a specified and numbered protocol and each segment of that study—animal or human subject, test product, treatment or procedure—is uniquely identified to ensure the protocol was executed flawlessly and that each component was performed completely and correctly.

Today, the quality function of ensuring identification and traceability increasingly relies on microprocessor-based systems. Quality electronic records and signatures are critical to performance requirements. Hence, the software and microprocessors themselves must be of the highest quality and suited to the tasks of identification and traceability. There are regulatory requirements for microprocessor-based systems, as described in Chapter 3.

Process Control

A phrase common to the pharmaceutical industry is "you cannot test quality into a product." This means quality must be built into the product at every phase of development and throughout the process, i.e., at each step in production. Biotechnology products are no exception to this rule and, therefore, process controls are used to ensure quality throughout the process. Process control relies on clear and concise written documents to guide operators at each step of biopharmaceutical production. These are standard operating procedures (SOP) and batch production records (BPR) or work instructions. For nonclinical studies or clinical trials, protocols and SOPs are used to guide processes.

Protocols, defined by the Oxford English Reference Dictionary as "official formality and etiquette," are formal written and approved documents that describe, step-by-step, how studies and trials are performed. The BPR, a formal written and approved document, guides manufacturing processes. BPRs also are used to collect information or data as they are generated. Typically, a protocol or BPR gives the reader broad step-wise guidance and references to SOPs, documents providing more exact technical instructions and describing exactly how the process is performed. In a quality environment, such as those mandated by cGMP, cGLP or cGCP, SOPs are used to provide a controlled work environment, thus ensuring products are made in the appropriate atmosphere, to instruct the use of equipment and utilities, and to mandate exactly how other activities are undertaken in the manufacturing environment.

The technical staff of a biotechnology firm prepare BPRs, SOPs protocols and related documents, but each one must be reviewed and approved by both supervisors and quality assurance staff. As approved documents, BPRs, SOPs and protocols are highly controlled and may be changed only using formal processes and with approval of all responsible individuals. Documentation control and change responsibilities, a quality assurance function, are described later in this chapter.

Environmental Controls

A product is only as good as the environment in which it is produced and biotechnology products are manufactured or tested in a wide range of environments. Genetically engineered plants are grown in greenhouses or open fields under controlled conditions. Sterile, recombinant proteins are manufactured and then formulated in highly controlled, indoor environments (see Chapter 6). Clearly there is a significant difference between the controlled environment in a corn or tobacco field and the environment within a biopharmaceutical manufacturing facility; yet, each environment meets specifications suited for its product's stage of development and intended use. We think of biomanufacturing endeavors as happening in clean rooms and, in fact, this is by far the most common. Environmental controls bring into play the issue of the quality of the facility, the air, the water and the personnel. For genetically engineered plants grown outdoors, the quality of the soil may also be considered. What goes on in the facility is also important. The flow of raw materials, product and waste in a biopharmaceutical manufacturing plant can have a major impact on the quality of the product. The quality of the people is no exception because their performance is part of the overall environment in a biotechnology operation. We discuss in Chapter 7 the technical efforts that go into ensuring the quality of all aspects of biotechnology operational environments. Quality is intricately involved in these efforts and ensures, through inspection, audit, validation, review, approval and documentation, that the environment meets preset specifications and is suitable for the intended operation.

Inspection or Testing (Quality Control)

Inspection and testing are integral to manufacture of biopharmaceutical products. As described in Chapter 7, quality control efforts are technical, performed primarily in laboratories. Quality assurance has responsibilities for ensuring that testing is fully qualified or validated for the intended purpose. Unlike discovery research, quality control testing is a formal process, performed under SOPs and using highly developed, qualified or validated procedures. A distinguishing feature of quality control is the use of specifications, normally a range of acceptable test values, against which sample test results are compared. Quality control testing is performed on raw materials upon arrival to ensure that they are, in fact, identical to claims made on the label or Certificate of Analysis. In-process testing is performed on samples taken during the manufacture of biotechnology products. While quality cannot be "tested into a product," in-process testing in part ensures the quality of products. And, once finished and filled into a container, final biopharmaceutical product is inspected and tested. Stability testing demonstrates that a product remains pure and potent once it has been stored for a designated time, hence the need to provide a stated shelf life for each biopharmaceutical. Quality control also involves the calibration or certification of test and measuring equipment to ensure that it meets standards or specifications. Just as manufacturing processes must be validated, so too must analytical tests and measuring devices (see Chapter 7). Quality assurance professionals work closely with quality control to ensure that adequate test methods are established, that test results are compared to specifications and that all testing is fully and exactly documented.

Release of Material, Service or Product

Before it is released for use, a manufactured product must conform to all specifications. This means that raw materials, processes, environments, identification labels and results of inspections and testing must meet predetermined specifications. Variance from any one specification can, in theory, lead to the rejection of that product so that it cannot be released for public use. The jargon used in the biotechnology industry to describe an acceptable product is "conform" or "pass," while an unacceptable product, one that does not meet specifications, is "nonconform" or "fail." Even though it is other operational groups—manufacturing and quality control—that produce and test product, it is the quality assurance function that reviews the data and decides whether a product conforms or does not conform, i.e., passes or fails. Product that does not conform is usually placed into quarantine while the data are reviewed and the situation investigated. However, if failure is the product's eventual fate, it is destroyed or completely reworked. Quality assurance has the task of reviewing all records and, ultimately, deciding whether or not a service or product is released to the market.

Release applies to services as well as product. For example, written reports of clinical and nonclinical studies are released to the client only after they are reviewed, approved and released by quality assurance. Not only is the report reviewed, but the performance and compliance of the complete study, from protocol through data collection, are closely scrutinized to ensure integrity and correct translation of results to the report.

Change Control and Corrective or Preventive Actions

Not everything goes as planned in a biopharmaceutical operation. Product or reports do not always pass scrutiny by quality professionals, making them unacceptable for release. Such products or reports are considered nonconforming because they do not meet specifications or were not made according to written procedures. Quality audits also may uncover hidden defects in a process, raw material or test program. Nonconforming product may be identified by customers who complain about a product even after it passes and reaches the market. In cases of nonconformance, corrective and preventive actions must be taken immediately. These decisions, and the investigations that typically precede them, are the purview of quality assurance.

A quality system includes procedures to collect and, when appropriate, review nonconformance issues as they arise. Quality assurance is tasked with ensuring that timely collection of information, including customer complaints, identifies problems with a product, study or other service. Quality professionals then make management aware of the issues and ensure that appropriate corrective action is taken to resolve the problem and prevent it from recurring. Corrective plans, instructions for investigating failures or deviations and initiating corrective action, are written and executed for this purpose. Investigations are prioritized based on the risk a problem poses to the user. A serious problem with a biopharmaceutical, one that might put patients' lives or health at risk, is addressed immediately and decisively, perhaps with a product recall and halt to production. Perceived or unproven problems are investigated over time as quality assurance staff follow trends and discuss the situation with professionals, such as functional area managers and consultants or vendors. Corrective actions follow identification of the root cause and lead to application of controls and additional monitoring to confirm effective resolution. Quality professionals pass information on to senior management because, as noted in Chapter 4, executives are ultimately held responsible for resolving issues. Another result of an investigation and identification of root cause of a problem is preventive action, a process also guided by quality assurance. Preventive actions address the root cause of a problem.

Change is normal, often good, for all aspects of a biotechnology operation and earlier in this chapter it was discussed in relation to design control. Corrective and preventive actions may lead to change in a manufacturing process or clinical or nonclinical study. As manufacturing progresses through the development life cycle, changes are made to improve product yield and quality. Clinical or nonclinical protocols require change, midstudy, to correct omissions or errors that threaten the integrity of the study itself.

Change is a carefully managed process, referred to as *change control*. For biopharmaceutical development activities, the controlled change process is mandated by regulatory agencies and is the ultimate responsibility of the quality assurance unit. Executed properly, change in a biotechnology operation, no matter how seemingly insignificant, requires forethought, extensive discussion, focused decision and follow through in action and documentation. Specifically, these six steps characterize the change control process:

1. Recommend specific changes.
2. Identify potential impact of proposed changes on the process and product or the study or outcome.
3. Plan procedures for making changes and documenting changes.
4. Complete changes to process or study documents.
5. Test the changed process or protocol, monitoring carefully to ensure no negative impact on product or outcomes.
6. Ensure complete documentation.

Packaging and Labeling

Each biotechnology product has a container and, adhered to the outside, a label that exactly identifies its contents. Containers must be appropriate to hold a given product and to maintain its identity, purity and potency. A product label contains very important product information, such as lot number, exact name, strength or dose, formulation and warnings or critical instructions to users. Labeling, as described in Chapter 6, provides additional information on a product as printed matter that is inserted into the packaging (hence, the term *package insert*). If any of this information is incorrect, then the product itself is compromised, *misbranded* in parlance of the biopharmaceutical industry, because it is not correctly packaged or labeled. Production and use of packaging and labeling are highly controlled processes and their quality must be perfect to prevent omission or error, otherwise the product is considered misbranded. Therefore, quality criteria for these processes are every bit as stringent as they are for making the product itself. Whenever a package or label is generated, quality assurance approves the printed material before it is used and they also ensure that the packaging and labeling processes are performed correctly.

Preservation, Storage and Handling

A biotechnology product may be perfectly manufactured and labeled, but if it is improperly preserved, stored or delivered, then it is of no use. In fact, improper storage renders a product adulterated. Therefore, a quality system ensures that product is properly handled postmanufacture. For example, if a protein solution is, by specification, to be kept frozen, but inadvertently warms to room temperature on the loading dock, then it is no longer a quality product. Biopharmaceutical manufacturers make every effort to ensure adequate controls are instituted and followed so that only pure and potent product reaches the customer. Quality procedures and records are used for all aspects of transportation, storage and delivery of a biotechnology product.

Servicing

This aspect of a quality system is seen largely in the medical device industry, where it is essential to provide service, metrology and technical support to customers. However, in other areas of biotechnology, such as production and sales of research reagents, manufacturing equipment or analytical instruments, servicing is also important. While the quality assurance unit does not itself provide servicing to a customer, it ensures that a servicing program is instituted and the unit monitors the quality of these efforts.

Customer Concerns and Adverse Event Reports

Every biotechnology firm with an investigational or marketed product collects and reviews comments from customers and, in the case of medical products, collects safety data. Trend analysis is often used to prioritize complaints and identify problem areas. Management and quality assurance review complaints and establish and maintain programs to address customer concerns.

Document Control

Also referred to as record control, documentation is a major element of any quality system. Keeping quality records is a major endeavor in a biotechnology firm and even a small operation generates thousands of critical documents each month. These records are used to support regulatory applications, to provide data used in business development, to verify compliance with regulations, to demonstrate the application of appropriate quality systems, to record technical proficiency and to document changes, trends and issues. Records (written, printed or electronic) are often "legal documents," meaning they can be requested by a court of law. Each record is reviewed for accuracy and completeness and, in most cases, signed and dated. Then it is archived, where it is available on short notice. Document management, as performed by the quality assurance unit, is discussed later in this chapter.

Training

Employees have the appropriate training before they begin work and that training is kept current during the course of their employment at a biotechnology firm. Training is directly relevant to an employee's duties and it is performed "to assigned procedures." For example, it is correct to employ an individual in a biopharmaceutical manufacturing operation if he/she has training and experience in a given skill area, but it is not correct to employ a laboratory technician in manufacturing if he/she does not have training and experience in that skill area or is not properly supervised during a training period. A quality system ensures that employees are fully trained and also have on file current job descriptions and training and education records before they perform a particular job function.

Auditing

Auditing is very important to maintaining a quality system in biotechnology. The ISO defines a quality audit as: "A systematic and independent examination to determine whether quality activities and related results comply with planned arrangements and whether these arrangements are implemented effectively and are suitable to achieve objectives." There are several important phrases included in this definition. First, auditing is systematic because it is a carefully planned and executed activity. Secondly, the audit is independent of the entity being audited, so there is no conflict of interest. Third, audits focus on quality aspects and not on highly technical activities. In other words, an auditor examines whether or not the appropriate quality system is in place to support a particular technology and functional area and has limited regard for scientific or technological details. Compliance, described in Chapter 4, is of utmost importance in a quality audit. This may seem counterintuitive for a high-technology industry, but scientists are responsible for technical issues while quality professionals work with technical staff to ensure the quality aspects of that same operation. Fourth, an audit compares what was actually performed against "planned arrangements." It determines if performance matches instructions. Finally, an audit examines whether or not these planned arrangements were really appropriate to achieve the stated objective.

For internal audits, a biotechnology firm's quality assurance auditors inspect the records of a functional department, such as manufacturing, within that firm. External audits are performed on contractors, vendors or collaborators, operations external to the firm performing the audit. It would be impossible to audit every function, internal or external, of a biotechnology firm or that firm's contractors and vendors. Restraints of time and resources allow only for the more critical functions of vendors to be audited and, therefore, priorities are established. Also, there are many ways to perform an audit; one method does not fit each entity or situation. Quality assurance procedures for conducting audits are explained later in this chapter.

Quality Assurance Unit

Under the Hallmarks of Quality section, above, we discussed the attributes of a quality system. In most cases, these attributes are managed, at least to some degree, by a group of professionals within a biotechnology firm referred to as quality assurance (QA) or the quality assurance unit (QAU). FDA regulations (21 CFR) refer to this entity as the quality control unit, but within the industry QA is widely used. The QAU serves several important roles at the biotechnology firm. First, it maintains a compliance posture to help the firm meet regulations. This means that QA works closely with functional area supervisors and coordinates frequently with regulatory affairs staff. Second, QA serves the users of product or clients of services by ensuring they receive products or services of the very highest quality. Third, the QAU is the gatekeeper of the quality plan, quality manual and the various hallmarks of quality for the firm.

Five aspects of quality operations—management, documents, training, auditing and change control and investigations—occupy much of each quality professional's time and effort. They also are quite important to the success of a biotechnology firm, especially for those involved in biopharmaceutical development. These five aspects also garner quite a lot of attention during inspections performed during due diligence or by regulatory agencies. This section provides more detail on those five important functions, expanding on descriptions provided in the Hallmarks of Quality section and describing how each is handled in a QAU.

Manage the Quality Assurance Function

A QAU is designated under the quality policy and the unit's responsibilities and authority are described in a quality plan. A primary function of the QAU is management of the quality system or systems instituted by the biotechnology firm under the policy and plan. The QAU is managed by a trained and experienced quality professional. Indeed, and as noted earlier, this individual at a small biotechnology firm may write the policy and plan and initiate the QAU. In addition to possessing quality assurance skills, the head of the QAU understands the technology being developed by the firm and technical aspects of operational areas falling under the proscribed quality systems. They also have knowledge of regulatory guidelines, especially as they apply to the quality systems. In some firms, the quality manager will be responsible for implementing one or more quality systems. To work effectively with the product development team and interact with upper management, the head of the QAU possesses leadership and negotiating skills as well.

As with other operational units in the biotechnology firm, the QAU is managed in all respects. Budgets, personnel, communication, serving on teams and establishing priorities are but a few of the routine management tasks. Two other quality requirements, communication and coordination, stand out and, if they are managed effectively, distinguish the excellent QAU from the mediocre department. These requirements are important because quality functions and, therefore, quality professionals are highly integrated into the daily operations of a biotechnology firm. Quality assurance is not limited by physical location as is a quality control laboratory. Quality professionals provide information, notably advice, to many individuals and they coordinate their activities with many others at the biotechnology firm. Functional area managers depend on communication and coordination with quality assurance professionals if they are to achieve their objectives. This means that quality assurance leaders must themselves understand complex technological and regulatory systems and then integrate their quality processes in a timely and effective manner to meet the objectives of the team. This makes quality assurance management a unique endeavor and provides many internal and external challenges to biotechnology development teams.

Control Documents and Manage the Documentation System

Earlier in this chapter, we briefly mentioned documentation as a hallmark of quality and stated that everything, no matter how seemingly insignificant, that happens in a biopharmaceutical operation is recorded. The QAU manages and controls each of these entries as an official record. In this section, we summarize documents and review QA responsibilities and procedures used to ensure an effective and compliant documentation system.

The process of documentation provides a means to generate, review, use and store documents. There is a saying in the pharmaceutical industry, "If it isn't written down, then it was not done." Every aspect of biotechnology operations, including plans and processes, generate piles of records, a variety of documents. While written records are the norm in many firms, the trend is to move to electronic records and signatures. Most small biotechnology operations begin using paper document systems, but, with growth of a firm, the electronic systems provide certain advantages. Transitioning from paper to electronic records does not in itself simplify the documentation challenges. Capture, review and storage of documents also must be done exactly and these tasks fall to the QAU. Regulatory agencies expect nothing less. Further challenging documentation efforts for growing biotechnology firms is operational growth beyond the capacity of the existing documentation system. Finally, senior management never sees all of these documents and, therefore, seldom appreciates their volume or the complexity of properly reviewing, approving then organizing and maintaining all these files.

So, then, who is responsible for writing, reviewing and approving each of these documents? Plans and strategies are written by high-level management with the assistance of technical and administrative staff. Protocols are prepared by investigators or study directors, individuals responsible for designing and completing a nonclinical or clinical investigation or validation. SOP, BPR and work instructions are written by technical staff, those who know exactly how a technical procedure must be performed. Finally, each document has considerable detail. Many instruct what is to be done, by whom and when. So-called forms record what was actually done, who did it and when it was performed. Corrections, reviews and approvals are exactly recorded on each.

Documents are prepared by operational staff and reviewed by their supervisors and managers and by QA specialists. Approval, that is, formal signing of each document, is the responsibility of three or more individuals: the writer, the reviewer and a representative of quality assurance. Therefore, in a biotechnology firm, QA has the major task of reviewing and approving each document generated by the development team and, in many cases, by each contractor, consultant and service provider. This is further complicated by the fact that documents are often amended, and each change or amendment must be reviewed and approved by quality assurance.

To guide documentation activities, the QAU has, you guessed it, its own SOPs to guide quality activities. Notable are procedures for writing, reviewing, approving and changing each type of document. QA establishes and maintains an archive in which to place, in an organized manner, those important documents that customers, regulatory agencies and investors might wish to review at a later date. Regulations clearly spell out the length of time a sponsor must keep operational documents.

Research activities are generally not under the prevue of a QAU. However, QA may be asked to review and maintain laboratory notebooks prepared in research laboratories, despite the fact they do not come from development operations. Well kept, accurate and detailed research notebooks are important to any biotechnology firm because they are used for intellectual property claims and because they are valuable to a product development team as the basis for planning early development activities. For example (and as noted in Chapter 7), many quality control assays used to support product manufacture begin as research tools in the laboratory and these records are helpful if not essential to the establishment of early specifications. Also, the exact history of genetic constructs, as produced in a firm's research laboratory, are important documents to support safety claims of products derived from that research. A prudent biotechnology firm gives research documents the same care provided to documents from development.

Plans, protocols, SOPs and BPRs instructing biotechnology operations, defined in an earlier section, are used in each functional area, ensuring consistency of operations, and are required by regulatory agencies. Each of these documents must be reviewed, approved and distributed to users. Most are revised periodically, meaning the process is repeated at least annually. Data captured on forms or in BPRs are also reviewed, approved and archived. Even a small operation generates a large amount of data, further complicating the documentation task.

The list of operational documents does not end here. Many others are identified in chapters throughout this book. But, there is some relief for QA as it does not manage every record generated by the biotechnology firm. Notably absent from its responsibilities are financial, human resources (excluding training), business and marketing records.

Investigate Situations: Manage and Control Change

Change is normal in biotechnology; changes are made to plans, protocols, and procedures. Any document or process or study can be changed, but change is in a controlled, proscribed manner; hence, the term *change control.* Earlier in this chapter, changes to product designs were discussed. However, downstream to design and in all functional areas change is to be expected. Thus, all change processes within a biotechnology operation are managed and ultimately approved by QA to ensure the integrity of each study, process or procedure and to communicate a proposed or completed change to all parties involved. The QAU maintains procedures for changes. If a document, study or process is to be changed, then the individual with responsibility for that functional area recommends the exact proposed change to the product development team. The various departments involved review the proposal and they then discuss the risks and benefits as well as the technical issues or challenges. Discussion and agreement to an intended change improves the likelihood of making the correct decisions and thus preventing subsequent problems with a product or study, as might be the case if change was made without proper consideration of all factors. A proposal for change may go through several iterations and extensive discussion before it is finally approved by each member of the team and by QA. The change control process is complex, often lengthy and, thus, involves careful documentation and, finally, modification of a document, such as a protocol or SOP.

QA also manages documents and approves activities, such as deviations, the retrospective discovery that a procedure had not been correctly performed. Variance, change that must be made without proper discussion, review or approval, is another type of change that is handled by the QAU.

Ensure Qualified and Trained Staff

Individuals working in biotechnology must have the training, education and experience commensurate with their assigned duties. Operational area supervisors are responsible for ensuring this is the case for their employees. However, because adequate training of personnel is critical to the safety of employees and the quality of biotechnology products, certain training responsibilities are given to the QAU. In addition to maintaining training records for each employee, the QAU provides training on subjects related to quality, compliance and documentation or record keeping, ensures that supervisors are qualified to provide technical training and confirms that each trainer maintains and follows a training plan and schedule. The QAU staff members also coordinate training activities with senior management and evaluate training programs to ensure they are effective and compliant.

Perform Audits

Earlier we noted auditing as a hallmark of quality and its importance to a quality system. Here, we discuss the performance of quality audits, internal and external, as managed and conducted by the QAU.

Quality audits support quality operations by applying two types of audits to compare actual performance and conditions to stated requirements (e.g., in SOPs). The external audit is conducted by a company engaged in an agreement or wishing to do business with another company. The vendor audit is an example. The internal audit is a firm's own audit of its internal operations. Virtually everything in the operational arena of a biotechnology firm is audited internally on a periodic basis or when problems are identified in a particular area. An example of an internal audit includes the review of study protocols, study records and study reports at a nonclinical or clinical study site to ensure compliance with cGMP or cGCP, respectively. In a manufacturing plant, an internal audit reviews SOPs, BPRs, equipment and validation records, quality control testing and records related to raw materials.

External audits may, for a reputable vendor, be performed by checking credentials, reviewing certificates provided by the vendor or performing telephone interviews and reference checks. However, for key materials and services, a quality auditor visits a vendor's manufacturing facility, nonclinical laboratory or clinical site and carefully inspects to ensure compliance with expected quality criteria. Reputable product and service providers to the biotechnology industry are frequently audited by their clients. Therefore, a vendor or contractor who denies or evades a quality audit without good reason is suspect.

Several rules apply to the auditing process. The auditor must be "independent" of the entity being audited. For internal audits, the auditor and the audited department typically have parallel but independent reporting schemes in the corporate structure. In a practical sense, this means that QA performs the audit and reports directly to a senior executive in the biotechnology firm. Even small firms may have a person trained to perform audits. Hence, audits and audit reports have become a major means of ensuring quality of a product or service. Careful preparation is important for success of any audit. The format, length and organization of the audit report should be considered before the audit is initiated. The auditing firm must address several questions. What is the purpose of the audit? Are there specific issues or problems with the system being audited? What is being audited and who are the individuals involved? The auditor should be carefully selected by QA. It would not be correct to send a very thorough and detail-oriented individual to perform an audit that was intended as a superficial overview of a vendor's quality system. Alternatively, if there was a need to examine in detail any aspect of that quality system, for example, a biopharmaceutical aseptic fill operation, then an expert in this area should perform the audit.

Performance of the audit is important to maintaining validity of the outcome. Most audits begin with an introductory meeting at which agenda, participants, purpose and scope are confirmed. The facilities, such as laboratories or a manufacturing area, are inspected by the auditor. Most audits focus on documents, both prospective and instructional documents, such as SOPs and performance or data records. This is critical because a quality audit is performed largely to demonstrate that procedures, in fact, were performed according to requirements. Notes are taken by the auditor. A closing meeting is held so both parties have an opportunity to discuss the findings and perhaps resolve any apparent discrepancies or misunderstandings.

Every audit results in a written report, prepared by the QAU's auditor, to relate important and relevant findings or discoveries. The report states the facts and cites regulations or guidelines, but without being judgmental or finely interpreting regulations. The audit report also may make recommendations when deficiencies are found, but it should not mandate exact procedures to resolve these issues. Functional area supervisors at the audited entity are left to take corrective action, but the auditor follows up to ensure that corrective action has adequately addressed the original issues.

There is another side to the audit coin and every biotechnology firm will, at some time, be audited by an outside party. This may happen as "due diligence," the result of pending business arrangements, as an inspection by an interested party, such as a regulatory agency, a client or a collaborator, or it may be a customer, someone wishing to purchase materials or services. When audited, employees should make full disclosure of any records requested by the auditor, but they are not responsible for volunteering materials or information that is not requested. It is in everyone's interest to be ready with complete, unambiguous and well-organized materials for the auditor to examine. The entity being audited typically sets the tone for the audit and a positive, cooperative tone is especially important because, like a dental appointment, getting it completed quickly and painlessly is important to all parties.

Initiate a Quality System for a Biotechnology Operation

Once a biotechnology firm decides to develop a product and enters the regulated or quality compliance arena, it finds an effective means to build quality systems into the proposed development operation. As described earlier, quality planning, specifically the quality plan and quality manual, are key to initiating an appropriate and effective quality system, one that will grow with the operation. Quality manuals and quality plans are living documents and may be changed as operations increase in development phase or scope. Thus, while QA is seldom the first operational element at any biotechnology firm, it must be adopted early and then it soon becomes an integral part of the development effort. The excitement and expense of entering into product development sometimes obscures this immediate need for quality function; if this happens, quality efforts may suffer. Nonetheless, senior management support of the QAU allows it to be established and then to support the other operational areas.

Where does one begin? The need for a quality assurance role at the biotechnology firm is often spelled out by a consultant or by a new employee who has worked in a regulated biotechnology environment and is familiar with quality and compliance. Or, it may be stimulated by recurring problems in operations. However it begins, the growing biotechnology firm may retain an experienced quality consultant, someone who has built a quality department within a growing operation, or they may elect to hire a quality professional with the same experience. Few succeed in establishing any quality assurance function by tasking an inexperienced or untrained staff member with such responsibilities.

The next step is to generate management support and then quality policies and plans that focus on the mission and operations of the organization. Budget must be considered, of course. The quality function is tailored to fit the PDP and it is written in parallel to, or as part of, that overall plan. Establishing a QAU is especially challenging for the virtual biotechnology company, the firm with less than 10 full-time employees managing a full product development program. Here it may be necessary to delegate, over a long period, quality functions or oversight to a contractor or consultant.

Many quality issues face a maturing or mature biotechnology firm, defined here as 100 or more employees working in a fully operational environment. First is the inability to sustain growth of the quality function. Unlike the virtual or start-up operation, the maturing firm already has a QAU, but it may be woefully understaffed, with management emphasizing rapid product development over quality programs. Indeed, midsized firms can, for various reasons, have less concern for quality functions than do smaller, younger companies. Functional managers may ignore quality or quality staff may experience burn-out and lose interest. A weak QAU is easily detected during business due diligence or inspections by regulatory authorities and this lowers the value of a biotechnology firm in the eyes of potential business investment, partnership or purchase.

Operational growth requires additional resources for quality efforts and so growth in quality requirements is planned and budgeted. Growth requires addition of specific quality skill sets. A documentation specialist may be needed or an experienced individual is required to perform audits. Mundane issues, such as space to work and secure files in which to archive all those documents, face the growing quality operation. The key to successful growth of the QAU in a midsized biotechnology firm is effective and timely quality planning through revision of the quality plan.

Unfortunately, some biotechnology firms have, for whatever reason, a largely dysfunctional QAU. This situation often results from inadequate planning, poor management support and, unfortunately, ineffective leadership.

Such QAUs first require immediate management involvement and support. This does not mean simply throwing money at the problem, but more often requires investigation followed by organizational change or resolution of issues, for example, resolution of interdepartment squabbles, by upper management. Inability for any one department to operate effectively is often a reflection of an ineffective project team. Once the root cause has been identified, then senior management begins to repair the quality system and QAU.

Unique and Effective Approaches to Quality Management

Risk-Based Approaches to Quality Systems

Risk-based approaches are a popular and effective means of ensuring quality in development. The U.S. Food and Drug Administration recommends risk assessment and management as a means to enhance and modernize pharmaceutical and biotechnology manufacturing and product quality. This initiative uses a scientific framework to find ways to mitigate risk posed by medical devices, drugs and biotechnology products.

Risk-based approaches in quality assurance examine a biopharmaceutical operation as a process and then identify those areas within that process that pose the greatest risk to the product. They also examine issues or problems associated with a product or particular type of study, such as number of failures. It focuses efforts early in the life cycle of product development. This naturally fits with quality development efforts for any product. Risk management then uses the scientific method to examine the risks and to address and lessen those risks using appropriate quality systems. Continuous, "real time" quality is a hallmark of this approach.

Risk analysis and management procedures are described further in Chapter 2, because they typically involve multiple operational areas and often fall largely in the purview of project managers. Nonetheless, the QAU has critical functions and plays an important role in this area, often identifying risks and recommending that risk approaches be initiated or completed.

Total Quality Management (TQM)

TQM aims at customer satisfaction. It has been adopted by many firms, including large biotechnology companies, and is especially popular with sales and marketing groups. It is a structured system for satisfying internal and external customers and suppliers by integrating the business environment, stressing continuous improvement, refining development processes, encouraging maintenance cycles, and changing, for the better, organizational culture. The term *structured system* relates to the fact that TQM relies on principles of quality systems and an environment that fosters such systems. TQM has three cornerstones:

1. Everyone and Everything: Total quality involves every individual and all activities in the company.
2. Quality: Conformance to requirements (meeting customer requirements).
3. Management: Quality can and must be managed. As one might imagine, TQM must be driven from the top.

Six Sigma

This program has been adopted by many firms worldwide as an avenue to produce quality products and reduce customer complaints. A major objective is to incorporate a quality system that is so effective that a firm claims less than 5% of revenues are used to address and repair quality issues. Specifically, it aims to reduce product or service failure rates. The Six Sigma process encompasses all aspects of a business, including management, service delivery, design, production and customer satisfaction. As compared to an operational quality system, such as current Good Manufacturing Practices, in which only the manufacturing and control departments are directly targeted, Six Sigma involves every aspect of a firm.

Quality Systems for Research

What quality considerations should be given to (or imposed on) biotechnology research laboratories? Is there a compelling business reason to establish a quality system for research efforts in any context? Application of a full quality system may be helpful for research quality, but, in many cases, it may actually hinder research endeavors. Research laboratories are for discovery and not for structured development. Research results must be of high quality, but discovery research does not directly lead to products or to users. The standards applied to research and development or commercial applications are different. Particularly frustrating are attempts to impose on a discovery research laboratory a formal quality system, such as cGLP, when it is neither needed, by regulation, nor helpful to achieving objectives.

However, some hallmarks of quality, mentioned earlier in this chapter, can be very effective at improving productivity and the research environment, at least if they are applied correctly. For example, vendor control is an excellent way to save time and money. An effective documentation system can be very supportive of patent applications and improve records on which future product development efforts depend. Hence, the small firm, engaged exclusively in research, is encouraged to institute quality hallmarks that help their research laboratories achieve success without burdening research efforts through establishment of a fully compliant quality system.

Resolving Quality Issues or Problems

The quality assurance professional takes on huge responsibilities and has great authority within a firm. He or she reports to upper management and may be more influential than many other functional area managers. One example is in a growing biotechnology firm, where the QAU questions whether a clinical study site is fully qualified, under cGCP, while the clinical manager strives to meet an already challenging schedule to begin a study. Another example, this time in manufacture and control, is whether or not to release product for a clinical study because a specification was not fully met by analytical results. Specifications for product in early development can be ambiguous in certain respects and individuals on the product development team disagree on whether it passes or fails. In another example, the QAU may question the validity of an important aspect of an expensive nonclinical study and suggest that it be repeated to be considered valid. These quality opinions have a tremendous effect on day-to-day operations, expenditures of time and money and, in the end, the success or failure of a biotechnology firm. Such situations often leave the quality assurance professional in the "hot seat." Because disagreements are not infrequent and because operational department directors disagree on important matters, tempers may flare or disagreements linger and fester.

Well-led project teams are perhaps the best means of resolving differences while ensuring correct decisions in a timely manner. Several guidelines must be considered. In the biopharmaceutical industry, FDA regulations make it clear that the QAU has, under cGMP, cGLP and cGCP, the final word on matters relating to quality. Secondly, individuals with different backgrounds often perceive the same data quite differently. We see this in the media and at scientific meetings quite frequently. These differences can lead to animosity and disregard for the other's opinion. Therefore, great quality professionals are good at negotiation, which is based on understanding the other person's point of view and then trying to work within that opinion to reach a solution or common understanding. This, in turn, requires them to listen carefully and to be patient. They also must clearly explain the reason for a judgment and they are well advised to seek the opinion of regulatory professionals. Again, a good project management team with a strong project manager is wonderful at facilitating negotiations, if only by setting a positive environment.

Upper management plays a major role in preventing hostilities and resolving disagreements between the QAU and other operational areas. Misunderstandings in biotechnology operations are often the result of poorly established or undercommunicated corporate and product development objectives. In these all-to-frequent instances, upper management bears responsibility. Further, upper management must recognize when communication has broken down between quality assurance and another department manager and then make every effort to resolve the differences and have each faction work together toward a common objective. To recognize developing issues, upper management must always be involved and alert. Project managers and team members ensure that management is engaged in development activities, including professional roles and disagreements.

Why is it so vitally important to identify quality issues in a biotechnology firm? First, these small firms are so fragile, very susceptible to failure for a number of reasons. Second, they have little depth—fiscal resources, product line, facilities—to rely upon in times of trouble. Third, the team has worked together for a brief period, as most firms are relatively new and development may have just begun. Differences related to quality aspects of the operation could spell the difference between success and failure, especially in biopharmaceutical endeavors. Often, quality issues are, for a variety of reasons, invisible to insiders, but most obvious to outsiders, such as consultants and auditors. Indeed, seasoned professionals in biotechnology have said that by examining the roles, authorities and responsibilities of a QAU, one can quickly surmise a key indicator of success versus failure at a young biotechnology firm.

This section on resolving quality issues has provided no magic solutions for problems one might encounter in a biotechnology operation. It aimed instead to summarize but a few of the situations one might encounter in the operational environment of a biotechnology firm. By understanding quality systems and through careful planning and effective management, quality functions and the quality professionals who manage them, the quality endeavors can be a valuable asset to any biotechnology development team and the firm they represent.

Summary of Quality Systems

Quality assurance is a planned and structured function designed to ensure a product or service provided by a biopharmaceutical firm meets established requirements and user expectations. Quality planning and institution of a quality system early in development is critical to the success of any biotechnology operation. The functional area that manages a quality system is referred to as the QAU. Hallmarks of quality are distinctive features of excellence comprising any quality system and include features such as management responsibility, definition of the quality system, design and design control, contractor control, product identification and traceability, process control, environmental control, quality control or testing and release, change control and corrective or preventive actions, packaging and labeling, preservation, storage and handling, servicing and customer concerns. The QAU focuses on the quality attributes of quality management, documentation, investigation and change management, training and auditing. Effective quality systems are developed specifically for each process and product or service and established quality systems, such as cGMP or ISO 9001, are often required of biotechnology firms, especially for biopharmaceutical development.

6

Biomanufacture

Overview of Biomanufacturing Requirements

The biotechnology operation focuses on development of a specific product. This concept carries with it the need to plan and then develop a biomanufacturing process to produce a biological substance of high quality and in amounts required for testing and marketing. Further, the biomanufacturing processes and the resulting product must be compliant, that is, they must satisfy regulatory agencies through application of good science and a quality system: current Good Manufacturing Practices (cGMPs). To achieve these objectives, the biotechnology operation must develop a biomanufacturing plan.

Even for the simplest product, biomanufacturing is a demanding endeavor and requires considerable time and financial and human resources. False starts in the biomanufacturing pathway, usually the result of inadequate planning, often lead to project failure and termination.

Hence, biomanufacturing planning begins early in the product development cycle and is based on an exact understanding of the product's nature and intended use, or indication. The overall Product Development Plan (PDP) (see Chapter 1) coordinates the manufacturing plan with plans for the quality assurance, quality control, regulatory, clinical and nonclinical aspects of product development. To ensure this integration, the biomanufacturing planning process requires leadership from biomanufacturing experts, considerable time and effort and frequent interactions between individuals from various departments.

This chapter on biomanufacturing considers design and planning, production technologies, compliance and quality, major stages and steps of manufacturing for various types of biotechnology products and the manufacturing facility.

Design in Biomanufacture

At the heart of a biomanufacturing plan is the manufacturing design or scheme, pictured from beginning to end, and with the various control testing, quality and regulatory elements that impact product production and release. The objective of biomanufacture is to produce a product that has attributes—strength, identity, purity, potency and safety—commensurate with the intended use. (Product attributes are further defined in Chapter 7.) Each biotechnology product is unique and is, or will be, produced using both well-characterized and well-known commercial processes, and special methods, these developed for that particular product or class of products. The flow diagram shown in Figure 6.1 is a general format or template used to design a product-specific biomanufacturing scheme. Three stages of biomanufacture, (1) upstream processing, (2) downstream processing and (3) formulation fill and finish, are the backbone of a biomanufacturing design. In upstream processing, or the first stage, the product is produced from raw materials using process technologies, such as cell culture, fermentation and synthesis. The second stage, downstream processing, involves purification of the desired product by separation from impurities and contaminants. In biopharmaceutical processing, the output is referred to as the bulk (drug) substance (BS). For a biopharmaceutical, the BS is also the active pharmaceutical ingredient (API), meaning it is the material that has the therapeutic activity. Stage 3 processing ensures the product is fit for use by applying the processes of formulation, filling into a container, packaging and labeling. The result is a final product (FP) ready for use.

Quality by design (QbD) is a concept applied to all product development endeavors (see Chapter 5) in biotechnology and it is a critical and early aspect of biomanufacture. It evolved in part from regulatory and quality initiatives in the medical device industry late in the twentieth century; more recently, the principles and practices of QbD have been adapted to biopharmaceutical development. Product development or manufacturing QbD is driven in part because regulatory agencies have provided evidence that, when followed, QbD consistently leads to high-quality products. The biotechnology industry recognizes also that QbD makes good business sense. Janet Woodcock* of the U.S. Food and Drug Administration (FDA) defined QbD in 2005 as "a maximally efficient, agile, flexible pharmaceutical manufacturing sector that reliably produces high quality drug products without extensive regulatory oversight."

In its simplest form and as now provided in International Conference for Harmonization (ICH) quality guidelines (see Chapter 4), the concept of QbD instructs the developer to design a product so it consistently meets the

* American Association of Pharmaceutical Sciences–FDA–ISPE Workshop, North Bethesda, MD, October 5, 2005.

FIGURE 6.1 (*Continued*)
This flowchart traces the biomanufacturing scheme applied to many biotechnology products. Boxes in the upper row define inputs, the resources required to begin biomanufacture of a product. The flowchart below is divided into three stages typical of a complete manufacturing process and describes outputs, results from each of the three stages. Process elements, shown in the shaded flags, are typical for production of a recombinant protein product.

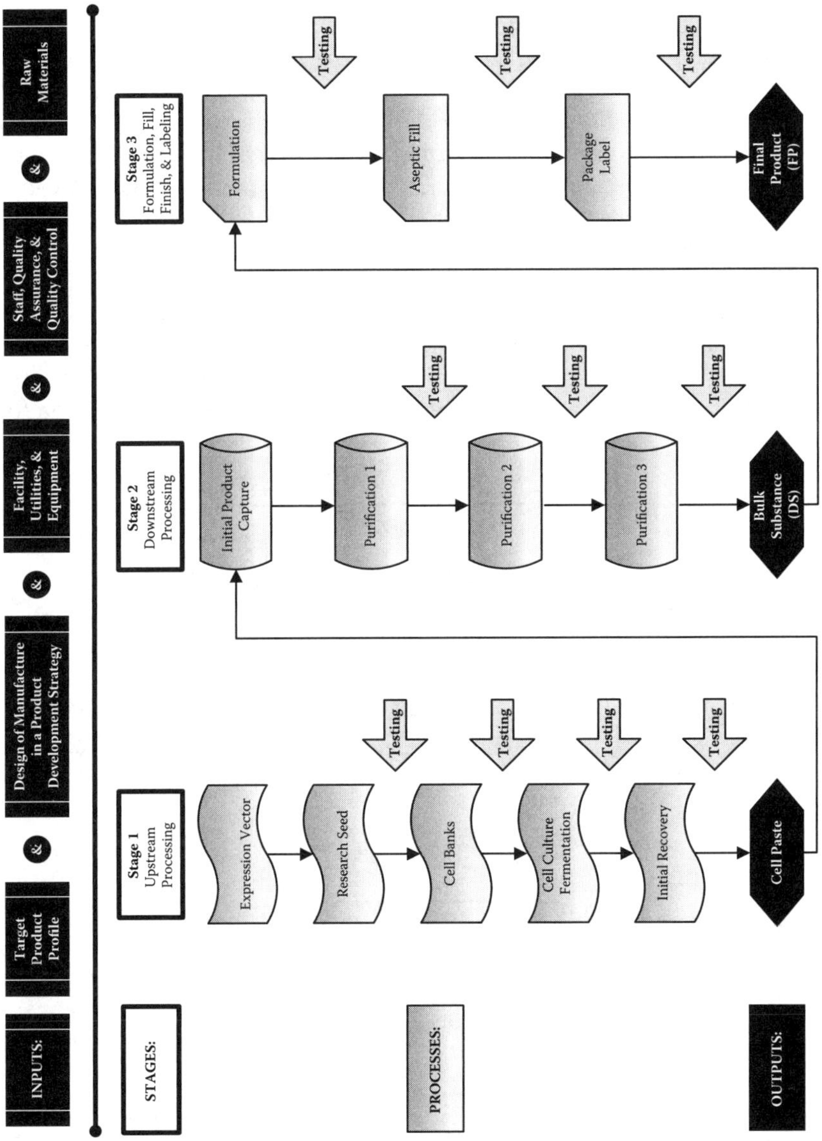

FIGURE 6.1
General outline of biomanufacturing activities by stages or steps.

Continued on facing page.

desired performance criteria and always meets expected quality attributes. This definition demands much of a product's sponsor. First, it identifies a need to integrate a manufacturing plan into the overall PDP, described in Chapter 1. It also directs the use of formal manufacturing design process in which the product designer considers and documents both the expected performance and quality attributes of the product.

QbD as it applies to biomanufacturing and the biomanufacturing portion of the PDP is shown in Figure 6.2. The nature of a biotechnology product, as provided in a Targeted Product Profile (TPP), must be carefully considered in the manufacturing design. QbD requires that a manufacturing process be designed using scientific approaches, quality criteria risk management and design space. QbD prompts the need for application of design concepts: input, output, reviews, design space and specifications, ranges of acceptable values or limits. These are discussed in Chapter 1. Design space, as it refers to biomanufacturing and control activities, is the requirement to design within limits or boundaries. For planning biomanufacture, the limits are imposed by constraints of manufacturing technology and by product specifications. Specifications, further defined in Chapter 7, are measurable quality criteria for

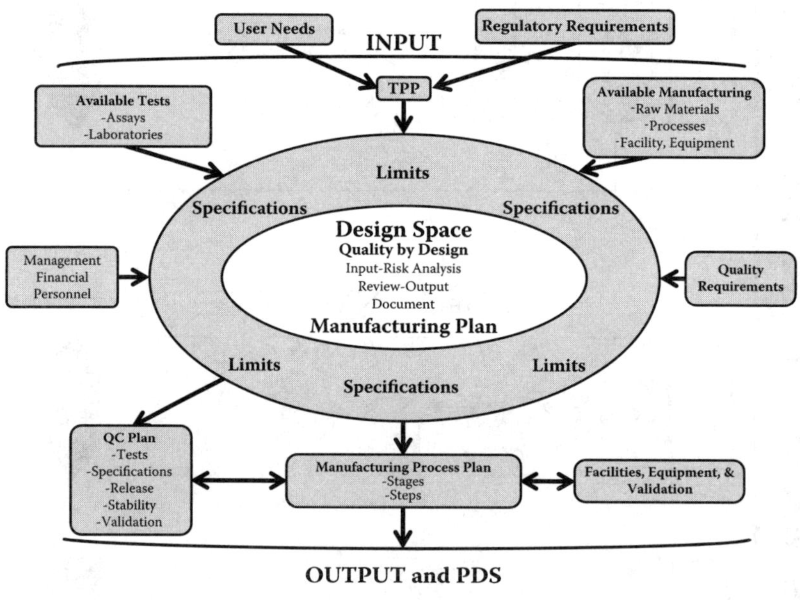

FIGURE 6.2
Quality by design in biomanufacturing. Manufacturing design begins with input, notably user needs and regulatory requirements that are synthesized into a Targeted Product Profile (TPP). A design space has specifications and limits as boundaries and quality requirements, available manufacturing resources, tests, and management and resources as inputs. This allows the manufacturing process to be designed within a design space, represented in the center of the figure. Outputs of the design process include a quality control plan, the manufacturing process plan and a strategy for facilities, validation and other requirements.

a product. While these boundaries on design space restrict initial manufacturing design, they also allow later changes in manufacturing processes, as long as the changes are implemented within the design space and consider the specifications. This allowance is based on the premise that changes to a biomanufacturing process made within a reasoned design space will most likely not change the quality of the biopharmaceutical product. Planning under QbD helps to ensure that a biomanufacturing process will be robust and the product predictably and consistently of high quality throughout the life cycle.

As shown in Figure 6.2, QbD also applies the concepts of input and output to biomanufacture. Input, notably user needs, the nature and profile of the product, and manufacturing and control technologies, are considered within the confines of design space. Within the design space, active manufacturing design and review leads to output, notably plans for the appropriate processes, facility and quality control tests.

In manufacturing design, the concept of risk analysis, elaborated upon in Chapter 1, is considered in light of how the manufacturing process and the design space might impact the user of the biopharmaceutical product. Any biotechnology product carries some inherent risk to the user or to the public. Other risks are imparted through the product's manufacture. Design and manufacturing planning activities both identify and attempt to mitigate all risks. QbD considers these risks in a logical manner and demands that design take into account any possible risk.

Biomanufacturing design and planning are greatly impacted by available and affordable biomanufacturing technologies available for a given type of product and process. For a given product, the biomanufacturing planner has a variety of process methods from which to choose. The process technologies chosen during design are added to the plan and referred to as input. The results of these technological applications are output and also become embodied in the manufacturing plan. Consider, throughout the remainder of this chapter, the need to design a biomanufacturing scheme, applying product limits and specifications, the inputs and outputs of design and the need for understanding process and product risk at every point. These concepts help us to better understand why products are manufactured in a certain manner. Finally, some biomanufacturing process change is inevitable for even the best conceived biomanufacturing plan. Consideration is given for making changes in a manufacturing plan as long as change is kept within boundaries of design space and risk is carefully considered.

Technical Considerations for Biomanufacture

Biomanufacturing is a relatively new field having expanded, by quantitative and qualitative measures, rapidly. Prior to 1970, the manufacture of biological products was accomplished largely by purification of biologically active molecules from various natural sources. For example, albumin was precipitated from human plasma and then separated from other blood proteins. Some years ago, vaccines were strictly natural products, such as subunits of viruses or protein toxins, derived from native or genetically unaltered microbes grown in culture. While these endeavors are still considered biological technologies, the advent of genetic engineering led to the endeavors we now refer to as biotechnology. Genetic engineering made possible the transfer of genes from one organism to another and this science allowed us to genetically modify bacteria or mammalian cells, which in turn led to biomanufacture, the production of small amounts of recombinant proteins or nucleic acids. To make large amounts of these recombinant products, first for further evaluation and then for commercial use, biomanufacturing protocols and technologies or methods were expanded. And biomanufacturers soon discovered that product quality mattered greatly. If a molecule was of poor quality, then it would not perform as intended when used in critical test protocols, such as in animal or clinical studies. Notably, when a manufactured biotechnology product failed to perform consistently, the sponsor was left with little product value and doubts about the product's utility and marketability.

It was recognized as well that quality control testing of a manufactured biotechnology product did not, alone, ensure product quality. Indeed, product quality reflected both the processes and the technology applied to biomanufacture, such as facilities, utilities and equipment used in biomanufacture. Consistency of manufacture also was critical to achieving the desired attribute; the biopharmaceutical product had to be exactly the same every time it was manufactured. Hence, commercial biomanufacture demanded attention to product consistency.

Rapid advances in biotechnology have challenged the young field of biomanufacture in other ways. The number and types of organisms capable of expressing recombinant proteins expanded greatly in just three decades and the types and classes of biotechnology products that must be manufactured by our industry continue to both expand and diversify. Somatic or stem cell engineering are examples of rapidly growing scientific endeavors that have brought about the need to develop and apply new biomanufacturing technologies. Transgenic plants and animals are becoming commonplace and biologically active molecules must be processed from these sources. Synthesis of biologically active molecules is a field that continues to expand. Other challenges include producing very novel products, refining old processes to more economically yield currently marketed products, improving the quality or consistency of investigational and marketed products, and engineering production of generic or "follow-on" biopharmaceuticals, new products that are safe and effective, exactly like a predicate product.

To meet these challenges, careful planning and development of new manufacturing technologies, analytical tools and processes continues unabated. Biomanufacturing scientists have invented, applied, adopted or adapted processes, procedures and skills. Facilities and equipment have been designed or redesigned, built, validated and commissioned to house and support these processes. In summary, there is much activity in the field of biomanufacture leading to excellent marketed products and each success is based on proper manufacturing planning and design and the ingenious application of existing and novel technologies.

For the remainder of this chapter, we present an overview of the stages and steps used in biomanufacture, identify technical considerations for various processes and integrate quality and compliance into biomanufacturing schemes. We then apply biomanufacturing criteria and technologies to several classes of biotechnology products, highlighting differences and similarities of various products. At the end of the chapter, we describe the design, use and validation of biomanufacturing facilities, utilities and equipment. Quality control and quality assurance activities are closely associated with biomanufacture and these are discussed in Chapter 7 and Chapter 5, respectively.

Phases and Scale Up: The Biomanufacturing Life Cycle.

Biomanufacturing is performed throughout the life cycle of a product. We identify phases in the life cycle and further ask the biomanufacturer to ensure a product possess particular qualitative and quantitative attributes or traits in each phase of development, with process and product specifications becoming increasingly stringent as the cycle progresses. For biopharmaceuticals, manufacturing phases of development follow those applied to clinical studies (Chapter 9): Phase 1 (early phase), Phase 2 (middle phase), Phase 3 (late phase) and Phase 4. This approach makes sense because we use product in greater amounts as the numbers of human subjects increase at each clinical phase. At Phase 1, requirements are in the hundreds of doses, but, as product approaches the marketplace, product might be needed for millions of users. Compliance issues, specifically adherence to current cGMPs, also increase in intensity and importance as biomanufacturing development increases through the phases. Both total amount of product and quality criteria undergo change as the product is used in a greater number of individuals. These relationships between clinical phase, biomanufacturing phase, product quality and product quantity are shown in Box 6.1 and they must each be considered in a manufacturing plan. A greater amount of product is produced in each subsequent phase and along with this comes the need to better characterize the product and to meet ever greater compliance standards through improved quality and production systems. Phased product biomanufacturing development is a dynamic process and change is desirable and normal. How this change is anticipated, planned and executed in the manufacturing plan is critically important to overall success.

BOX 6.1 BIOMANUFACTURING ACTIVITIES BY PHASES OF BIOPHARMACEUTICAL DEVELOPMENT

Phase	Design and Plan	Manufacturing Processes	Quality Control Laboratory	Quality and Compliance
Planning (0)	Targeted Product Profile Product Development Strategy Technology	Develop constructs Technology transfer from laboratory Cell bank development	Research laboratory development of critical analytical tools QC constructs, cell banks	Identify regulatory guidance Establish quality plan and basis for quality system Ensure quality assurance activities
Early Phase (1)	Implement design and process development schedule	Accept constraints Produce clones, cell banks Perform process two or more times Produce product for nonclinical and Phase 1 studies	Establish Certificate of Analysis with product attributes, tests and specifications Test Phase 1 products	Institute Phase 1 cGMP compliance
Midphase (2)	Refine plan based on findings	Scale up for multiple batches and lots and Phase 3 requirements Adjust process, refine steps	Further develop assays, qualify critical tests, refine specifications, add new tests Test Phase 2 products	Increase scope and depth of cGMP compliance
Late Phase (3)	Plan commercial process and validation activities	Execute multiple lots at or near commercial scale Validate process, facility	Validate or verify each assay. Test product at scale up and for Phase 3	Come to full cGMP compliance as applied to commercial production Approve validation
Postlicense/ Commercial (4)	Plan, document all change	Manufacture for commercial market	QC for commercial product	Approve change, maintain full cGMP

The quality of product required at Phase 1 clinical studies is, to some extent, mandated by cGMP, but it is also a function of the indication, intended use and proposed manufacturing process. In planning and developing the process, it is important to consider that greater amounts of product will be required later in development; therefore, a biomanufacturing process must be amenable to change due to scale up and more stringent quality and compliance criteria. Process control, quality control testing and consistency of manufacture are important measurements that can demonstrate product quality at Phase 1. Hence, the process is well defined and some quality control assays are established prior to initiation of biomanufacturing. In early phase development, it is best to produce multiple batches of bulk substance and multiple lots of final product to ensure consistency of manufacture—amount and quality—from batch-to-batch and from lot-to-lot.

At midphase development, the objectives are to confirm and extend the findings of early phase biomanufacture. Biomanufacturing scale up at midstage further tests application of quality criteria to the process and end product. Product manufactured at midstage is used in those clinical studies and also it is applied to refinement or qualification of analytical tests. Process improvements are often implemented and tested to ensure that changes yield a product with quality the same or better than that seen at early stage development. The ranges of acceptable values for product specifications, both process and quality control, are often narrowed at Phase 2. Midphase biomanufacture provides product that is used for the qualification or validation of analytical tests and also under additional stability protocols. It is not uncommon for a biomanufacturer to miss critical midphase manufacturing objectives. Failure to achieve consistent manufacture or the need to abort processes are examples. In such situations, it is important to review the manufacturing plan, make changes and then repeat midphase biomanufacturing before progressing to scale up or late-stage manufacture. Unfortunately, such advice is too often ignored, leading to biomanufacturing failures at Phase 3 and leaving the biotechnology operation with only the options to repeat manufacturing development and then repeat both Phase 2 and Phase 3 biomanufacturing and clinical studies, both very expensive propositions.

Midphase is the best time to make significant and necessary process changes. Process changes often result in changes to purity or potency of the product and these may negate the validity of nonclinical and clinical data generated during Phase 1 and Phase 2 studies. Every effort is made at this time to improve the process without changing the molecular or cellular nature or the quality profile of the biotechnology product.

Late phase biomanufacturing development focuses on preparing material for Phase 3 clinical studies. Production also ensures a robust process at greater scale, and the late-phase product is used for further assay development or validation and stability studies. Biomanufacturing process validation, described later in this chapter, is another objective of the late-stage biomanufacturing program. Manufacturing changes, in addition to scale

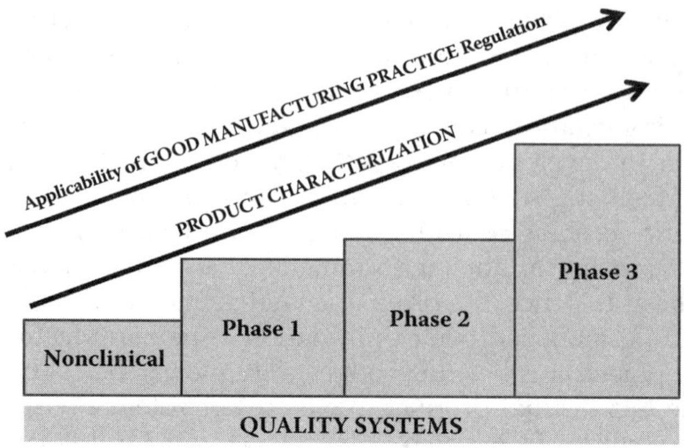

FIGURE 6.3
Manufacturing by phase of development. Simultaneous increases in product quantity, quality, characterization and regulatory compliance through phases of biomanufacture. (Graphic courtesy of Anthony Clemento, 2008.)

up, can be instituted late stage, but they must be thoroughly analyzed to ensure the product remains consistent in quality with material made in earlier phases and used in clinical and nonclinical studies.

To this point, we have discussed changes to biomanufacturing processes as qualitative. We now cover the subject of quantitative manufacturing changes, or "scale up." Early phases of biomanufacture yield limited amounts of product, certainly not the quantities required for a market. Biomanufacturing scale up, depicted in Figure 6.3, is used to gradually increase the total amount of product that is available from each production run. A production run is a distinct series of processes with defined beginning and ending. Each run results in one batch of bulk substance and/or in one lot of final product. Scale up must be considered within the confines of design space. One should, at the outset, have an idea of commercial requirements and then base a scale up plan on these requirements. Scale up may follow several pathways. It may be achieved by increasing the yield, that is, by increasing the amount of product produced within a batch or lot. Or it may up the scale of manufacture for a batch or lot. Or it may increase the number of manufacturing runs. Often, a scale up plan considers two or even all three methods to increase the total amount of bulk substance and final product. This is particularly important with therapeutic biopharmaceuticals, as small changes in molecular structure may have significant physiological impact for the patient.

Scale up is an expensive endeavor and is not instituted until success has been demonstrated in clinical or field studies. Scale up has a greater impact on the production of bulk substance than it does on final product. Scale up of final drug product manufacture often involves increasing formulation batch size by using larger vessels, by building another facility or by identifying contract capacity to perform a greater volume of fill, finish and labeling. In contrast, scale up to greater amounts of bulk substance most often cannot be economically accomplished by simply multiplying the number of bioreactors or fermenters or by installing many rows of chromatography columns. Instead, scale up involves developing new or modifying existing technologies to produce much greater volumes in a single batch (i.e., a single, large fermentation vessel). Biomanufacturing systems are often uncooperative when it come to scale up; hence the need for imagination and extensive experimentation. At large scale, these experiments require investments in equipment and are costly to perform. It sometimes forces the operator to devise bulk substance production systems that appear to be quite different from the smaller systems used in early phase biomanufacture. Changes to yield may impact product quality and so process and laboratory controls are continually applied during scale up. And, as always, changes to a process must be kept within the design space.

Raw Material Considerations

Like any other manufacturing endeavor, biomanufacturing requires raw materials, sometimes referred to as components. These provide the structural building blocks, the biomanufacturing environment and the sources of energy for every product and process. Raw materials, such as water, gases, salts and nutrients, are employed at every phase of biomanufacturing. The quality of each raw material is unchanged through the manufacturing cycle but amounts increase with scale up. Raw material requirements and specifications are included in a manufacturing plan.

Box 6.2 presents a list, albeit incomplete, of raw materials that might be used in "upstream" production, such as fermentation of a yeast to yield recombinant protein. Box 6.3 is a list of raw materials that might be used downstream, that is, in that product's purification. Because the quality product, output, is in part a reflection of quality of input, raw materials, biomanufacturers and regulatory authorities take very seriously the source and quality of each raw material, no matter how insignificant it is to the process or application. Special consideration is given to raw materials that contact or are incorporated into a final product. Raw material specifications and acceptance criteria are important to consistently meeting standards set forth in manufacturing plans and procedures. The possibility of raw materials containing toxins or adventitious agents is especially noteworthy because they present risk to the user and because they might enter the product stream and are difficult to detect and remove.

A raw material for biomanufacturing may be purchased from a vendor or it could be produced in-house by the product manufacturer. For example,

BOX 6.2 EXAMPLE OF A MATERIAL LIST: UPSTREAM FERMENTATION

Material Number	Description and Attribute	Source	Specification	Comment
H2-115	Working Cell Bank	TA Biotechnical	CoA	Manuf. 1/11/10
145621	Yeast Nitrogen Base without Amino Acids	DB/Fidco	CoA	No animal product
RX001	Glycerol	Spectarm	USP	No animal product
LC1121	8 N Ammonium Hydroxide	TJ Booker	USP	
31772	Glucose/ Dextrose	TJ Booker	USP	
32274	TSA Plates	Ramal	SOP QC-1181	Passed
32371	YPD Plates	Ramal	SOP QC-1181	Passed
16-2010	Water for Injection	TA Biotechnical	SOPs MF-1141 & QC-1832	WFI passed
4-115	Biotin	Spectarm	CoA	

Note: Manufacturer's material number; CoA = manufacturer provided Certificate of Analysis; SOP refers to internal testing by QC laboratory standard operating procedure with specification and passed by QC and QA; USP = U.S. Pharmacopeia-grade material; animal product = manufacturer provided certificate ensuring no animal product was used in this material.

sodium chloride is typically purchased, while highly purified water, water for injection (WFI), is often produced in the sponsor's facility. As you might expect, raw materials are highly controlled. To prevent misidentification or contamination, vendor-supplied raw materials are inspected, clearly labeled, sometimes retested and then kept in controlled storage areas of the manufacturing facility until consumed. (Quality of raw materials is further discussed in Chapter 5, Quality Systems, and in Chapter 7, Quality Control.)

Compliance and Quality in Biomanufacture: Current Good Manufacturing Practices

Quality considerations for biomanufacturing begin with design and planning and continue through the life cycle of a product. In the United States,

BOX 6.3 EXAMPLE OF A MATERIAL LIST: DOWNSTREAM PURIFICATION

Material Number	Description and Attribute	Source	Specification	Comment
1-110	Clarified fermentation Supernatant	TA Technology	BPR-661-00	Manuf. 1/10/11
040721	Water for Injection, WFI	TA Technology	SOPs MF-1141 and QC-1832	Passed
SF-1418	Sodium phosphate, (Monohydrate)	TJ Booker	USP	
TM0012	Sodium hydroxide, pellets (NaOH)	Spectarm	USP NF	
SF-1416	Sodium phosphate (dibasic) Heptahydrate	TJ Booker	USP	
C3HN5-9990	Millipak-20 filter units (0.22μm)	Milepour	CoA	Meet specifications
65 SD105	Superdex 200 Chromatography gel	EG Healthcare	CoA	cGMP Grade
30 SO672	Sepharose HP Chromatography gel	EG Healthcare	CoA	cGMP Grade

Note: Manufacturer's material number; CoA = manufacturer provided Certificate of Analysis; SOP refers to internal testing by QC laboratory standard operating procedure with specification and passed by QC and QA; USP = U.S. Pharmacopeia grade material; NF = National Formulary.

biopharmaceutical manufacturing quality is promulgated in a set of regulations known as current Good Manufacturing Practices (cGMPs). Other countries also have manufacturing guidelines and, for the biotechnology firm intending to export a biopharmaceutical, attention must be paid to directives from European, Japanese, Canadian, World Health Organization (WHO) and other national and international agencies or organizations. In addition, the ICH has guidelines on manufacturing quality. These references are further identified and discussed in Chapter 4.

Biotechnology products and raw materials, those other than biopharmaceuticals, also have manufacturing and product quality criteria, either known as an "industry standard" or established by industry trade organizations, national or international bodies and regulatory authorities. For example, the International Standards Organization (ISO) guides activities and establishes standards for biomanufacturing and thousands of other industrial endeavors.

Good Manufacturing Practice guidelines, no matter what the source, are followed by biomanufacturers both because they are regulatory requirements and because the guidance in cGMP makes sense from the standpoints of business development, product marketing, financial stability and product liability for pharmaceuticals. Recalls of product are expensive for a biotechnology firm and adverse events due to substandard product can be devastating even for a large company.

cGMP has the objective of consistently delivering the highest quality product to the user. Today, cGMPs apply beyond production activities in a biomanufacturing plant. They encompass the concept of biopharmaceutical design, risk analysis and manufacturing planning, functions that begin well before the product even enters the plant, and extend to warehousing, an activity found at the far end of the biomanufacturing development. The full embracement of cGMP, the U.S. FDA regulation, is phased into the manufacturing plan, as shown in Figure 6.3. Phase 1 manufacturing would, under cGMP, ensure that raw material and process hazards are identified and steps taken to ensure they do not endanger human subjects enrolled in a clinical study. However, the FDA recognizes that not all aspects of cGMPs apply to a given product, especially in early development and the agency offers additional guidance for Phase 1 manufacture of biopharmaceutical products. Application of cGMP requirements is considered in a manufacturing plan by focusing on product-specific attributes and quality issues that might impact biopharmaceutical manufacture for nonclinical and Phase 1 clinical studies. Risk analysis of the manufacturing plan is one way for quality issues to be identified and cGMPs applied to production processes in a rational manner.

The plan also considers manufacturing compliance as product moves through subsequent phases of manufacture and greater numbers of patients are exposed to a product. Now, cGMP application is broadened and becomes increasingly stringent for each stage of biomanufacture. There is an important disclaimer for certain processes such as aseptic technique or sterile fill because these are critical to product safety. With sterility and several other manufacturing areas, there is but a single interpretation of cGMP and it applies from Phase 1, or early phase biomanufacture, through commercial manufacture. Yet, where safety is of lesser concern, the concept of cGMP application is a gradient, beginning with cGMP at Phase 1 biomanufacturing and increasing through commercial production. The cGMP regulations are outlined in Chapter 4. Specific examples of quality criteria and application of cGMP are provided in subsequent discussions of biotechnology products and biomanufacturing technologies in this chapter.

Biomanufacturing Processes for Biotechnology Products

The discussion on biomanufacture now shifts from general subject matter to focus on various biotechnology products and the technologies used to manufacture these products. We begin by reviewing standard production methods used to biomanufacture recombinant proteins in living cell-based systems and follow with a discussion on the use of transgenic organisms. The text then moves to the field of stem-cell or somatic cell and tissue production; delves into technologies, such as the synthesis of biologically active molecules; and introduces the growing field of combination products, where biopharmaceuticals are merged with medical devices or pharmaceuticals (drugs). There is a great diversity of biological products, so it is impossible to mention each one, or even to discuss each class. However, the examples presented in this chapter should provide the reader with an idea of what has been achieved and, in a few instances, what could be done in the future to manufacture biotechnology products.

Expression of Recombinant Proteins and Nucleic Acids

Production of Recombinant Molecules from Expression Vectors

Laboratory methods to manipulate living organisms and the biological molecules they produce are at the heart of biotechnology. Operational endeavors, including biomanufacturing, flow from discoveries made in basic research laboratories where tools or methods are devised and first applied in discovery research. Hence, it is no wonder that the first step in biomanufacturing is discovery or invention of an organism or molecule that expresses a desirable trait or otherwise serves a useful function. To date, thousands of discoveries or inventions have enabled genetic engineering of nucleic acids and living organisms. Research laboratories have yielded a wealth of information regarding recombinant DNA, cell metabolism and the basis for life itself. However, there is a caveat in all of this. While it is these scientific findings that form the foundations on which we base product development, it is biomanufacture, the production of large amounts of biotechnology product of high quality, that brings the product to market and the user.

The production of recombinant molecules, notably proteins and nucleic acids, represents, by volume, the bulk of biomanufacturing capacity today. A variety of active recombinant molecules—proteins, such as insulin, human or bovine growth factor, monoclonal antibodies, and vaccine antigens and nucleic acids for genetic therapy and diagnostic purposes—are nearing or have entered the marketplace. Some are sold in large quantities and represent blockbuster products in the marketplace. Today, biomanufacture of recombinant proteins and nucleic acids meets growing demand and it represents an important economic sector of the biotechnology industry.

Genes, Vectors and Host Cells

The first stage in biomanufacture of a recombinant product involves three processes: gene cloning, development of an expression vector and production of cell banks. The process and controls of this first stage are outlined in Figure 6.4. First, a gene of interest is identified and isolated through molecular cloning, most often using polymerase chain reaction (PCR) and other technologies. Alternatively, the gene may be selected from an established library of cloned DNA. The gene is characterized by molecular weight determination and DNA sequencing. A vector, available from public or private vector libraries, is selected, this decision based on suitability for biomanufacture of the designated product. These attributes include ability to adapt and function in a suitable host, replication, promotion of protein expression, protein chain termination and absence of undesirable characteristics, among others. The gene is then inserted into a selected vector using methods such as recombinase-based cloning or restriction-ligase cloning.

Next, the vector is transformed into a host cell: bacterial, yeast, insect or mammalian. The host also must be carefully chosen so as to be compatible with the vector. Each type of cell has particular attributes and no cell is universally well suited for expression of every recombinant DNA or protein molecule. Following transformation, the vector must be stable, i.e., held within the host, and be maintained as one or more copies of the vector over many generations, as the host divides. Methods are applied to increase chances of successful transformation, but many attempts may be required before a stable and fruitful match between host and vector is achieved. Ultimately, transformed hosts are produced and one is selected, but only after extensive testing is performed to ensure that all objectives have been met. Once the transformed host is deemed acceptable, it is cloned by limiting dilution to ensure that all future transformed cells are derived from a single cell. The progeny of this host cell is considered a "research seed" and this seed is characterized for purity of cell line, retention of vector and other traits or attributes (see Chapter 7).

Selection of the host cell is worth additional mention. There are many species to choose from and within a species there are several strains, each well characterized. Bacteria, notably *Escherichia coli* and yeasts, such as *Pichia pastoris*, are common choices.

Host cell lines are purchased from a reputable source, such as the American Type Culture Collection or a biological supply house. These vendors maintain several strains or lines of cells, each with a full genetic history, and the buyer expects and should receive only top quality, highly characterized cell lines. Much like pedigreed horses, cell lines and strains are noted for various attributes, such as a posttranslational capacity or large yields of recombinant protein, under specified conditions. Also, they may have known limitations or deficiencies, such as slow growth or stringent nutrient requirements. Although several species and strains of host may be able to express a given recombinant protein, there are caveats, and care is taken in selection of any expression system.

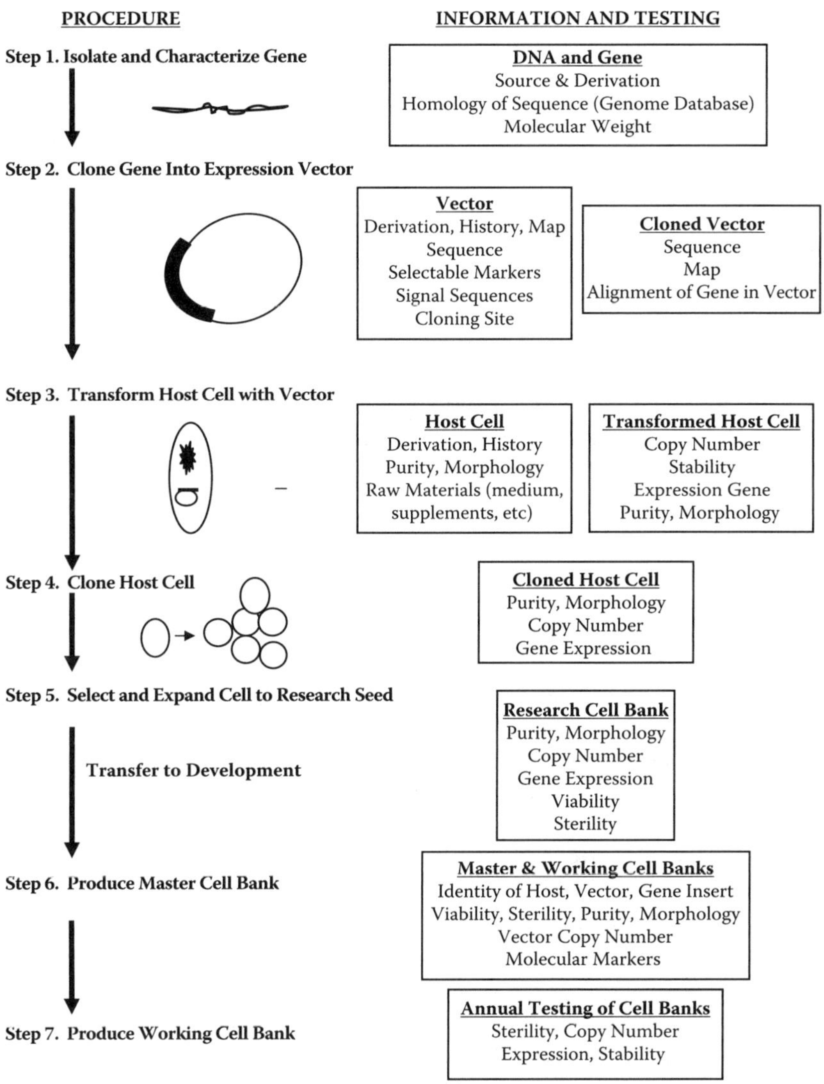

FIGURE 6.4
Production and testing of a recombinant molecule in an expression system and production of cell banks. This flowchart serves as an example of the steps that are taken in early development and biomanufacture of a recombinant molecule in an expression vector. The expression system is constructed by genetic engineering and then it is produced as a cell bank. Quality testing is performed throughout the process.

Bacterial Cell Expression Systems

Bacteria are often chosen as host cells because they express large quantities of a wide variety of proteins very economically. *E. coli* has been used for decades; the genome has been sequenced, laboratory strains are plentiful and very safe, and this bacterium is receptive to accepting, holding and expressing recombinant genes from vector plasmids. *E. coli* is often the first choice when biomanufacturing is considered. Within the species *E. coli* there are many strains to choose from and each has particular attributes and advantages as well as disadvantages. For example, some strains secrete the desired recombinant protein into the culture media during fermentation and this can simplify downstream processing. However, production by *E. coli* can have drawbacks as well. With certain proteins, *E. coli* does not secrete but instead harbors protein internally, within inclusion bodies. To obtain recombinant product from inclusion bodies, cells must be split open, lysed, adding extra steps, complicating purification and possibly adding unwanted impurities to the product stream. Protein from inclusion bodies may not be properly folded, necessitating refolding steps. However, purification from inclusion bodies may be easier and more productive than purification from cytoplasm for some recombinant molecules.

Another disadvantage to bacterial expressions systems is the inability to make or correctly complete certain posttranslational modifications to a recombinant protein. Bacteria do not add carbohydrates to proteins as do eukaryotic cells. Hence, if gylcosylation is required for bioactivity, a bacterial host cell may not be the best choice. Another issue with bacteria is the production of undesirable contaminants, these released into the process stream. Gram-negative bacteria have certain molecules, such as the cell wall component endotoxin, and if molecules like endotoxin cannot be readily separated from the desired protein, then such organisms are not good candidate hosts. These examples demonstrate the importance of identifying the proper host prior to beginning experimentation. Alone, this aspect of planning may save considerable time and resources.

Yeast Cell Expression Systems

Yeasts are eukaryotic, unicellular organisms offering both advantages and disadvantages as host cells. Two species are commonly employed, but other species are available. *Saccharomyces cervasiae*, brewer's yeast, is well characterized as an expression host as is *Pichia pastoris*, which has the advantage of secreting recombinant proteins into the culture medium. Yeast cells grow rapidly and economically in commonly defined medium, even in large vessels up to 10,000 or more liters and this can enhance protein expression scale up.

Yeasts are very efficient at producing some recombinant proteins and the fermentation of yeast cells is usually inexpensive. Both yeast and bacteria can be grown in the same types of fermentation vessel and equipment is standard, reusable and comparatively inexpensive. Yeast host cells are available in many strains, allowing selection based on attributes. As with bacterial host cells, and unlike mammalian cells, yeast growth medium is very well defined, so there is little concern about co-production of adventitious agents, such as human or animal retroviruses, with yeast cells. In contrast to bacterial cells, yeasts have the capability to correctly add and process many posttranslational modifications. Yeast strains commonly used in fermentation are genetically engineered to be inducible. This highly desirable trait means the yeast cells begin to produce greater amounts of a recombinant protein once a simple chemical, such as glycerol, is added to the fermentation chamber or when an exact environmental condition is established in the chamber. It allows for greater control of fermentation. Hence, production in a yeast cell system provides opportunity for production of recombinant molecules.

Mammalian or Insect Cell Expression Systems

These expression systems are increasingly selected by sponsors for biomanufacture, especially for production of high value human recombinant proteins, such as monoclonal antibodies. Although transformation of a mammalian or insect cell line with a genetic construct can prove more difficult, as compared to bacteria and yeast cells, these cell systems have the advantages of accepting and expressing a large gene and completing most posttranslational modifications. Because large proteins with gylcosylation (such as monoclonal antibodies) are common to the world of biopharmaceuticals, mammalian cells are frequently chosen as an expression and production system. However, as compared to yeast or bacterial cells, mammalian or insect cells are often less robust and more fragile, grow more slowly, may be more fastidious and, thus, require more stringent environmental controls, may require a continuous flow of complex medium to deliver nutrients and may demand continuous waste removal. Concerns regarding the presence of latent virus in mammalian cell lines, specifically cells from a new and poorly characterized cell clone that might harbor and then shed viral particles into the product stream, have slowed the introduction of new cell lines. However, advances in mammalian and insect culture techniques and extraordinary characterization efforts have overcome some of these difficulties and today there are several effective cell bioprocessing systems on the market.

Several factors enter into the choice of a mammalian cell line intended for protein expression. The ease of transfection with a particular gene or transfection technology, cell growth and protein secretion profile, and environmental requirements all enter into the decision. Hence, the key to selecting the correct cell line for expression of a given gene is experimentation with several of the highly regarded lines, which allows comparison with a particular construct. Cells are named by their derivation; most used for biomanufacture are cells of epithelial origin. High on the list of choices is the mammalian Chinese hamster ovary (CHO) cell, a cell line in use for over 50 years. This was widely used first in virology and cancer research laboratories and later adapted to biomanufacturing. CHO cells are very well characterized and certified free of adventitious agents (with the exception of endogenous retrovirus-like particles); products derived from CHO cell production have been used for decades without safety problems. Other cell lines chosen for biomanufacturing are Vero, MDCK, embryonic human kidney (HEK-293), baby hamster kidney (BHK), Per C6 and NSO.

The process of establishing an expression vector, referred to as transfection, is outlined in Figure 6.5. The expression gene is isolated and cloned into an expression vector, using methods described earlier in this chapter and technically in the manner described for yeast and bacterial hosts. The method of delivering that gene to mammalian cells, transfection, differs noticeably from methods applied to bacteria or yeast cells. One of several transfection methods, most commercially available, may be used to transfect a gene to a mammalian cell. Notably, the method must deliver the intended gene directly into the nucleus of the target cell. Most transfection methods rely on chance, the probability that a gene will enter into the nucleus of a mammalian cell and result in a stable and productive transformed cell. In practice, this means treating a large number of cells and using cloning and selection methods to determine which cells are stably transformed. Additional experimentation then characterizes cells and the best is chosen, cloned and expanded to become the transformed cell research seed (see Figure 6.5).

Transfected insect cells also are used to produce recombinant proteins, oftentimes large quantities of proteins that could not be well expressed in other systems. The process of gene transfer to insect cells is quite different from that applied to mammalian cells. The gene of interest is first inserted into the genome of baculovirus, a virus that normally infects insects, which then acts as the delivery vector. The cell lines used as targets are derived from insects and, thus, are free of potentially harmful human viruses, but capable of hosting baculovirus. Insect cells are desirable as well because, like the mammalian cells, they perform complex yet accurate posttranslational modifications. Upon infecting an immortalized insect cell of a well-characterized cell line, genetic information is transferred from the virus to the cell nucleus; some insect cells are stably transfected. The transformed cells are then identified, selected, characterized, cloned and expanded to produce a research cell seed. As the transfected insect cell grows and multiplies, it expresses the recombinant gene of interest and, in ever greater amounts, the gene product, a recombinant protein, is produced and it can then be harvested.

Production of Master Cell Banks and Working Cell Banks

Research seed, described earlier, is transferred from the research laboratory where it was produced, to a development laboratory where it will undergo additional examination and characterization. The complete history of the construct, to include descriptions and sources of all raw materials and

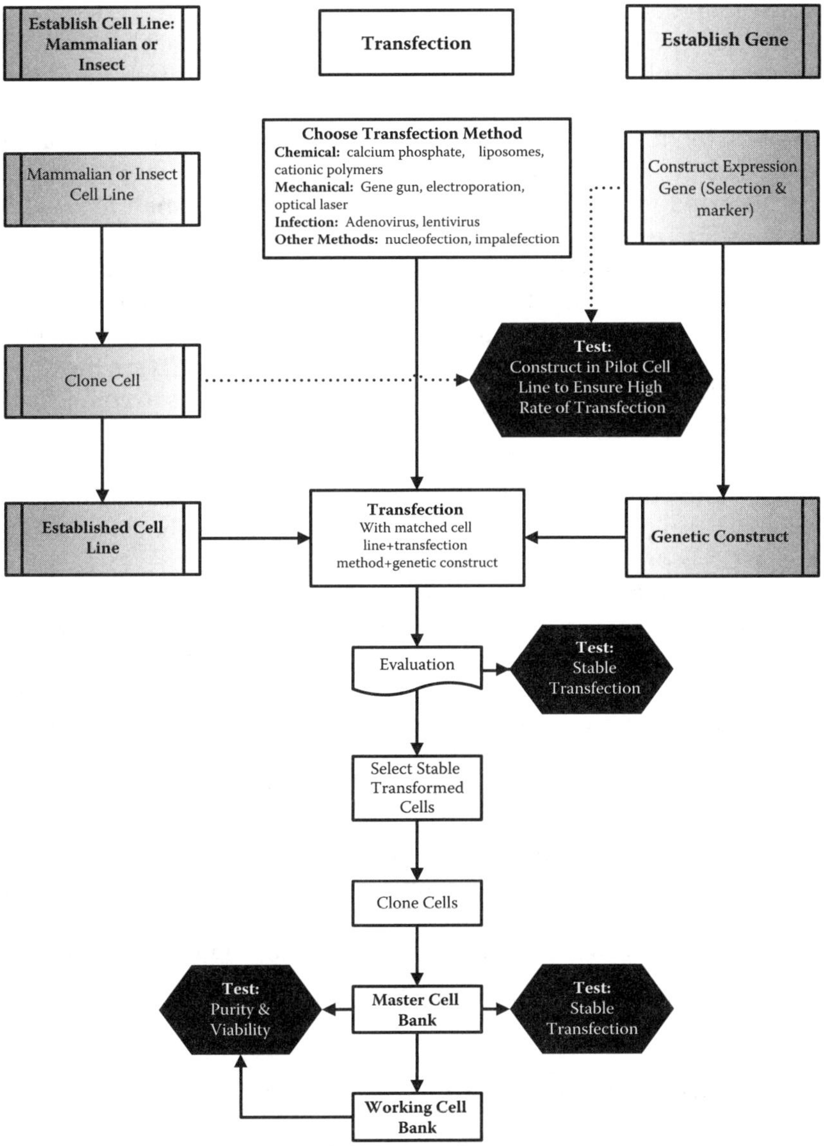

FIGURE 6.5
Process of gene transfection for mammalian cellular expression. An established cell line is selected and tested in the left panel and a gene is engineered in the right panel. Using a carefully chosen transfection method, the cells are transfected with the gene and following evaluation it is used to produce the expression product.

detailed summaries of procedures used to derive the seed, is archived as the research seed history. Once development scientists are satisfied the seed is adequate for production of the intended product, master and working cell banks are manufactured.

Cell banks provide a stock of cells with genetic inserts that are available to production for future use. Cell banks include the Master Cell Bank (MCB), derived from the research seed, and the Working or Production Cell Bank (WCB or PCB), this derived from the MCB. To produce an MCB for bacterial or yeast cells, a single clone of research seed is expanded in culture and then transferred for further growth in a shake flask or small fermentor, as outlined in Figure 6.4. A limiting dilution step may be employed prior to transfer, to ensure a single cell is indeed the ancestor of the MCB. The cells are harvested, counted and specified numbers are aliquoted into each of dozens of vials; these are labeled, frozen and placed into secure storage, usually divided between two or more sites. The vials of an MCB are the ultimate source from which the product will be derived for decades to come. Because the MCB is limited in number of vials produced while demand for an MCB could be great, a WCB is produced from a vial of an MCB. Procedures used to produce and control a WCB are very similar to those used for an MCB. Biomanufacturing uses stock from the WCB until it is exhausted, and then another WCB is produced from a vial of MCB.

To ensure identity, purity and safety of the MCB and the WCB, samples taken immediately after production from both an MCB and a WCB are extensively tested, as outlined in Figure 6.4 and described in Chapter 7. Tests of MCB and WCB samples are repeated at specified intervals (e.g., annually). This stringent test regimen once again emphasizes the need to develop analytical tools early in development, even before biopharmaceutical production begins It is important as well to note the precious nature of the construct, the MCB and WCB because they represent much time, effort and expense. Secure storage is critical. The processes of producing research seed, the MCB and WCB are often both rewarding and instructive to a new biotechnology operation and often represent their first introduction to biomanufacture under cGMPs. There is great satisfaction in having completed the first stage of biomanufacturing by having produced the foundation for later production efforts.

Biomanufacture of Recombinant Proteins

Planning Production of a Recombinant Protein

In a product-specific manufacturing plan, most processes have at least three stages and each stage is further divided into several technical steps, as outlined in Figure 6.1. We previously covered the steps of Stage 1 and will now consider Stage 2, or upstream processing, production of recombinant protein in the expression system and Stage 3, downstream processing, or purification of recombinant protein product as bulk substance. A good biomanufacturing plan goes beyond the initial process outline and also considers facility, utilities, equipment, raw materials, quality control testing, staff requirements and compliance, or cGMPs, for both upstream and downstream processing.

The earliest attempts at biomanufacturing under a new process or for a new product is referred to as pilot production. Pilot production involves performing defined, sequential runs in an attempt to develop the process and to eventually "get it right," i.e., to make a safe, pure and potent product. Indeed, pilot production is much like research experimentation because it involves trial and error, tweaking various systems, even making significant changes in process protocols and procedures. Pilot production may precede Phase 1 production, described above, or it may overlap or be synonymous with Phase 1 production. The term *run* is used in biomanufacturing to describe performance of one defined process, such as all steps in the upstream fermentation stage, or a full set of process steps, fermentation followed by purification from beginning to end. It also demands repeatability to confirm the system is performing properly. It is not unusual for a biomanufacturing operation to attempt a new process in 5 or even 10 runs before it is considered reproducible and robust. Hence, no matter what the biotechnology product is, pilot production can be a long, arduous and expensive endeavor stretching over several phases of development.

Upstream Process: Production by Bacterial or Yeast Cell Fermentation

Fermentation is an ancient process, best exemplified by brewing of beer in the presence of yeast, a skill developed over the ages. Substrates for fermentation of biomolecules remain simple and use well-characterized materials, such as water, salts and sugars. In some instances of biomanufacture, more complex nutrients, such as soy extracts or vitamins, may be added to the fermentation vessel. Only in special circumstances, where they are essential to the success of a process, are animal materials, such as liver powder or serum supplement, used in fermentation of a biopharmaceutical product. Such materials may harbor adventitious agents that can contaminate final product and their use is discouraged.

Fermentation to produce biotechnology products is performed in a fermentor, a closed and sealed glass or stainless steel vessel with a series of portals, stirring devices and tubes entering the chamber (Figure 6.6). To begin the fermentation process, one must have raw materials of the highest quality, including a growth medium, gasses, a seed of recombinant bacterial or fungal cells and a means to control the process. Seed material, billions of cells capable of active division, is derived from a vial of a WCB that has been expanded in a flask containing the defined medium. This is called the inoculum. The environment inside the vessel is controlled by human intervention or, when on "autopilot," by a microprocessor. The fermentation process is initiated once all ingredients and the seed have been added together in the vessel and the fermentor has been closed and sealed. Once the environment

FIGURE 6.6
Equipment for microbial fermentation. This picture shows fermentation equipment in a biomanufacturing suite. The operator in the center is programming the microprocessor controller in the square unit. On either side of the controller stand two fermentation vessels, a small one in the background on the bench top and a medium one behind the controller. In the foreground is a large, cylindrical storage vessel also made of stainless steel. (From Waisman Clinical Biomanufacturing Facility, University of Wisconsin/Madison. http://gmpbio.org. With permission.)

inside the chamber is optimal, the cells divide and are active manufacturers of the recombinant product. The chamber is typically stirred or otherwise agitated in an effort to evenly distribute gasses (particularly oxygen), nutrients, cells and, if secreted, product. Because stirring and movement of gasses may cause foaming, addition of antifoaming agents, chemicals designed to reduce microbubble formation, is often helpful. Cell growth and product production are monitored by taking samples from the chamber and critical measurements, such as pH, gas tension and osmolality, are measured by probes placed directly in the chamber. This in-process testing allows the operator to follow progress and correct variables if they should deviate from specified limits. For example, if pH should drop beyond operational limits, sodium hydroxide could be added to raise pH.

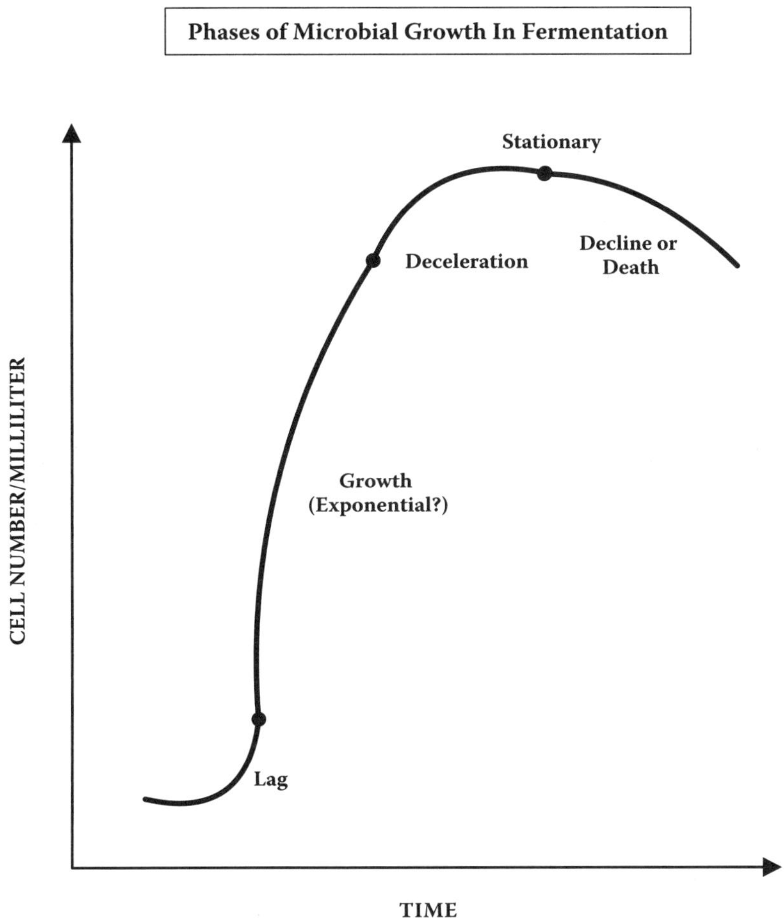

FIGURE 6.7
Phases of microbial growth in fermentation.

Cell division and growth are divided into several phases, shown in Figure 6.7. First, is the brief lag phase during which cells adjust to their environment. Next is the exponential growth phase during which cells divide rapidly and hopefully produce the intended product in great amounts. The deceleration phase represents a slowing in growth and the fourth phase is stationary, with little cell growth or even cell death. The final phase, decline, represents decline and cell death. Once measurements determine that the cells have grown to the required optical density, that cell death is extensive or that sufficient product has been produced, then the reaction is halted by radically changing the pH, rapidly cooling the chamber or by another intervention that is not destructive to the product. It is important to terminate cell growth at the right phase of growth, as there is a time beyond which little more product will be produced and yet dying cells still release enzymes or impurities into the medium; continuing a controlled fermentation beyond that point can interfere with product purity or complicate downstream purification. The result of successful microbial fermentation is a vessel filled with a slurry of cells, cell debris, expended medium and, hopefully, the intended product.

Certain cells, notably yeast used in some fermentation systems, may be cued or induced to begin production of recombinant protein. To engineer an induction system, a gene or genes are inserted into the expression vector for the purpose of controlling protein production by a cell. These genes are active in the presence of certain environmental cues. An example is selective induction of recombinant protein production by *P. pastoris* upon addition of glycerol to the fermentation vessel.

Upstream Process: Production by Mammalian or Insect Cell Culture

Mammalian or insect cells are cultured in a sealed chamber referred to as a bioreactor. While mammalian or insect cell culture has superficial resemblance to fermentation, the processes are, overall, quite distinct. The objective in both systems is to produce a recombinant protein product that is either stored within the cells or secreted into the medium. Both bioreactors and fermentors are closed and sealed systems with a high level of monitoring and environmental control. Mammalian or insect cells typically demand more complex substrates than do bacteria and yeast. Hence, cell culture media used in a bioreactor contains a complex mixture of nutrients and vitamins. When animal products are used for biopharmaceutical production, in fermentation or cell culture, they must be carefully tested and controlled so as to ensure that microbes and impurities do not contaminate the cells or the product. Typically, mammalian or insect cells must be grown as anchored cells. This is because, unlike bacteria or yeast, these cells in nature exist in a tissue or organ where cells are interconnected and held firmly to a basement membrane or other connective tissue protein. Alternatively, hollow fibers, convoluted vessel surfaces and microcarrier beads, sometimes coated with collagen or other connective tissue matrices, may be used to create optimal microenvironments, facilitating cell anchoring or adaption to culture conditions yet confining cells to a small area in the absence of mechanical stresses. Mammalian cells engineered to grow without anchorage are grown in suspension, but agitation or stirring is exceptionally gentle, as insect and mammalian cell membranes are fragile. Gentle air movement or wave action is used in some systems to maintain the requisite movement of the cell medium for suspension cultures. Mammalian cells are particularly susceptible to reduced growth due to low oxygen tension, high carbon dioxide tension, buildup of waste, changes in pH, and other metabolic–environmental influences. Cell bioreactors are closely monitored and the addition of gases, buffers and nutrients is highly controlled, both by the operator and by microprocessors. The vessel environment of a cell bioreactor is monitored using measurement probes and microprocessors and by operators, and it may be adjusted by allowing materials to enter through aseptic ports.

Mammalian and insect cells demonstrate growth curves representing cell growth as a logarithmic or semilogarithmic phase, followed by a plateau phase and finally a decline phase. While growth and protein secretion of mammalian cells is typically slower than is seen in bacterial or fungal cell fermentation, under well controlled operating conditions, the growth and secretion of mammalian cells may be sustained for much longer periods. A cell culture is terminated at an exact point in the growth and protein production cycle to maximize protein production and minimize contaminants. The result of successful cell culture is a bioreactor vessel filled with a slurry, this composed of cells, cell debris, expended medium and, hopefully, the intended product.

Upstream Process: Recovery

Immediately upon stopping cell growth in a fermentor or bioreactor, the material is harvested, chilled and the cells and other large solids separated from the liquid phase. This is done by moving, with pumps, the contents of the chamber into a capture vessel. In the case of cells anchored to substrate, or when product is contained within the cells, it may be necessary to dislodge cells by mechanical or enzymatic means. If most of the product is harbored in the cells, as would be the case with protein that is not secreted, the cell paste is retained and the supernatant discarded. Whole cells containing product are then lysed, using chemical shock, osmotic shock or mechanical pressure, and soluble proteins are harvested, again using centrifugations and filtrations. Primary clarification, to remove any remaining cells, cell debris and other large solids, is performed by centrifugation or tangential flow filtration (Figure 6.8). The resultant filtrate or supernatant with the recombinant protein is now kept in a storage tank under controlled conditions until it is purified. The storage step is referred to as a *hold*.

Downstream Process: Purification

No matter what the source—fermentation, cell bioreactor, transgenic animals or plants—recombinant proteins and other biological molecules must be purified from what is typically a complex milieu of cellular debris, impurities and contaminants. By way of definition, impurities are undesirable materials, both particulate solids and soluble molecules, remaining with product following production. Common impurities derived from biological processing are host cell proteins or DNA, endotoxin or other microbial toxins, cellular debris and organelles, and materials from culture media. Contaminants are substances that enter the product stream, often during purification, and frequently are shed or leached from the process materials or equipment. Small particles from tubing, glass or metal containers or chromatography gels or heavy metal ions, leached from metal containers,

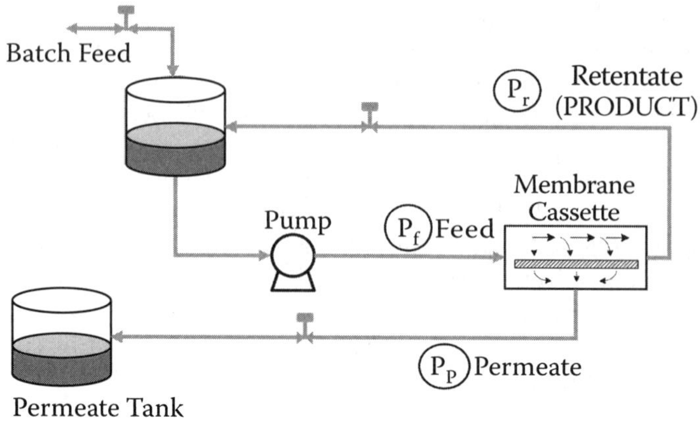

FIGURE 6.8
Tangential flow filtration. Material enters the filtration scheme from the batch feed, usually a storage vessel (holding tank), and is pumped under pressure (feed) across a membrane cassette where some material of the correct molecular weight passes through the membrane cassette as permeate and is then held in a tank. Material that does not pass through the cassette reenters the batch feed tank and is again pumped across and, in some cases, through the membrane cassette. Continuous flow across the membrane cassette deters clogging the cassette selective filter.

are examples of contaminants. Impurities or contaminants may be debris, suspended particulate or soluble in nature. Both are considered necessary evils because their presence reflects the contents and environment of a culture that yielded the product. However, impurity and contaminant levels are greatly reduced during purification and, in the end, biotechnology products are tested for common impurities and contaminants (see Chapter 7) to ensure the levels meet specifications and the product is safe.

Purification steps are designed to remove one or more impurities or contaminants, at least to the greatest extent possible, and yet retain the desired recombinant protein, or other biologically active molecule, thus maximizing the yield of product. Yield is especially important because a low yield of very pure product is as unacceptable as a high yield of product with significant levels of impurities. Hence, in-process testing (see Chapter 7) is applied throughout purification to ensure improvements in purity and maintenance of yield at every step. The field of biomolecular purification has progressed rapidly in recent decades and dozens of methods, some simple and others quite complex, have been developed and entered the marketplace. We will mention but a few of the most commonly used methods.

To plan downstream processing, a purification scheme is produced (Figure 6.9). To do this, it is first necessary to understand the biophysical and biochemical properties of the recombinant molecule because purification methods take advantage of those properties. This understanding is based on experimental data derived from the research laboratory about the product and the nature of that product as it enters purification. Knowledge is also needed

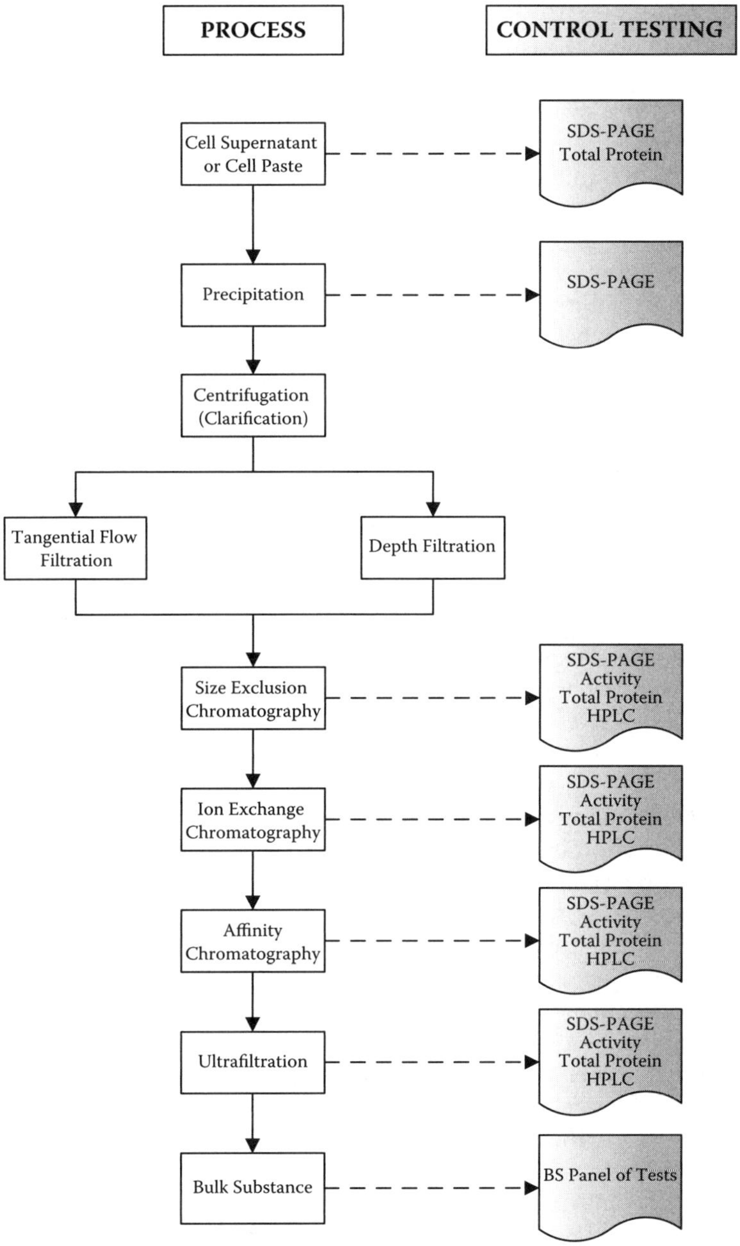

FIGURE 6.9
Purification scheme for a recombinant molecule. This is a classic purification scheme, or downstream process, for a recombinant protein and includes precipitation, centrifugation, filtration and multiple chromatography steps, yielding bulk substance (BS). (Graphic courtesy of Waisman Clinical Biomanufacturing Facility, University of Wisconsin/Madison. http://gmp-bio.org. With permission.)

of the possible contaminants and impurities and their properties. For example, it is critical to know the isoelectric point of the desired molecule under given conditions, the pH or salt concentration at which the molecule precipitates from solution, the size and shape of the molecule, the glycoslyation profile or any propensity to bind to other molecules or inert substrates. Purification schemes, as shown in Figure 6.9, take advantage of these attributes.

Many technical methods, or purification tools, are available for downstream processing. Choice and application of a method is based on the nature of the molecule and knowledge that certain tools have, in the past, been used successfully to purify similar molecules. Some purification tools are quite simple and inexpensive, while others require significant investment. The sequence in which methods are applied is as important as the choice of the tools themselves and the downstream plan must choreograph their use very carefully. It is often necessary to test each purification method alone and the full sequence of methods, at small scale in the laboratory, in an effort to derive the optimal sequence of events, and before beginning operational purification. In-process tests are another critical component of a purification plan. These are developed to ensure that materials, such as solutions, are of the correct composition, pH or strength and can be effectively used to measure levels of the desired molecule and of impurities throughout the purification process. These assays, which can require expensive instrumentation and extensive development efforts, must be available from the outset of purification process development because they are essential to an understanding of purification outcomes.

Early in downstream processing, and often at other stages in biomanufacture, the liquid fraction must be clarified, without significant loss of the desired protein, using combinations of precipitation, centrifugation and filtration. Precipitation is a simple and inexpensive application because many methods for precipitation, such as changing pH or adding simple salts, may neither degrade the protein of interest nor add contaminants to the product. Precipitation is based on the knowledge that a desirable protein or an undesirable impurity becomes insoluble under certain conditions, such as low pH or high salt concentrations. Once a precipitate is formed, it is separated from the undesirable materials by centrifugation or filtration. Desirable molecules in the precipitate are recovered by diluting the precipitate with a physiologic buffer. Undesirable proteins in the precipitate can then be discarded. An example is purification by precipitation of an immunoglobulin upon addition of a buffer with a high salt concentration. Immunoglobulin precipitates in this environment, leaving impurities in the supernatant. This is centrifuged and the pellet is recovered and diluted with physiological saline solution to again solubilize immunoglobulin protein. Another example is application of a polycationic agent, such as polyethyleneimine, to precipitate undesirable nucleic acids, while the desired recombinant protein remains in solution. Following centrifugation, the pellet with impurities is discarded and the supernatant is retained or visa versa, depending on which fraction holds the desired product.

Centrifugation, a relatively simple and often effective method, is employed whenever possible and often constitutes the first step in a purification scheme. It separates materials based on density, shape and other physical properties that impact their gravitational movement in a fluid. It can be used without other treatments, such as in effective separation of large impurities, like cell walls or nuclei, from smaller particles and soluble proteins. Centrifugation also is used in a step-wise manner to sequentially remove matter of different density. It is often applied in conjunction with other applications, such as precipitation. Centrifugation equipment is available in many designs, from small instruments to large, continuous-flow machines that can process large amounts of product.

Flow filtration or tangential flow filtration are methods used in purification schemes as well. Both methods remove debris and clarify a solution in which the recombinant protein is suspended. Flow filtration involves passing the material through a selective membrane filter, a synthetic sheet that has holes of a specific size. As solution is pushed against a filter, solids of that size or less move through the membrane and larger particles are trapped atop the membrane. But a disadvantage to flow filtration is buildup of material on the membrane surface that can clog and foul filters. Some filtration protocols, therefore, use a series of flow filtration filters. The fluid stream is fed first through filters with larger membrane holes. Then, in a series, it is fed through filters with smaller holes, thus distributing particles over many filters and avoiding fouling and clogging of a single filter.

Tangential flow filtration (TFF) is a more expensive, but often more effective, method and is also a choice for processing larger volumes. In TFF, solute passes over the filter in a horizontal stream even as filtration is happening in a vertical plane (Figure 6.8). This horizontal movement constantly sweeps debris off the filter surface and prevents clogging. Because it can be performed rapidly, TFF is often used to exchange solutions, such as one buffer for another, and to selectively remove low molecular weight impurities, all in a single step.

While extremely useful, filtration must be applied judiciously because, under some circumstances, it also may destroy molecular integrity. Filtration causes shearing forces as the fluids move under pressure across or against a membrane and shear can destroy cells or molecules, with each product having a unique tolerance for shear. The only way to fully realize the effects of shear on a given molecule is through experimentation followed by characterization of the desired molecule. Movement of fluids rich in proteins may create foaming, an undesirable outcome of many biomanufacturing processes, and this must be avoided or countered when it does happen. A third cautionary note is avoidance of adsorption of the desired molecule to surfaces of equipment, transfer tubes, filtration membranes, and even to impurities. Again, any purification step must provide consistent and useful yield and adsorption can greatly reduce the amount of desirable cells or proteins left in the product stream.

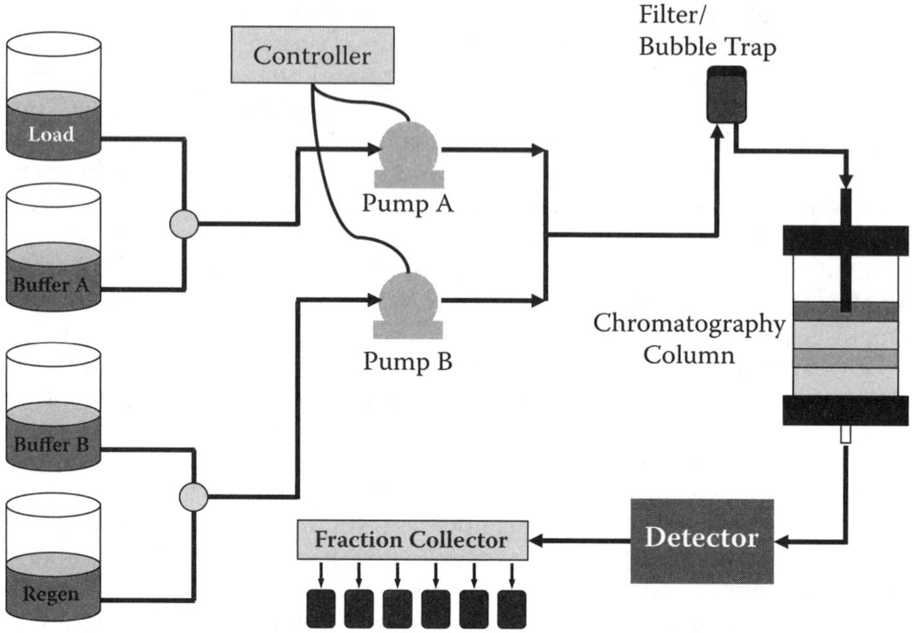

FIGURE 6.10
Flow diagram for preparative chromatography. This scheme depicts the equipment and flow for a chromatography system used as one step in purification of a recombinant protein. The product (load) and Buffer A are pumped into the column via pump A where the gel matrix of the column binds or otherwise slows progress of the molecule of interest (e.g., through affinity binding, size exclusion, ionic interaction). Other molecules pass through the column and are detected and collected into fractions. Once this has been completed, Buffer B is pumped onto the column with the intention of releasing the bound molecule. The desired product is detected as it comes off the column and it flows into later fractions, where it is collected.(Graphic courtesy of Waisman Clinical Biomanufacturing Facility, University of Wisconsin/Madison. http://gmpbio.org. With permission.)

Chromatographic methods are used in most biomanufacturing schemes involving molecular purification. Chromatography is based on various properties of proteins and other macromolecules: charge, size, shape or affinity to a substrate. Preparative chromatography is used to purify significant amounts of materials while analytical chromatography, described in Chapter 7, is used to characterize macromolecules. Preparative chromatography, the subject of this discussion, is performed using aqueous suspension of resins or gels packed into a vertical column (Figure 6.10). Preparative chromatography columns come in various sizes and shapes (Figure 6.11) to suite the intended purpose with some exceeding the volume of household refrigerators. They are controlled with pumps, valves and microprocessors. Each column-and-resin chromatography system has a unique set of properties that allow for the differential separation of molecules based on physical or chemical properties of the molecules to be separated. In the simplest

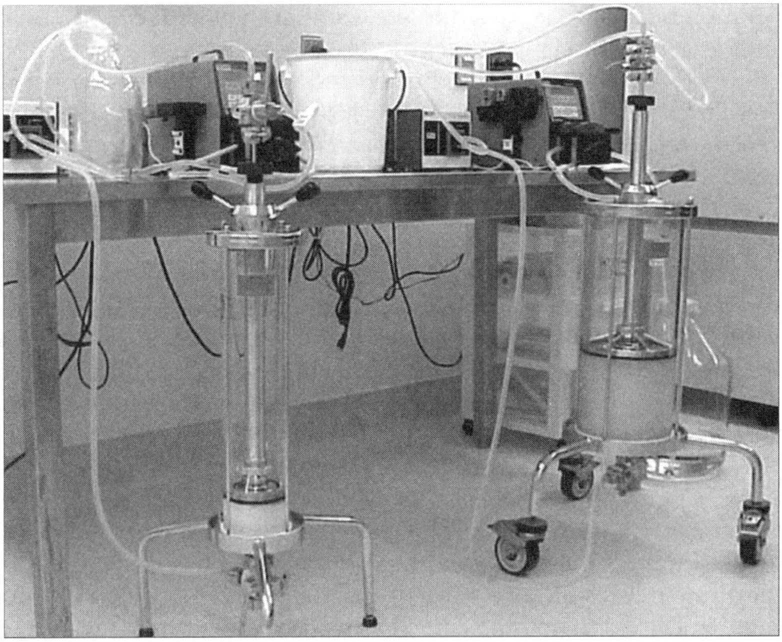

FIGURE 6.11
Equipment for preparative chromatography. Two preparative chromatography columns rest on tripod supports. On the table top are controller units and pumps with tubing leading to the glass columns. This is performed within a chromatography suite of the biomanufacturing facility. (From Waisman Clinical Biomanufacturing Facility, University of Wisconsin/Madison. http://gmpbio.org. With permission.)

chromatography protocol, clarified supernatant containing both product of interest and impurities is placed at the top of this column; then, using gravity or a pumped stream of buffer, it is passed through the column. As fluid passes over the column, the molecules contact the resin and they may or may not bind to the column. As it passes out of the column, fluid is collected in a series of tubes, each a fraction representing a specific volume and time of collection. By eluting with various buffers, a gradient is established on the column and desirable proteins leave the column in one fraction, while undesirable proteins exit the column in another fraction. This process is shown by simplified format in Figure 6.10.

Chromatography uses distinct molecular properties to separate one molecule from another. For example, ion-exchange chromatography uses the charge properties of various molecules to separate desirable from undesirable proteins. Here, the resin has a known electrical charge at a given pH and ionic strength (salt concentration). Elution buffers added to the chromatography column may be changed by the operator over time. Each buffer has a given salt concentration and pH. Taking advantage of ionic properties of both the gel matrix and the desired protein, the buffer strength and pH of the elution buffer are adjusted to ensure that the target protein binds, by ionic interaction, to the ion exchange resin. For example, at pH 7.0 and ionic strength of 100 mM (Figure 6.10, Buffer and Pump A), a recombinant protein might bind to the resin while impurities pass through the column to be collected in the early factions. Next, a second buffer (buffer and pump B in Figure 6.10), of pH 6.9 and ionic strength of 150 mM, is added to the column to release the recombinant protein. And so on. Eluate is collected in later fractions, and some of these contain product, largely free of impurities.

Many other types of chromatography are available to the biomanufacturing operator. Hydrophobic interaction chromatography takes advantage of a molecule's affinity for or, alternatively, rejection of water. To purify hydrophobic proteins, a gradient with high-to-low gradient of salt concentrations is established in a column containing hydrophobic interaction resins. Size exclusion chromatography takes advantage of the size and/or the shape of a molecule. It is particularly useful to purify proteins of interest if they are particularly large or small or have an unusual shape. Affinity chromatography methods employ ligands, attached to chromatography resins, to capture the desired protein as it passes through the column in the presence of physiological buffer. For example, Protein A resins are commonly used to retain monoclonal antibodies to a column. The protein A molecule, derived from bacteria, naturally sticks to antibody molecules. When Protein A is immobilized on a resin and placed into a column, any antibody passing over the resin will, in physiological buffer, adhere to the Protein A while impurities pass through the column. In the second step, a buffer solution, known to force Protein A to release the antibody by molecular or ionic competition, is passed over the resin. Now, monoclonal antibody, without impurities, elutes into the subsequent fractions.

Another purification method is the use of "tags" in affinity chromatography, with the polyhistidine tag being quite popular. Here, the protein of interest must, in research or early development, be genetically engineered to have at the C- or N-terminus a series of nucleotides that repeatedly code for histidine. Polyhistidine tag chromatography is an affinity method in which the chromatography resin has immobilized nickel ions. As the protein harboring the polyhistidine tag passes over the column, it binds to the nickel, while other impurities, without the tag, pass through the column. Subsequently, selective buffers passed over the column trigger release of protein from the nickel. While providing an efficient purification method, the use of tags is discouraged by regulatory agencies since they might impart undesirable properties to a biopharmaceutical.

More chromatographic methods are available and still others have been developed to purify specific biomolecules. For some molecules and even for living cells, in situations where readily available purification methods are not useful, a new and very product-specific chromatographic method, such as affinity chromatography, is often developed out of necessity.

Purification is a lengthy process as the tools are applied over days or even weeks. Pauses, referred to as holds, in a series of events are commonly incorporated into process schemes to allow for in-process testing and allow operating staff breaks in schedule. However, pauses require product storage and storage can result in product degradation. So pauses must be carefully planned and monitored. During a hold, the product may degrade, which can be due to impurities in the material, such as proteases, to influences of the hold environment, like oxygen or pH, or even to the container surface acting as a catalytic agent. In general, greater lengths of storage time and higher storage temperatures accelerate product degradation. To prevent degradation, it is important to understand the cause and to plan holds to prevent degradation of given product. For example, protease degradation is reduced by storage in various buffers or by addition of protease inhibitors, substances that are inherently safe and can later be separated from the product. Oxygen tensions can be adjusted, antioxidants added or containers may be lined with inert materials to prevent product breakdown. Planning each process and hold step is based on a knowledge of the product, possible impurities and contaminants and the product's stability profile.

The end result of purification efforts is bulk substance, a pure, potent and stable product within the proper bulk container. For a biopharmaceutical this is referred to as bulk (drug) substance or BS.

In summary, purification processes are planned based on the properties of the product and possible impurities or contaminants. Success at purification frequently involves experimentation and trial-and-error in the research laboratory. Purification is first attempted at small or model scale, to better understand the attributes of each application. The biomanufacturing operator does not expect to "get it right" the first time through. Indeed, it may be shown that a purification tool or a series of methods negatively impacts the molecule of interest or it may be discovered that product yields are unacceptably low. If the negative impact is irreversible and the protein cannot be recovered to the native form, then the operator might drop that step and try another. Alternatively, the method can be modified. A third possibility is application of a recovery step intended to return the molecule to its native or desired state. In reality, it may be necessary to apply several tools and determine, experimentally and by trial and error, the impact of each, before the correct process or formula is discovered. Again, success in this endeavor is based on knowledge of the protein and all tools available to the operator.

In-Process Testing and Analysis of Bulk Substance

In-process testing is a hallmark of product purification. The operator needs to know, at each step, whether his/her purification scheme is achieving the intended objectives of removing impurities and contaminants while enriching the desired product, and without significant product loss. Hence, quality control (in-process testing) is applied to samples taken at the completion of each step. More is written in Chapter 7 about individual analytical tools commonly applied for in-process testing of biopharmaceuticals.

Examples are measurements of product, particles, contaminants or impurities. At each step, the operator is interested in learning if product remains in the stream and, if so, to identify its molecular integrity. Relatively rapid methods, for example, examination and measurement of protein bands following SDS (sodium dodecyl sulfate)-polyacrylamide gel electrophoresis of a sample, quickly provide information, qualitative and quantitative, about yield of both the intended protein and contaminants and impurities at each step in the process. These tests must be readily available to a biomanufacturing operator.

Both quantity and quality of a bulk substance matter greatly to the sponsor. Quality control test results of bulk substance must demonstrate that the product, at this stage, possess intended attributes. In Chapter 7, we discuss the tests used to measure these attributes. Each test is classified under the attribute it measures: identity, safety, purity or potency. Purity is of particular importance because this is a key objective of downstream processing. But how, in a general sense, do we define purity of a molecule such as a recombinant protein? One guideline often applied to biopharmaceuticals is that over 95% of the bulk substance is the intended and intact ingredient and less than 5% is known or unknown impurities and contaminants. Most biomanufacturing operations strive for over 98% or greater purity, certainly for commercial manufacture. However, there are caveats to this purity guideline. First, the balance of material in bulk substance, the remaining 2 or 5% if you will, must be known, indeed be characterized, for it cannot be toxic or potentially toxic to the user and it must consist of various materials without a predominant molecular entity: impurity or contaminant. Second, it is not always possible to reach the 95% purity level and, in such instances, it may be acceptable to identify impurities and show that they cannot be harmful to the product or the user.

Knowing what could be or should not be in bulk substance is helpful in making these determinations. For example, a recombinant protein product derived from bacterial host cells might be expected to have very small or trace amounts of bacterial chromosomal or plasmid DNA and it would be tested for such impurities and measured against a specification for host cell DNA. Also, in-process testing focuses on materials that are there and those that could be there, but not those that are highly unlikely to be there. For example, the bacterial product would not be expected to have mammalian or yeast cell DNA and so the operator would not test for DNA from eukaryotes. In-process testing will, in part, help the operator to understand the makeup of material and to pinpoint the step where it either entered into, or was not fully eliminated from, the product stream. Understanding the potency of product is another critical step in characterizing bulk substance and at this stage of manufacture there must be either an indirect or direct measure of potency.

Recombinant proteins are not the only biotechnology products produced by biomanufacturing technologies. The following paragraphs provide by way of example an overview of possible manufacturing approaches for a few of the many other biotechnology products.

Production of Bacterial Plasmid DNA

Bacterial plasmid DNA, used in DNA vaccines, genetic (DNA) therapeutics, diagnostic tests or as research laboratory chemical, is produced by bacterial fermentation. Once purified, biomanufacturing processes may yield up to 1 g of plasmid DNA per 10 L fermentation. RNA also can be produced, albeit in milligram quantities, by *in vivo* transcription in *E. coli.* To produce DNA, a plasmid vector is constructed in the research laboratory and tested for the intended biological effect. An appropriate cell line of *E. coli* is transfected with the DNA plasmid and cell banks are produced from research seed. Beginning with a WCB, cells are grown in a fermentation vessel and then harvested and lysed to release supercoiled plasmid DNA, the intended product for most purposes. The plasmid DNA is then purified using physical separation and chromatographic methods. Separation of DNA from impurities, such as chromosomal DNA, RNA and bacterial host cell proteins must be considered during purification. During all processing, the plasmid product must remain supercoiled, that is, in the closed circular form that coils about itself. Quality control testing focuses on identity, purity and identification of minor impurities, and potency in a relevant biological assay for all DNA products. The form of the plasmid (linear, circular, relaxed or supercoiled) is also determined by analytical testing.

Production of Live Recombinant Organisms: Bacteria and Virus

Live virus, bacteria and even protozoa are used as biopharmaceuticals or as diagnostic or laboratory reagents. Virus, e.g., retrovirus, may be used to transfer therapeutic genes to patients in gene therapies. Live virus, such as vaccinia, adenovirus or alpha-virus, are often employed as vaccines. Live bacteria are used as investigational biopharmaceuticals, both therapeutically and as vaccines. Even protozoa, such as attenuated malarial sporozoites, serve as live vaccines.

There is considerable experience in bacterial and viral culture and purification, largely for vaccine production, though there are potential issues with some viral or bacterial constructs. One concern is possible release into the environment, or spread by close contacts, of live recombinant virus or bacteria. Another is the question of whether these microbes retain the capacity to cause disease in humans, plants or animals, even a very few individuals. To address these issues, live organisms are carefully designed and endowed with redundant systems engineered into their genome. Some systems ensure they cannot survive outside the environment provided by cell culture or a living host organism. Other systems delete genes responsible for pathogenicity, making them unable to cause disease. Further, safety issues related to exposure and release are considered in manufacturing plans and process controls. Extensive safety testing is performed to measure attributes of organism in research seeds and stringent specifications are applied.

These recombinant organisms are constructed in the research laboratory using well characterized, attenuated strains of virus or bacteria. Live bacteria or virus (Figure 6.12) are grown by fermentation or cell culture, respectively. Because live virus is propagated in cultured cell lines, it is necessary to identify a virus host cell, often of human origin, which both yields large quantities of viral particles and is free of adventitious agents or other undesirable traits. This requires significant resources because the number of satisfactory choices is limited and the processes of adapting virus to the host, cell banking (both MCB and WCB) and testing for identity and purity is lengthy and expensive. Cells are first grown to optimal cell density in a bioreactor or in large flasks. Cell growth medium is both well-defined and -characterized to prevent contamination of cultures with adventitious agents. Raw materials from animal sources, such as serum, are seldom acceptable as media supplements. Virus from a WCB is inoculated onto cells and, after a period of incubation and viral replication, the cells are harvested, lysed and then virus is separated from large debris by centrifugation or filtration of the medium. It is then purified, using density gradient centrifugation or selective filtrations to separate viral particles from impurities, such as cell debris and culture medium. Virus of high purity, bulk substance, results from the process and this is extensively tested for identity, safety, sterility, purity and potency. Potency tests are developed to reflect the intended use of each product. Virus may be examined for desirable traits, those that enhance the intended mechanism of action. For example, if a vaccine must first attach to epithelial cells to be immunogenic, testing might examine virus for that receptor. Hence, both potency and safety tests are often complex assays, many immunological or molecular, and others performed in tissue culture or with live animals.

Live bacteria are propagated from a bacterial WCB as recombinant biotechnology products. These are grown in defined medium, preferably without animal-derived supplements and in the manner described earlier for fermentation of recombinant proteins. Bacterial cell growth and expression are monitored for amount and for desirable traits or attributes, these often identified in the research laboratory. The cells are harvested in a manner that retains viability and are processed further to remove impurities, such as dead cells, cell debris and components of the medium. Purification of live bacteria is based largely on physical separation with centrifugations, washes and filtrations. Bulk substance consists of live bacterial cells held in a physiological medium and perhaps with a cryo-preservative because they will likely be stored frozen. Testing protocols are developed to ensure that traits are retained throughout the processes. Identity, safety, purity and potency testing are completed as well. Potency testing may include the ability to express an antigen or attach to a cell substrate.

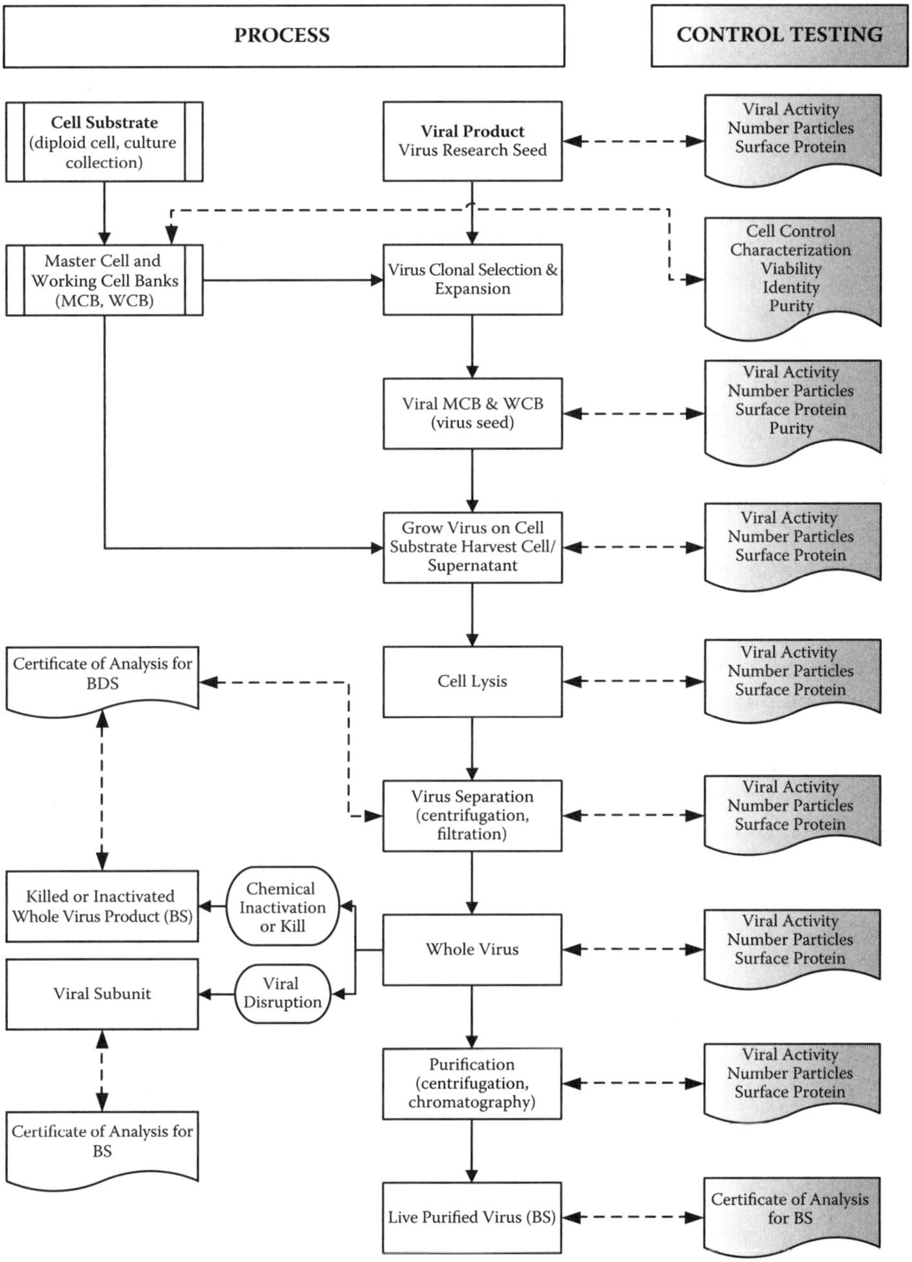

FIGURE 6.12
Production and preparation of virus (live, killed or subparticle).

Production of Products Composed of Mammalian Somatic Cells or Tissues

With the advent of tissue and cell replacement therapies, biomanufacturing operations developed methods to expand somatic tissues and cells. While selected individual cell types, such as those used in laboratory tissue culture, had been produced for decades, the growth of somatic cells and tissues intended as replacement therapies in human subjects presents new manufacturing challenges. Today, autologous tissue regeneration and replacement is a small but growing biotechnology industry and a quality system suited for this technology, Good Tissue Practices (GTPs), has been developed as regulatory guidance.

Replacement of knee joint cartilage provides one example. The objective is to grow, *in vitro,* healthy autologous cartilage that can be used to replace diseased cartilage in a joint. To begin, a piece of cartilage is surgically removed from a healthy joint of the patient and the cells from this normal tissue are transferred to a biopharmaceutical production facility. Here the cartilage cells are expanded on an inert biological matrix to confluence a sheet of cultured cartilage cells. This sheet of cells is returned to the surgeon who then implants it into the damaged joint. While this product does not utilize recombinant technology, it does apply the biotechnology practices of cell growth production, cell and tissue purification and quality control testing to a complex biomanufacturing scheme. Many technical and quality hurdles were overcome, largely through planning, proper application of existing technologies and invention of new methods.

This approach also has been used to illustrate another example—growth of autologous epidermis, new skin for patients subjected to severe burns. A flow diagram for skin production is shown in Figure 6.13 where major attributes and production steps are highlighted. The seed tissue is taken from an unaffected area of skin on a burn patient. This critical raw material must be of high quality. Culture methods encourage rapid and consistent growth of skin tissues on a matrix or artificial membrane to the stage of confluence. Scrupulous aseptic technique is applied at every step to ensure a safe and sterile product. Unique quality control methods focus on attributes of this product. For example, measurements of skin tissue tensile strength ensure the tissue could be transported and then surgically implanted without tearing. Identity testing confirms that donor skin sample matches exactly the skin tissue yielded at the end of manufacturing. The outcome is a biomanufacturing system producing a high-quality skin or tissue product able to alleviate patient suffering and provide new market opportunities.

Production of Cellular Products Derived from Pluripotent (Stem) Cells

The biomanufacturing community is just now beginning to produce products from pluripotent cells (i.e., stem cells), thus, there remain many unknowns

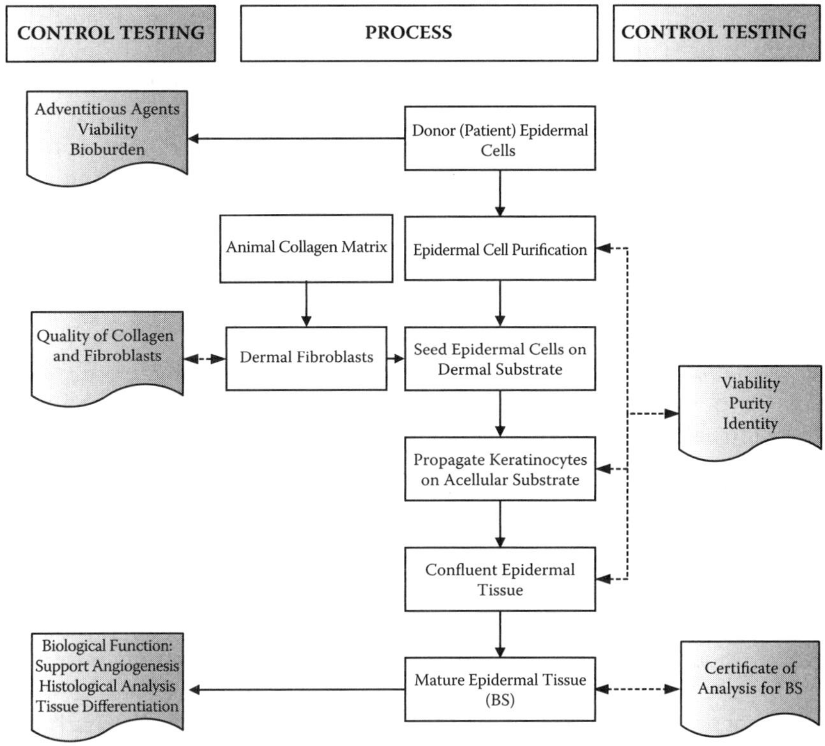

FIGURE 6.13
Production of epidermal somatic cells and skin tissue.

regarding this technology. The following overview of technologies available for production and testing of differentiated cells from pluripotent cells gives one an idea of how biomanufacture might be performed in the future.

A biomanufacturing plan for products derived from stem cells is based on a cell therapy indication and a product composed of differentiated but growing cells of a particular lineage. As a new technology, regulatory guidance, even general information from the FDA, is important to developing a compliant product. Fortunately, the FDA has announced general guidelines for development from pluripotent cell sources and established helpful information on related products in the GTP and other regulatory guidelines. These recommendations, while still quite general and largely unproven for pluripotent cell-derived products, still provide a foundation for production schemes.

However, many questions regarding application of technologies to ensure production of safe and effective cells or tissue remain to be answered. First, do the proposed manufacture and control methods result in a safe product and how exactly do we demonstrate product safety using scientific methods currently available? Second, can any biomanufacturing scheme actually generate differentiated cells and tissues from pluripotent cells and, as BS, FP and after growing in the patient, will these cells demonstrate the attributes of identity, purity and potency? Third, is it possible to apply to pluripotent cell-derived biomanufacturing protocols those methods and quality criteria used for somatic cell and tissue production or will it be necessary to begin anew and develop unique schemes for these cell types. And, when produced in great numbers, will differentiated cells, derived from pluripotent cells, remain differentiated or will they revert to an undifferentiated status or even to a malignant state of differentiation? And how do we ensure that pluripotent stem cells derived from an unknown source not carry adventitious agents; do traditional methods applied to somatic cells provide adequate safeguards for pluripotent cell-derived products?

The biomanufacturing design and subsequent plan must address these issues and answer questions using novel manufacturing methods and applying novel quality control tests. A hypothetical scheme for biomanufacture and control of a pluripotent cell-derived product is proposed in Figure 6.14. Despite the novel technology and source of the product, issues that confront the manufacturing team are very similar to those experienced several decades ago by teams of biomanufacturing scientists intent on producing a recombinant protein in *E. coli*. Then and now, application of precedent, good planning and careful and thorough experimentation are keys to preparing a biomanufacturing plan, moving a novel product through the manufacturing cycle and bringing it to market. Indeed, by application of good scientific and manufacturing practices, it is conceivable that any biotechnology concept can be taken from the research laboratory and be successfully produced and marketed for the benefit of mankind.

Production of Biological Molecules by Transgenic Animals or Transgenic Plants

A host of alternative production systems for recombinant biopharmaceutical molecules are currently in development. Most purport to make a product that is of equal or greater quality when compared to biomanufacture by traditional methods, such as fermentation or cell bioreactor production. Most animal and plant systems promise expediency, higher quality and lower cost. However, today few plant or animal biomanufacturing concepts are technically mature or proven and many have already fallen by the wayside as they have proved difficult to manage, give small yields or produce an unacceptably impure or impotent product. Nonetheless and based on some successes, notably production of recombinant protein in transgenic goats, there are a host of new plant- or animal-based biomanufacturing technologies in development. *Biopharming* is a casual name given to the application of transgenic plants or animals to produce biopharmaceuticals.

Two examples are mentioned. First is the production of proteins by domestic, recombinant animals. As one might imagine, this requires a bioengineered domestic animal to carry the engineered gene and, within that healthy animal, transgenic cells multiplying and expressing the genetic trait. Further, to

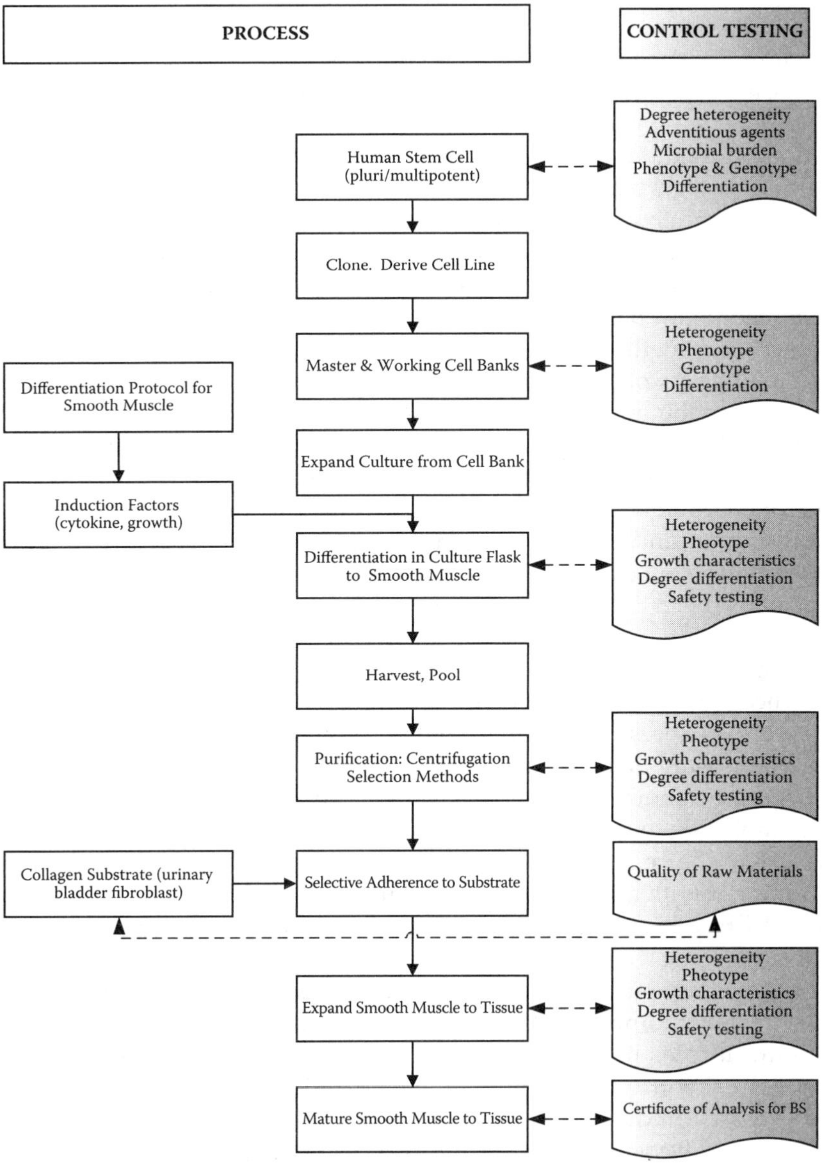

FIGURE 6.14
Hypothetical scheme for biomanufacture and control of human cells and tissue (bladder smooth muscle) derived from pluripotent cells.

allow for collection and purification of the protein, the transgenic gene must express and secrete, into a harvestable body fluid, the protein or glycoprotein product. To achieve such design objectives, human proteins have been transferred to dairy animals, such as goats, for the purpose of gene expression in the mammary tissues with protein secretion into the animal's milk. Protein is then purified from milk using protocols that consider separation from cellular debris, other milk proteins and fat. This is accomplished using downstream purification methods, such as centrifugation, filtration and chromatography, but under design plans that consider the unique impurities in mammalian milk. Significant hurdles, such as developing the equivalent of cell banks and ensuring absence of adventitious agents in animals residing in decidedly septic environments, have been overcome. This technology reached the marketplace with a therapeutic blood clotting protein, demonstrating proof of principle.

The second method, producing biopharmaceuticals in plants, has been tested in a variety of plant species for over two decades. A scheme for production of a recombinant protein by a transgenic higher plant species is depicted in Figure 6.15. The ability to transform commercially useful plants, first with genes of prokaryotes or of other plant species and then with genes of animal origin, has provided the foundation for this technology. Cultured plant cells of both higher and lower order plants, e.g., maize or tobacco and algae or mosses, have been tested as biomanufacturing systems. Higher order plants are usually used, although systems are under development using lower order species. The first step is production of a transgenic plant, a process that is facilitated by several novel gene delivery methods well suited to transfection. Once mature, samples of various plant tissues are tested for the desired trait, such as expression of a mammalian protein. Next, these plants must be cross-bred, using methods applied to the development of hybrid plants; this is followed by another round of selection. The ultimate source of the recombinant molecule may be any plant tissue expressing the product, but seed is a preferred choice, given its ability to store large amounts of protein in an environment with a low microbial bioburden. The plant tissue must then be processed to release the desired recombinant protein from cells and tissue matrix and bring it into solution. The solution is clarified of plant debris and processed, using purification methods described above for other proteins, to derive the recombinant molecule as bulk substance.

Several hurdles had, or currently do, confounded the application of biomanufacture with transgenic plants. Glycosylation is one example. Plants have evolved unique methods for making posttranslational modifications to proteins. Glycosylation modifications of plants are unlike those in animals, yet they are encoded in the plant genome. Sometimes, expression of a mammalian gene by a transgenic plant unpredictably results in a molecule that has the expected protein backbone, but possesses a totally different and undesirable carbohydrate moiety. This is the result of a gylcosylation by plant enzymes, a process that would not happen to the protein molecule manufactured by an animal cell. A case in point is a monoclonal antibody produced

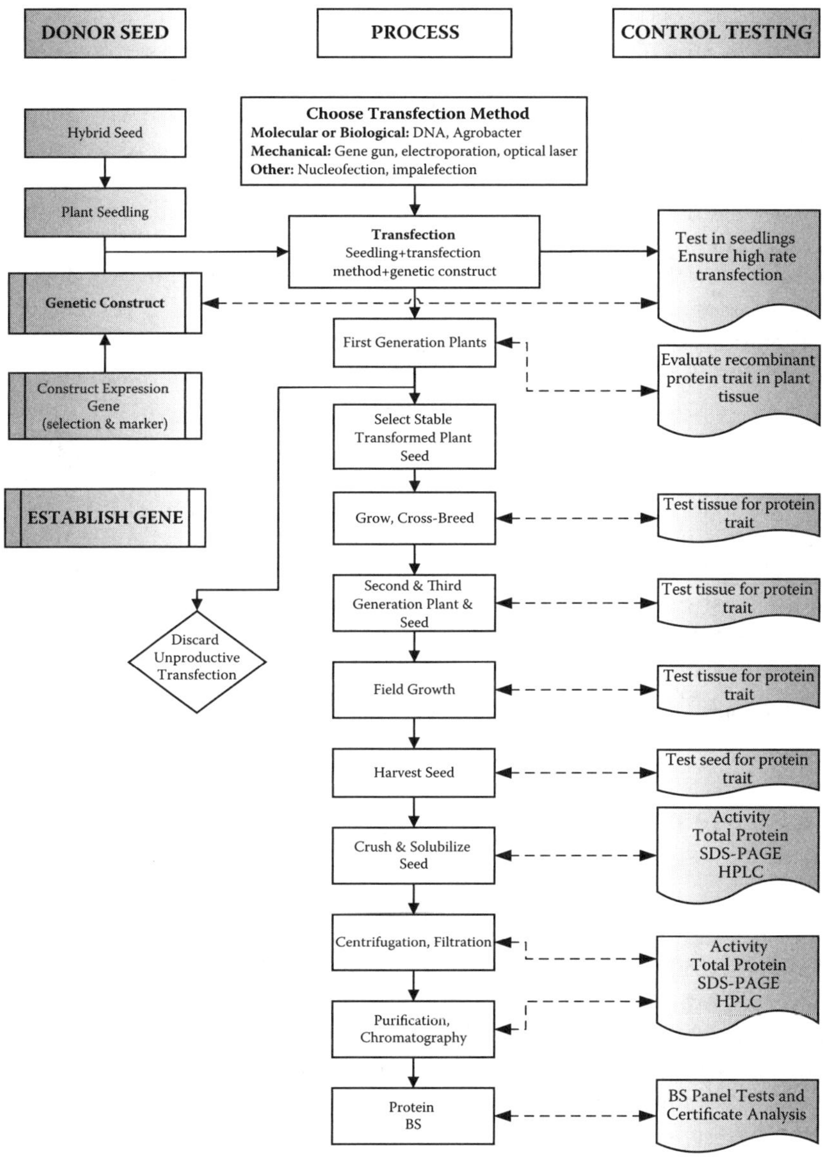

FIGURE 6.15
Production and control of a recombinant protein by a transgenic higher plant species.

in maize cells and glycosylated with a series of carbohydrates unique to plants. Such modifications change the properties, often the potency, of the recombinant molecule. Hence, careful planning, based on an understanding of the product and the transgenic host system, are essential to successfully using transgenic plants or animals for biomanufacture.

Another issue is the need to redefine cell banks to meet transgenic plant technology. For example, corn, unlike bacteria or immortalized cells, relies on cell banks composed of actual monocotyledon seeds. Due to sexual reproduction and other traits, higher-order plant seeds are quite heterogeneous in nature and genetic makeup. A corn seed bank is not derived from a clone and is not genetically pure or homogenous. This can cause problems with variable field growth, sexual reproductive capacity or protein expression. Also, the environment of field-grown plants is quite difficult to control. Weather in a corn field is highly variable and higher plants grow, quite literally, in dirty environments, certainly as compared to that of the fermentation vessel in an aseptic biomanufacturing facility. Thus, plant-derived recombinant molecules begin as septic entities, adding challenges to aseptic purification processes. Finally, purification of biopharmaceuticals also has presented new challenges, many unexpected, because extracts of plants have impurities and contaminants not found in bacteria, yeast or mammalian cell systems. Novel approaches, such as use of unicellular and asexual plant species (e.g., algal or moss species), growth of plants in controlled environments and derivation of recombinant proteins from selected plant tissues, have overcome some, but certainly not all, of the issues facing biopharming. Yet, there are potential advantages, notably large quantities of product at low cost, to be derived by biomanufacture from transgenic plants or animals.

Production of Biologically Active Lipids, Glycolipids and Complex Carbohydrates

Peptides, lipids, complex carbohydrates and glycolipids are potentially important biomolecules of economic and therapeutic value. This is perhaps best demonstrated by complex carbohydrates and glycoplids, which have been successfully used as vaccine antigens and by lipid molecules, which are used as vaccine adjuvants. Each has a biological derivation and function in nature and can be manufactured by man in the laboratory. Biomanufacture of these molecules in larger quantities is through one of two routes: (1) production by a live organism that is not recombinant, but naturally expresses the molecule in nature, followed by isolation and purification of the macromolecule; and (2) chemical synthesis, using processes analogous to, but often more complex than, those applied to production of small molecule drugs. An example of a natural product, derived from bacterial cultures and then purified using biomanufacturing methods, is outlined in Figure 6.16. This complex carbohydrate is used as a vaccine.

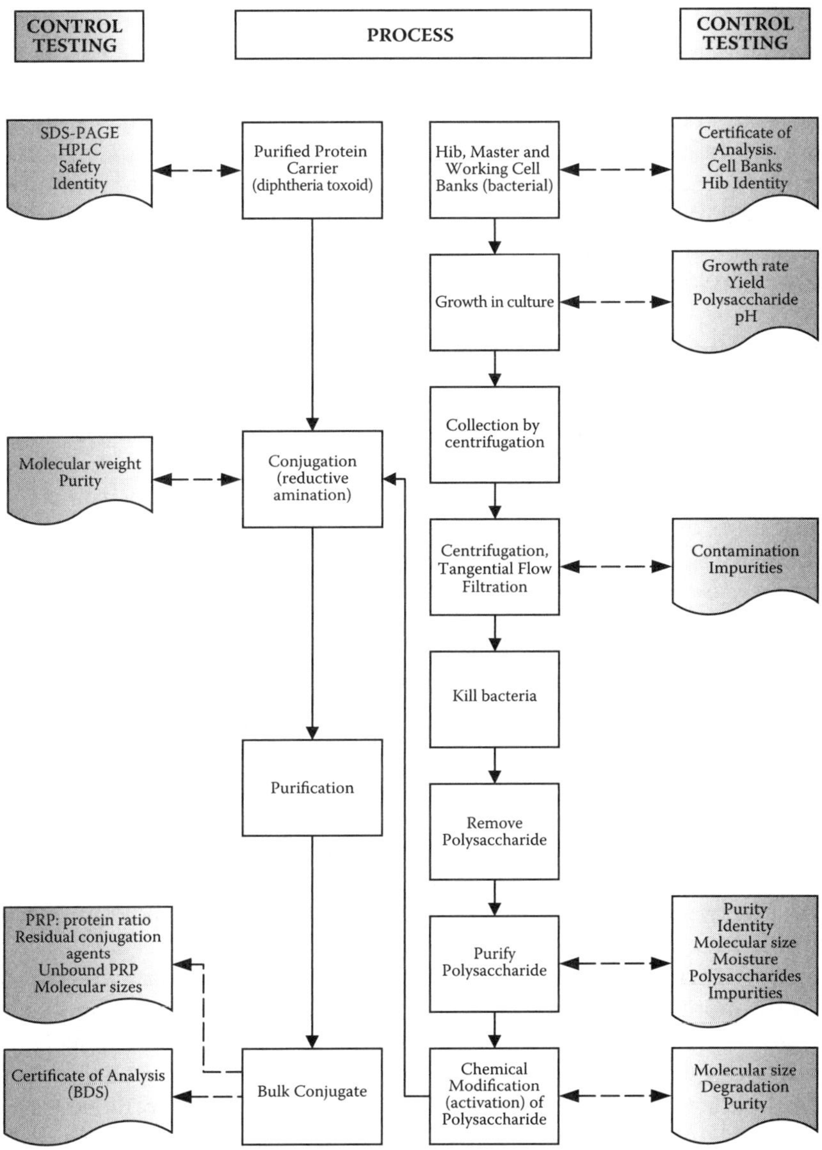

FIGURE 6.16
Production of a polysaccharide–protein conjugate product (Haemophilus influenzae type B vaccine).

The chemical synthesis of biologically active, nonproteinacious macromolecules is becoming more elegant and widely adopted. For example, instruments are used to synthesize complex carbohydrates on lipid-like backbones, a method that is dependable but still gives low yield.

The downstream or purification steps for lipid, carbohydrate or glycolipid biomolecules vary in some respects from those used to purify proteins. This is understandable because these lipid or carbohydrate molecules are quite unique in chemical structure and composition and physical characteristics. Manufacturing planning takes into consideration the molecular properties of these candidate products and matches them with technologies available for purification. Many of the methods applied to a manufacturing step may be borrowed from the research laboratory and adapted to biomanufacturing.

Production of Biologically Active Peptides

A peptide is a string of amino acids, 40 or less by most definitions, in a given sequence. Biologically active peptides are used or tested as biopharmaceuticals, enzyme inhibitors, laboratory chemicals and for other purposes. Because they are shorter than proteins, peptides typically have no posttranslational modification or tertiary or quaternary structure. Various technologies are available to manufacture peptides and choice of a method is based on the intended use and specifications, notably purity, the number and nature of amino acids in the string, the amount required and cost. Peptides are often manufactured using automated equipment, the peptide synthesizer, to build the string one amino acid at a time, beginning at a solid matrix, such as a plastic bead, and continuing to the end. Hence, most peptides are made using tools common to synthesis of other organic molecules.

Impurities—truncated peptides, fragments and free amino acids—remain in solution with the peptide product postsynthesis, so purification is required. Further, peptides can be unstable, perhaps because they represent incomplete or fragmented protein sequences. Formulation, fill and finish procedures must carefully consider postmanufacture hold and storage environments to ensure a potent and stable product.

Certain other biomolecules are peptide-based, in that they are peptides, but with another large molecule bound to them. Examples include glyco–peptide and protein–peptide combinations. The production of such molecules may require considerable planning and technical effort as there are no established large-scale or cGMP manufacturing methods available for these biopharmaceuticals.

Production of Combination Products: Biopharmaceutical with a Drug or Medical Device

Biomolecules are not infrequently used with material from another source to produce a combined effect. For example, a recombinant bacterium used to clean oil spills might be applied to the spill along with a short-chain organic molecule that disburses oil and facilitates metabolic activity of the bacterium. In biopharmaceuticals, the pairing of a biological product with a medical device, a drug or both is referred to as a combination product (see Chapter 3). A recombinant DNA vaccine that is delivered exclusively with a special injector device is one example. Another is an engineered retrovirus that must be given with a specific drug, one to enhance the potential therapeutic activity of the retrovirus. A case in point is the use of a drug to facilitate insertion of a therapeutic gene into the genome of the recipient. For the full intended therapeutic effect, both substances, drug and biologic, must be pure and potent under a defined schedule of usage and indication.

Biomanufacturing plans identify combination products and present a strategy for producing both products to specifications that will yield the desired, combined effect. Manufacturing specifications for combination products are often challenging for the operator because individual roles must be assigned to attributes of each product and synergistic effect also considered. For the example of a recombinant DNA vaccine (a biopharmaceutical composed of plasmid DNA), delivered by a needleless jet injector, a medical device, the vaccine plays the dominant role because it imparts the therapeutic effect through stimulating the immune response. However, the device must perform per established specifications, otherwise the DNA vaccine will not be delivered correctly and might not then exert the intended pharmacological effect: vaccination. Early considerations for a manufacturing plan takes into account this need for synergy in combination products. Specific concerns in the DNA example might be:

- DNA formulation must be compatible and stable with materials of the device.
- DNA must be concentrated so it can fit into the device chamber during storage, but it also must be capable of rapidly exiting via the device needle upon actuation.
- Device must consistently deliver an exact amount of DNA.
- Device must be easily and correctly used by medical staff.

From this example, it is clear that design, actually co-design of a biological and medical device, is critical to success whenever a biotechnology product is partnered with a drug, a device or even another biopharmaceutical. Manufacturing plans and technologies consider complex performance issues and ensure they have been addressed early in development. Experimentation is performed to test the effects of additional variables as each product brings into play a new set of issues.

Final Product: Formulation Fill, Finish and Labeling

A product is used commercially or in clinical investigations only after it has been properly formulated, placed into a protective container or delivery device and then labeled. Having reviewed the various types of biotechnology products presented earlier in this chapter, one might well imagine the need for a host of formulations, containers and labels as well the procedures to complete their final production. In this section, we review technologies used to bring a biopharmaceutical from bulk substance to the format intended for the user, i.e., the final product or FP.

The biomanufacturing plan for a final product focuses on the indication and the user. Biomanufacturing is a market- and user-driven process and, therefore, the formulation, container and label are designed for a specific purpose: needs of the user. For biopharmaceuticals, the user is the patient or a subject enrolled in a clinical investigation, but for many products, a full definition of the user includes the medical professional prescribing or administering the product. For a tissue product, such as the skin or cartilage mentioned in an earlier example, the surgeon is also a user of that product. Indeed, the patient may never see the cartilage tissue product as provided by the manufacturer because cartilage is transferred by the surgeon from the package and into the knee joint. In contrast, for the combination product of recombinant monoclonal antibody in an auto-injector, the patients, not the medical professional, use the product, receiving it directly from the pharmacy and then injecting it themselves. The medial professional is, however, familiar with the product and its attributes before a prescription is written and may train the patient in proper use. Most biopharmaceutical final products are considered by more than one person and each of their needs are considered in final product design.

The FP design considers various product attributes. Formulation decisions rest on the intended shelf life of a product, possible and desirable storage conditions and proposed dose. The target dose also allows planning for the amount of product that will be included in the final container and the packaging and labeling that are provided with the container.

A scheme for final product production of a typical biotechnology product is shown Figure 6.17. A first step is to select or devise a formulation that suits the nature of the product, intended use, route of delivery and container. There are many formulation choices available and others can be developed for a unique product. Common formulations for biotechnology products consist of salt solutions, buffers and a variety of excipients. Once a promising formulation is identified, experimentation and trial and error are the basis for selecting the one that is just right for that product; this means significant testing for product attributes and for product stability. Because a product may need to be matched with several formulations before the correct combination of ingredients is identified, the process can be extensive.

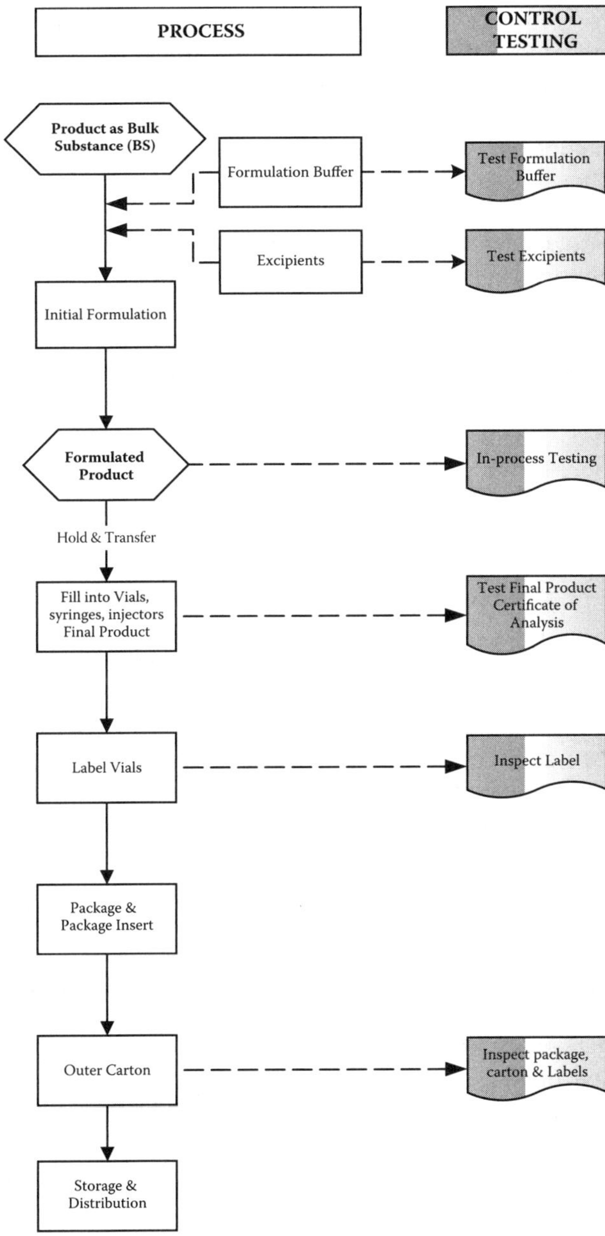

FIGURE 6.17
Manufacture of final product: formulation, fill and finish processes.

Once a formulation has been selected, the actual formulation process for a biopharmaceutical begins with final clarification and, for most products, sterile filtration of the bulk substance. Following purification and until formulation, biopharmaceuticals are held as bulk substance under refrigerated or frozen conditions. If the active ingredient in BS is stable, as a frozen recombinant protein might be, it may be stored for several months, and maybe even years. In contrast, a tissue product, like skin tissue, might withstand refrigerated storage for a few days. Hence, the time allowed between completion of bulk manufacturing and formulation, fill, finish and labeling varies greatly by product. (Product stability testing is discussed in Chapter 7.) No matter what the product, once the decision is made to begin formulation, the process is completed as rapidly as possible to reduce hold time and thus minimize product exposure to the environment. Because most biopharmaceuticals are sterile products, formulation must be a strictly aseptic process applying stringent techniques and environmental controls.

An example of formulation, fill and finish, is instructive and a recombinant protein, such as a monoclonal antibody, suits this purpose. After sterile filtration, the storage buffer for a monoclonal antibody, active ingredient in bulk substance, is exchanged for the final product buffer, here 0.9% sodium chloride in water (normal saline) for injection, with a small amount of detergent. After mixing, the formulated product is again sterile filtered and then it is transferred to the fill area, where it is dispensed into vials.

In contrast to a recombinant protein, a more complex biopharmaceutical product, such as skin tissue, may require more formulation effort. Here, it may be necessary to wash the skin tissue with various buffers and then place it into a nutrient solution enriched with oxygen and chemicals that maintain the live tissue. Terminal sterilization, even by filtration, is not possible with skin and other cellular (or certain other biotechnology) products and so strict aseptic technique is practiced throughout the formulation process. To meet stability profiles of some fastidious biotechnology products, such as live bacterial or viral vaccines, some products have unique additional steps in formulation. For example, dry powders may be prepared by spray drying the product once it is in the final salt solution. Freeze-drying, referred to as lyophilization, is also used to prepare powder from formulated liquid product, after it has been aliquoted into vials.

Fill is the next process step. Here, product is placed into a final container with fill procedures tailored for each product and container combination. For the skin tissue example, product might be placed into a pouch of some type. And, for the living tissue, the outer container might be a special transportation apparatus, replete with systems to provide sterile nutrients and oxygen and an external heat source to maintain a temperature suitable for this living product.

Returning to the example of a recombinant protein, the monoclonal antibody product formulated in saline with a detergent, container and handling requirements presents a less complicated fill and finish procedure. Standard

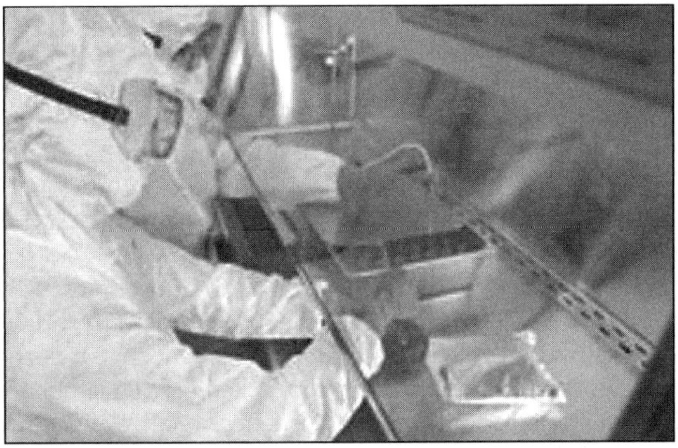

FIGURE 6.18
Fill and finish of final product (FP) in a clean room, Two operators, working under a Class 100 hood, manually fill and finish vials of a biopharmaceutical. They are carefully gowned and covered to prevent any possibility of contaminating the FP. The operator on the left is filling the vials, held in a box at the center of the hood, with a dispensing pipette while the operator on the right adds a cap to each vial. In the next step, an aluminum seal will be crimped to tightly close each vial and a label added. (From Waisman Clinical Biomanufacturing Facility, University of Wisconsin/Madison. With permission.)

pharmaceutical glass vials are chosen as the primary container, with butyl rubber caps and aluminum crimp to seal the cap to each vial. Vials and caps of the highest quality are scrupulously cleaned and heated to remove any contaminants before use. Vials are either filled and then capped and crimped (force a seal over the cap) manually by operators (Figure 6.18) or they are filled, capped and crimped using a machine. Prefilled syringes and injectors are becoming popular containers for biopharmaceutical products and these are always filled using automated pieces of equipment, which are fast and accurate.

Once containers have been filled, capped and crimped, then a permanent label is placed onto the container to provide the user with an abstracted description of the contents. Container labels include a description of the product, the dose, volume, total number of doses per container, source (manufacturer), expiration date, and special instructions or warnings. Imagine fitting this much information onto a label, 2.5 cm × 1 cm, for a small vial.

Packaging is the next step in producing the final product. Containers are placed into a protective inner package, a light cardboard box for example. A package insert is added to this box before an adhesive label is attached to identify the contents and then it is closed with a tamper-proof seal. The package insert, sometimes referred to as *labeling*, provides product information to both the medical professional and to the patient or user. Multiple product containers in their outer package are then placed into a larger carton, the outer container, and this, too, is labeled.

By way of a final example of "formulate, fill and finish," we consider a biotechnology product that is not a biopharmaceutical, but instead is a genetically engineered plant. It is a strain of genetically engineered, disease-resistant corn, grown in the field, harvested, cleaned and then placed into storage in 50-pound bags as bulk substance. The product will be planted in fields by farmers. Prior to packaging, seed is formulated by various treatments, perhaps drying to an established moisture level and followed by addition of powdered fungicide to prevent decay in storage. The seed is then filled, placed into moisture-proof containers, such as 10-gallon plastic buckets, and the buckets are sealed. These buckets are then labeled on the outside and several buckets, along with product information (the package insert), are placed into a cardboard outer carton and this, too, is labeled. The package may then be shipped to a distributor or the user. While technically quite different from formulation–fill–finish production methods used for a biopharmaceutical, the processes for this and many other biotechnology products follow the same general steps in production, protecting it from harmful environments and making it ready for the user.

Biomanufacturing Facilities, Utilities and Equipment

Facility Design Considerations

Biotechnology products, notably biopharmaceuticals, are manufactured in very special facilities. Biomanufacturing facilities are not only unique to our industry, but also custom designed for each class of product and even for a specific product. All the activities we have mentioned in this chapter, with the possible exception of upstream production in "biopharming," must be performed under a roof, above a floor and between four walls. If only it were that simple. Biomanufacturing facilities are complex and expensive to both build and operate. Staff are professional specialists and everything, from emptying the trash to filling the final containers, follows written procedures and results in permanent records. Because each biotechnology product and every biomanufacturing process for a product is unique, facilities are custom designed and have been built and equipped in every imaginable way. But, there are similarities, and our discussion of biomanufacturing facilities, utilities and equipment will provide basic principles of facility design and operation.

Biotechnology firms skilled and lucky enough to find themselves in development are faced with the need to manufacture their product. Following this realization, the first question to be posed by management to the product development team is this: Thanks for the opportunity to manufacture, but when, where and how will this be accomplished and how much will it cost? The answer must be in the manufacturing plan; therefore, options must be considered and choices made. A team, composed of individuals with business, financial, management, biomanufacturing and facility and process engineering experience, is chartered to reach a decision on manufacturing options. There are at least three possible ways to meet a biomanufacturing requirement and these are: (1) do not manufacture the product (and, hence, do not develop it further); (2) manufacture the product in-house; or (3) manufacture the product at a contract manufacturing operation (CMO). A fourth possibility is to split manufacture and do some processes in-house and have others completed at a CMO. Not surprisingly, few biotechnology firms select the first option and the rest seem evenly divided between choices numbered two, three and four. Many factors beyond the technical need or a desire to have an in-house operation often influence the final decision; these are resources, location or business and exit plans.

A biomanufacturing facility is designed to produce a specific class or type of product. One would not expect to see a recombinant seed corn operation co-located with a formulation–fill–finish operation and it would be unusual to find, in the same biomanufacturing building, both epithelial cell culture and bacterial cell fermentation. Yes, some facilities can process quite a number of different products, that is, manufacture on a campaign basis, but even then they are limited in scope. When considering the building itself, one must choose between erecting a new building or remodeling an existing structure. Many excellent biomanufacturing operations have been built within an empty warehouse, albeit one of high quality in a good location and with adequate services. Actual facility design is best left to professional engineers and architects, those with experience in biomanufacturing process engineering and building, equipping and maintaining facilities. Early drawings are produced by engineers and architects. These should be carefully examined by process engineers, individuals with an understanding of the process intended for the facility. General descriptions of utility and equipment requirements are added to the facility plan. Following discussions and one or two more rounds of review, a final facility plan is established. Cost estimates are made by professional biomanufacturing facility engineers.

Facility and Utilities: A Controlled Environment

Most biopharmaceuticals are sterile products of exceptionally high quality and reliability. Biopharmaceutical production requires special controls because products are intended for use as medicines by large numbers of people and many are given by the parenteral route. Biopharmaceutical production, and hence the facility producing a biopharmaceutical, is highly regulated and, for marketed product, must have a government license, this to ensure that only safe, pure and potent products reach users. Regulations and guidelines make clear exact standards and specifications for biopharmaceutical manufacturing facility design and operation. Further, good business practices also demand a quality manufacturing facility for biotechnology products. Not a month goes by without a news story covering an inadequate drug, biologics or medical device facility that produced and continued to provide substandard product to consumers from substandard facilities. These incidents result in expensive product recalls or even consumer illness or death and in regulatory actions against the sponsor (see Chapter 4). Such cases lead to negative publicity for the firm and loss of product sales, even for product not made in that facility. Indeed, a single incident at a manufacturing facility has led to the loss of a product line or even to the financial demise of a biopharmaceutical firm. The combined lesson is that a manufacturing facility must be first-rate, from the ground up and in all respects.

A biomanufacturing facility is, from the outside, a building that looks like any other commercial structure. Yet, inside it is established and equipped exclusively for the production of biopharmaceutical product. How then is the proper environment for biomanufacture established under the general facility plan? First, that building is designed to house a particular process or several similar processes. Secondly, the design considers the need for production by aseptic processes so as to reduce the incidence and spread of microbes and particulates through the use of segregated, clean work areas. The facility also is planned to house adequate utilities and equipment, these also properly designed. Biomanufacturing requires sufficient space for the various processes, utilities and equipment; further, regulations require and common sense suggests separation of various activities to prevent cross contamination of product or spread of microbes. The design must consider utilities, reliable sources of water, electricity, natural gas, and heat, ventilation and air conditioning (HVAC). Finally, a facility must be well managed by highly trained and experienced professionals and, just as with the manufacturing process, it must be run by strict and compliant written procedures and records. Sound rather expensive? It is.

Once a decision is reached to build a biomanufacturing facility according to a design, additional planning is in order. Detailed product manufacturing plans allow the facility planner to impose on the facility design a process map, a schematic in which the manufacturing process is drawn, stage-by-stage and step-by-step onto a general facility design. This allows one to determine whether or not all the pieces—processes, equipment, utilities, work flow—will work in harmony. Consulting engineers and architects and biopharmaceutical process engineers, those with experience in biopharmaceutical manufacturing, are retained to examine rough plans and refine the drawings to ensure compliance with local and state regulations. Plans and drawings are revised, discussed, changed and finally finished. The facility plan now supports business functions, such as accepting bids from building contractors and utility and equipment manufacturers. Now upper management can be given a firm estimate of the cost.

Operation of Clean Work Areas for Biomanufacture

Controlled processes and aseptic processing are crucial to biomanufacture. At the heart of aseptic processing is the need to keep viable particulate matter, specifically bacteria and yeast, from contaminating a product. The primary source of microbes is humans (skin, hair) and materials that enter into a clean area. Introduction and spread of microbes is controlled by facility, utility and equipment design, environmental awareness regarding microbial burdens and sound aseptic operational procedures. The facility floor plan is critical to maintaining a clean environment, as it allows for segregation and the logical flow of activities and products. Entry of people and raw materials and exit of waste are carefully controlled to reduce entry of contaminants into a clean work area. Prior to entering, individuals don special gowns and masks that cover much of the body and reduce the microbial burden shed by them into the clean room environment. Materials are sterilized or disinfected immediately before entering a clean area. Prior to entry, water used in a clean area is filtered to remove any viable particles. Because microbes and contaminated dirt particles move freely in air, the HVAC system is designed to constantly filter air through high efficiency particle air HEPA filters. The air is circulated rapidly and in great volume to ensure that any particles generated by process activities are swept from a room and filtered before they contaminate product. Doors and pass-through openings control airflow between rooms, moving air by an established pattern from the cleanest to less-clean rooms, thereby further limiting microbial movement. Because microbes adhere to and multiply on surfaces, all ceilings, walls and floors are finished with highly resistant epoxy surfaces to withstand repeated scrubbing and disinfection. All equipment surfaces are designed for harsh antimicrobial treatments.

Movement of everything—people, raw materials, trash and product—within a facility must flow in a predetermined direction. Highly controlled aseptic operations, such as fill and finish, are performed in highly classified, by air quality, areas segregated from other manufacturing and nonproduction areas so as to prevent contamination of final product. During and after biomanufacturing operations, measurements are taken of air quality and each room must meet an air quality standard, the established classification. There are two manners of classifying the air within a clean area and both are shown in Box 6.4. Everything in a clean area, including air, water, surfaces and the gloved hands of every staff member, is sampled or swabbed to test for microbial contamination.

Operations not requiring a clean environment are kept apart from clean areas dedicated to aseptic processing. For example, packaging and labeling are generally relegated to an area that is not highly controlled or classified. Also, quality control laboratories, offices and meeting rooms are established outside the clean area. In summary, a facility design facilitates the prevention of mix-ups, provides a product flow from cleaner to dirtier, allows segregation and isolates critical steps in an effort to ensure pure, potent and safe product.

BOX 6.4 MEASUREMENT OF PARTICLES IN A BIOMANUFACTURING CLEAN ROOM OR AREA

	Published Specifications	
	ISO 14644-1	FED STD 209E (US)
Room Classifications	ISO 3	1
	ISO 4	10
	ISO 5	100
	ISO 6	1,000
	ISO 7	10,000
	ISO 8	100,000

Note: The particle counter instrument is used to measure particles in air. An air sample is taken into the instrument and particles of a specific size (e.g., >0.5 μm) are counted and total air volume is measured. Particle counts are then given as number of particles per cubic foot (Federal Standard 209E; U.S.) or particles per cubic meter (ISO 14644-1; International and European). A manufacturing facility might have air classified as 100,000/ISO 8 in general preparation or laboratory areas, as 1,000/ISO 6 in clean work areas, or as 100/ISO 5 in areas for performing aseptic operations, such as filling vials.

A biopharmaceutical operation also considers the utilities in support of manufacturing operations. Temperature is always controlled by HVAC equipment and the production of some products requires humidity control as well. Gases to incubators, steam to sterilization equipment, and water to make solutions must enter the clean area. Water for injection, WFI, is the purest grade and may be produced by the manufacturer, in-house. However, due to the complexity and expense of producing WFI, many biotechnology firms simply purchase it in bottles, a USP (U.S. Pharmacopeia) reagent. Steam is used in most biomanufacturing facilities to sterilize raw materials or equipment and to clean and disinfect surfaces. In such cases, the steam is produced as "clean steam," which is generated with special equipment and from WFI. Mechanical equipment, such as HVAC and water purification systems, is always monitored and alarmed.

Biomanufacturing Equipment

Many pieces of equipment, including biological safety cabinets, centrifuges, filtration apparatus, fermentors, bioreactors, chromatography apparatus, controllers, microprocessors and incubators, to name but a few, are operated in a clean area. Special or unique pieces of manufacturing equipment (and there are many to manufacture biopharmaceuticals) must be of the highest quality and some are specially designed for the biomanufacture of one product. Equipment specifications ensure proper performance and each piece is periodically calibrated, controlled and monitored during operation. Rigorous cleaning is necessary and cleaning protocols are often complex procedures because all residual product and cleaning agent and microbes must be removed after use. Also, many pieces of equipment are sterilized prior to use or reuse. Once the facility has been built, facility, utilities and equipment are commissioned and validated, along with the process itself. In addition to product-specific requirements, there also are basic quality requirements for a cGMP operation.

Contract Manufacturing Options

It is no wonder that many biotechnology firms select a contract manufacturing operation (CMO) to produce their product. At first, many resist this option because they believe that they will lose some control of product manufacture, especially if the CMO is some distance away. However, renting capacity can be economically attractive and there are many ways to establish a partnership in which the sponsor retains adequate control of its biomanufacturing program after it is placed at a CMO. Indeed, procedures, reflected in Box 6.5, are recommended for selection of a CMO.

Even if a CMO is retained to perform critical biomanufacturing stages, many firms elect to perform some processes in-house. Using outside services to formulate, fill, finish and label is quite common in the biotechnology industry. Preferably, there are multiple CMOs capable of doing the work. Selection of a CMO requires considerable diligence by the sponsor. A history of previous projects provides assurance that a CMO can work with this product and is capable of performing the processes properly. Precontract site visits, frequent communications and a detailed contract, with a quality agreement, are the best means of preventing future misunderstandings or conflicts. A sponsor's efforts do not end with award of a contract and there must be frequent visits, communication and coordination between the sponsor and the contractor throughout the life of the contract. Including contractor representatives to sit on the project management team is an especially effective method.

Validation of Biomanufacturing Facilities, Utilities, Equipment and Processes

Validation is an expensive and time-consuming, but a very necessary, process that is completed during late phase development and only after a

BOX 6.5 CONSIDERATIONS FOR RETAINING A CONTRACT MANUFACTURING OPERATION (CMO)

Plans	List your needs from a CMO as regards type of product, phase of development and biomanufacturing by stage. Is this Phase 1, 2 or 3? What exactly must be done by the CMO: early development, cell banking, upstream, downstream, formulation, fill and finish, quality control testing or more than one of these functions? What are the deliverables: BS, FP, amount of product or number of containers, reports or records?
Competencies	Identify core competencies of CMO and match to your requirements: History; size; management philosophy; experience and willingness to work with small, medium or large biotechnology firms; experience by type of product, phase or capacity; references; location (region, country); profile in the CMO community; possibility of strategic partnership.
Equipment	Will it be necessary to purchase or lease specialized utilities or pieces of equipment or is everything already in-house at this CMO?
Microbiology	If necessary, could they work with live organisms or material that requires stringent or unusual aseptic technique or environment?
Design	Consider responsibilities for manufacturing design, planning and risk analysis and mitigation.
Quality	Note quality responsibilities and willingness to enter into an extensive quality agreement.
Schedule	Examine scheduling possibilities, typical slack and busy periods, opportunities on calendar and flexibility going forward.
Cost	What are the projected costs and unusual expenses associated with the CMO and will it be cost-effective, resource-effective and time-effective to retain this CMO?

manufacturing facility, with all equipment, utilities and staff and the exact process, have been established and commissioned. It is considered a regulatory requirement and good business practice; marketing approval is not possible unless a manufacturing plant and process are fully validated. Validation, which is defined as "to make sound, defensible" in common practice, carries a more complex definition in biopharmaceutical development. For biotechnology operations of any type, validation is a formal process of establishing documented evidence that a specific process (or test, equipment, facility or utility) consistently performs and will, with a high degree of assurance, continue to perform within determined specifications and quality criteria. Quite a mouthful, but once broken down this definition makes sense in light of the complexities of biomanufacturing and the importance of quality in the endeavor.

All manufacturing systems have inherent variation, much of it acceptable, and some undesirable. Validation is based on an understanding of the nature of that variation, its impact on the process and the ability of process controls to keep that variation within manageable levels. Hence, validation is an experimental endeavor based on deep knowledge of the process and under which evidence is generated using a protocol. Perhaps this is one reason that validation is only undertaken in later stages of manufacturing development. Careful planning is an absolute requirement and validation efforts are always included in a manufacturing plan.

A Validation Master Plan is prepared, usually in late phase development, for any major validation effort. The scope of the plan is broad and the level of detail is great. The overall philosophy and approach taken to validate the facility and process are provided in the validation plan. The master plan will outline validation activities related to a facility, to include the physical plant, environment and utilities, equipment (both installed and movable), computer systems, software and hardware, critical raw materials and biomanufacturing processes, with all stages and key steps to include aseptic, cleaning and monitoring processes or procedures. Many analytical tests are also validated and this is described in Chapter 7.

From the Validation Master Plan, validation protocols are written to experimentally examine each critical step of a manufacturing process. Validation of a facility might involve dozens or even hundreds of protocols. A protocol breaks down a system into simple parts and describes, for each critical parameter, the quality attributes and operational specifications. It ensures that attributes are measurable and testable and that each measurement is scientifically sound. Specifications are developed for results derived from each test or measurement. Validation activities are spelled out in detail, usually with standard operating procedures, production records and testing instructions, and these are identified in the protocol. The work is done by scientists and engineers working closely with quality assurance professionals, and it is normal to employ consultants and contractors to assist employees with these herculean efforts. Good statistical practices are also utilized in most protocols. Validation is fully documented, from the master plan to the final validation reports. And validation has a "pass versus fail" outcome; either one meets all the specifications or one fails validation. With this in mind, the prudent biomanufacturer ensures that all systems are performing as expected before executing validation protocols.

Technical steps are involved in the validation process. First, each piece of equipment, utility and facility component must undergo installation qualification (IQ), a process to confirm each item was correctly designed, built and installed. Limits-for-usage are confirmed in IQ as well. These are simply operational limits that could not be exceeded during normal operations. For example, instructions and specifications limit use of a 10 L fermentation vessel to 5 to 8 L of media volume. The next step is operational qualification (OQ), a verification process in which the operating ranges of each item are confirmed under operational conditions. Items may be stressed to furthest ranges of operational performance. For the fermentation vessel, it might be tested, against specifications, three times, with 5, 6.5 or 8 L of media. In

OQ, calibration is completed on mechanical or electrical systems and control systems are shown to work as designed. Process qualification (PQ) is the third step in which the manufacturing process is performed, typically three or more times, so as to demonstrate control, consistency and achievement of specifications. In PQ, critical systems are stressed to ensure they function properly at the limits of operating ranges. Two or three successful repetitions are the norm under validation protocols and each one must meet specifications. Validation is not a one-time endeavor and critical systems must undergo the process of revalidation at periodic intervals, postmarket approval. Any significant manufacturing change must be validated as well.

A primary outcome from validation efforts is confidence that the product will be successfully manufactured within specifications over a long period of time. Validation also ensures that product quality, safety and efficacy have been designed into the product as confirmed in early manufacturing endeavors and that ongoing process monitoring will be a part of the product manufacturing life cycle.

Summary of Biomanufacture

Biomanufacturing is a phased and often challenging process that occurs throughout the life cycle of product development. It aims to develop, improve and increase in scale the production and purification of the biological substance. Quality is important and this is achieved using various process and laboratory testing controls, high quality raw materials, application of cGMP and close coordination of all activities with quality assurance.

Because biotechnology products come in many types or classes, the manufacture of a given product is based on a custom design and a carefully crafted biomanufacturing plan. There exists precedence for many product types, but actual production of an individual product to the bulk substance stage is always based on trial and error and this involves much skill, experience, patience and resources. It also requires a facility in which the product can be made. Most early phase manufacturing is performed in small facilities following basic guidelines for aseptic manufacture and a proper environment. However, as the scale of manufacture increases, so do the quality criteria until, at the later stages, validation of the facility and process are required.

Production of the final product, exactly formulated biopharmaceutical in the proper container and with all labeling, is another challenge. The process must be scaled up during the development cycle. The processes of formulation, fill and finish must be scrupulously aseptic and produce only the highest quality product, further challenging the manufacturing operator.

7

Quality Control

Quality Control Overview

Quality control (QC) is a laboratory endeavor aimed at ensuring that the highest quality biotechnology products are manufactured and released to users. It requires careful planning and coordination with other functional areas, notably biomanufacturing. Quality control tests do not just happen, they are designed for a specific purpose following principles of Quality by Design or QbD (see Chapter 5). QbD promulgates concepts of a superior quality testing program and it is based on an understanding of the product and the analytical tools used to test it. Laboratory tests are designed to meet certain objectives and quality standards. This chapter reviews principles of quality control planning, describes the life cycle of product test and specification development, identifies analytical methods most often applied to biopharmaceutical development, discusses the qualification and validation of these methods and mentions the application of quality control tests for product release and stability.

In contrast to the management and administrative nature of quality assurance (QA) (see Chapter 5), QC is a technical endeavor, using analytical methods to achieve specific objectives in a biotechnology operation. In the past, the terms QA and QC were used interchangeably, particularly by regulatory agencies, and this was confusing. Today, the U.S. Food and Drug Administration (FDA) still refers to QC as a largely administrative function, which is not the same as that applied in most biotechnology operations and in this book, where QC is a laboratory or testing function and QA is a quality management and administrative function. As with any endeavor in biotechnology, QC has developed a jargon and commonly used terms are defined in the Glossary in the back of the book.

The technical objective of QC is to apply laboratory testing to measure the quality of product, be it in-process material, bulk substance (BS) or final product (FP), and whether the purpose of testing is at release or over time for stability evaluation. Quality control planning involves several steps, outlined in Chapter 1 and described in greater detail throughout this chapter.

Quality control planning leads to important documents, the Certificate of Analysis and the Stability Protocol, which are data reporting formats that identify product attributes, test methods, specifications and, following the completion of testing, the test results.

There is a second dimension to quality control planning, consideration of the biopharmaceutical quality control development cycle, shown in Figure 7.1. The cycle identifies various quality control functions and these processes are not strictly ordered, but instead are progressed, sequentially, in many small steps and in close coordination with biomanufacturing, clinical studies and nonclinical testing. Each step in this cycle is considered a generation in the life cycle of a given test method, involving the test, the specification and the product that is tested. Presumably, each generation of a test and specification is a slight improvement over the past and at each step there is a greater understanding of how, for a given biotechnology product, test, specification and overall product quality relate to each other.

Quality control planning gets more complex because a single biopharmaceutical product requires many quality control tests, each measuring an attribute. It is typical for an investigational product to identify 10 or more quality control tests and for a marketed product to have over 20 tests. Together, the tests developed to support one product are complementary, one test adding to knowledge gained from the others. Secondly, ongoing nonclinical and clinical studies often validate the usefulness or validity of a test by demonstrating biological activity of that product in animals or man. At some point in quality control development, midphase perhaps, some tests are qualified and later in development they all are validated. These processes add considerably to confidence in using the assays.

Finally, quality control testing is required for samples of BS and FP at release, and for product on stability protocol and at several points during manufacture (in-process testing). Multiply the number of individual quality control tests times the number each test is performed and again times the number of samples per test and it is clear why quality control is a critical, yet time-consuming, function in biopharmaceutical product development.

Quality control test development begins very early in biopharmaceutical development because operational and biomanufacturing requires excellent analytical support. Failure to have available analytical methods often delays other efforts, such as clinical and nonclinical studies or biomanufacturing process development. Time and again this point is proved as biotechnology firms ignore the need to plan and develop quality control technologies and soon find they are unable to evaluate product quality and, subsequently, the development program stalls.

We now describe each of the steps in developing a panel of tests, using quality control of BS as the first example, and then reviewing this process as it applies to FP and to stability testing.

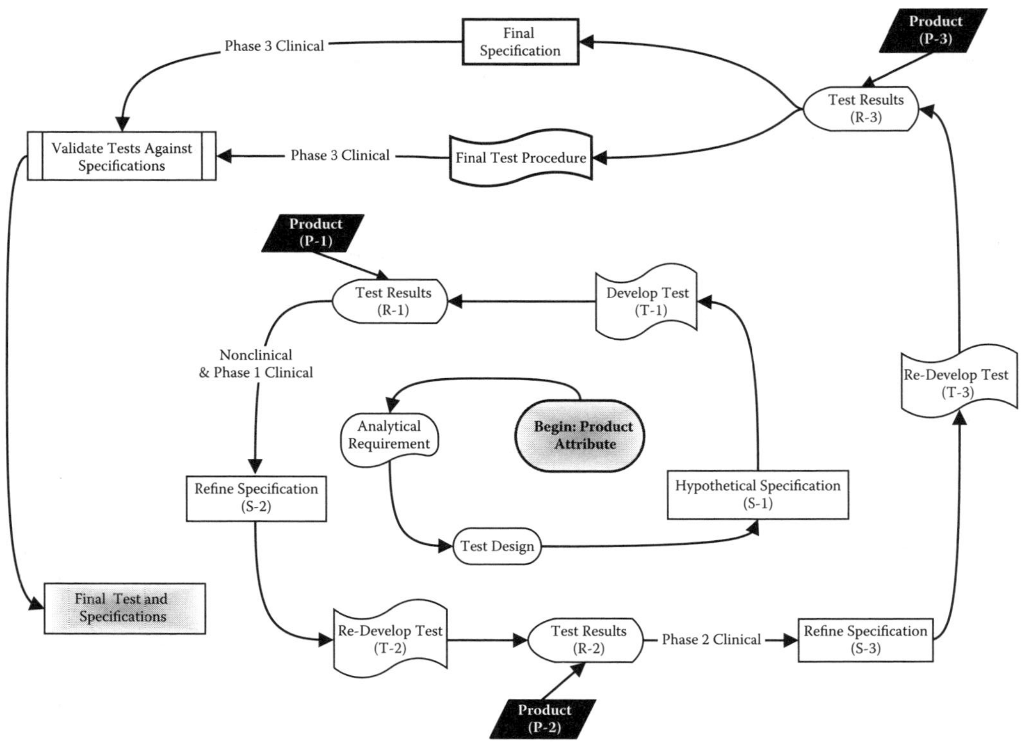

FIGURE 7.1
Life cycle for development of quality control tests. The test life cycle begins with an understanding of a product's attribute and the analytical requirement. An appropriate test is then chosen or designed and the development cycle then applies specifications (S), test development (T), input of manufactured product (P), BS or FP, upon which to perform the assay and results (R) from tests at that development step. It is a multistep process (e.g., S-1, S-2, S-3, final specification).

Define Product Attributes

The quality control development cycle begins with a QC plan, drafted only after there is an understanding of the intended product, a treatment indication, and at least an early or research version of the candidate biopharmaceutical (see Chapter 1). Also, some experimental work must precede quality control planning, as it is not possible to develop a test for a product that has not been at least slightly characterized in the research laboratory. This requirement is for a clear understanding of that product. If it is a protein, the primary, secondary and tertiary structure must be known, as well as the molecular, cellular or immunological basis, for its activity. Also, quality control planning cannot begin until one understands how the product will eventually be used; the intended treatment indication as well as the mechanism of action the product must have to achieve a desired endpoint.

An example product will be applied throughout this chapter to illustrate principles of biotechnology quality control. Consider a recombinant protein (r-protein) product, a biopharmaceutical that functions in man by neutralizing an undesirable molecule, perhaps a complex carbohydrate, located within a diseased cell. The therapeutic r-protein has the potential to ameliorate a disease. However, this r-protein must first enter the cell via a receptor on the cell surface. Further, it is known that r-protein binding to the cell receptor must be of a magnitude to trigger the cell to internalize the r-protein. Also, the r-protein must survive within the cell and bind to the target molecule, an undesirable carbohydrate associated with cell death and disease. Binding leads to the carbohydrate molecule's functional elimination from the cell. Further, to enter the cell, the r-protein must be an intact molecule with proper primary and secondary structure, i.e., not degraded or unfolded.

With such knowledge of a product and its mechanism of action, the quality control planner begins by developing a list of attributes for the product. An attribute is simply a desirable or necessary characteristic that a product must possess to be safe or effective. In biopharmaceutical development, the most commonly applied attributes are appearance, identity, strength, purity and impurities and potency. Returning to our example, several attributes of the r-protein, namely binding to a cell receptor, internalization into the cell, survival within the cell and binding to the target molecule, are easily identified. Each of these attributes is listed in a typical Certificate of Analysis (CoA) and some appear several times, as they are measured with different tests. The CoA, to be described later, is a formal document that lists each attribute, QC test, specification and test result.

Further to our example, the active ingredient in the product is the r-protein. It also is referred to in testing parlance as the *analyate* or *test substrate*, the material we wish to measure in whatever way. The r-protein also possesses attributes. When in solution, it has a distinct appearance. Our r-protein has an identity in the same way every human has a distinct fingerprint. In a given configuration, like a vial of FP, it has a particular strength or level of power (e.g., a concentration). Our product also has purity and, as with most products, there are impurities as well, even if in trace amounts. Finally, like all biopharmaceuticals, the r-protein has a biological potency, in that it affects a biological system. A biopharmaceutical may possess other attributes, but this list is adequate to begin planning QC tests for most biotechnology products.

Analytical Methods Measure Attributes

The heart of quality control testing is developing tests that measure a product's attributes. The quality control scientist maps a strategy, matching tests to particular attributes, sometimes applying multiple tests for each attribute. Further, he/she determines if an attribute requires, or deserves, a qualitative measurement or a quantitative measurement. This necessitates an understanding of what makes an assay a good means of measuring an attribute.

Nonetheless, a perfect match between a method and an attribute often is not possible. In such cases, the quality control scientist adapts a given analytical method to suit the exact nature and intended use or indication of his/her product. Fortunately, most analytical tools are quite adaptable. Even then, some attributes for a given product cannot be measured with "off-the-shelf" analytical methods or even with available, but slightly modified, tests. Here, the quality control scientist must be creative and design a new or unique analytical tool, right from the beginning. This is often the case for potency tests that measure complex biological functions, such as an immune or cellular physiological response.

Traits of Analytical Methods

Each assay has one or more unique traits, somewhat like a facial feature, distinguishing it from other analytical tests and making it attractive for a particular application. Some tests have the trait of qualitative analysis, others of quantitative analysis, and a few possess both. Tests are selected for their traits, a description of what they can do for us, really. The QC scientist must have a pool of analytical methods available, at least in theory, once he/she begins matching tests to measurement of attributes. Fortunately, a number of analytical tools, each with its own peculiar traits, are available for testing common attributes of many biopharmaceuticals.

One trait of any test is system *suitability*, which means that the chosen analytical method must be appropriate in all ways for the intended purpose and measurement. *Specificity* is a second trait and it means a test has the ability to measure the intended product, and nothing else that might be in the test material. *Precision* is the closeness of agreement between several measurements, much like precision in shooting an arrow means coming close to one point on the target with several arrows. The trait of *linearity* is applied when the assay must generate a linear curve. Linearity means that the results are directly proportional to the concentration of the analyate. *Range,* closely related to linearity, is the interval between the upper and lower concentration of analyate in the linear part of the curve. *Limit of detection* (LOD) is understanding how little of the analyate can be reliably detected in a sample. *Limit of quantitation* (LOQ) defines the lowest amount of analyate that can be quantitatively measured and not just simply detected. The trait of *robustness* carries many related meanings, but overall it means a test is reliable with respect to normal or expected variations in the analytical or testing environment. For example, three operators should obtain the same result if they each perform the test on three different dates, then an assay is robust and, one would think, reliable. Analytical tools are chosen to measure an attribute only if it possesses traits that allow them to complete a stated measurement. Hence, possession of the proper traits is a criterion for choosing the proper analytical tool. To paraphrase a saying of automobile mechanics, "You must have the right tool for the job."

Draft a Certificate of Analysis (Bulk Substance)

As attributes and methods are identified they are listed in tabular form, on a draft document referred to as the Certificate of Analysis (CoA), shown in Table 7.1 (for BS) and in Table 7.2 (for FP). Each attribute, and these are further defined below, is listed in the first column of the CoA. More than one test may be applied to an attribute, as each test measures a different parameter of that attribute.

- **Appearance**. Most products have a distinct appearance to the eye. Bad product often looks bad. Under appearance, traits may be further defined as color, clarity or opaqueness, or presence or absence of particulates or aggregates.
- **Identity.** This trait simply ensures that the product is what we believe it to be and have labeled it as, and not something else. Each biotechnology product is unique and possesses "fingerprints" that can be analyzed.
- **Safety.** A safety test cannot by itself tell us whether a product is safe or not, but it can provide some assurance that it is not overtly toxic or otherwise lethal or seriously harmful to the user. And multiple safety tests, each examining a specific aspect of the attribute, can additively increase the chances a product, in fact, is safe.
- **Purity and Impurities.** All biotechnology products are purported to be pure, to have only those molecules or cells or other active ingredients they are supposed to contain. Purity is a measure of that product claim and impurity testing informs us as to the nature of anything else in the product vial.
- **Strength.** It is important that each product have enough of the active ingredient so that it has the potential to cause the intended effect. If

TABLE 7.1

Certificate of Analysis for Biopharmaceutical Bulk Substance (r-Protein Example)

Attribute	Analytical Method	Reference to Method	Specification	Result
Appearance	Visual inspection of DS in clear glass tube	SOP# QC-01	Liquid, opaque, off-white to straw color; no particulates or aggregates	
Safety	Microbial Limits Test (MLT)	USP <61>	<1 viable organism/mL	
Identity	N-terminal sequence	SOP# H411B	Confirm known sequence	
Identity	SDS-PAGE	SOP#QC-02	Single band at 30 Kd	
Safety	Endotoxin. Gel Clot LAL	USP <85>	<1.0 EU/ mg protein	
Purity	SDS-PAGE	SOP#QC-03	Single band at 30 Kd; comparable to reference standard	
Purity	HPLC	SOP# QC-04	Single peak integrated >98% material in sample	
Purity	Aggregates by size exclusion chromatography	SOP#QC-05	>98% of material is not aggregate	
Purity	Peptide map	SOP#QC-06	Map equivalent to reference standard	
Purity	Host cell protein, ELISA	SOP #QC-118	<0.1 mg DNA/1 mg protein	
Purity	Host cell DNA, fluorescence probe	SOP #QC-120	<10 μg DNA/1 mg protein	
Purity	Silicon lubricant Atomic absorption	Contract Laboratory SOP#X147-1	<10 ng silicon per mg protein	
Purity	Aggregated protein Light Scatter	SOP#QC-08	<7 μg aggregated protein/mg total protein	
Strength	BCA. Concentration and amount of protein	SOP#QC-07	1.0 ± 0.1 ml containing 2.0 ± 0.3 mg/mL protein	

Continued

TABLE 7.1 (*Continued*)

Certificate of Analysis for Biopharmaceutical Bulk Substance (r-Protein Example)

Attribute	Analytical Method	Reference to Method	Specification	Result
Potency	Receptor binding	SOP #QC-111	0.60–1.05 μg r-protein/1.0 μg receptor	
Potency	Viability of cultured cells at 1, 2, 4, 6, 12 h	SOP#QC-09 SOP#QC-10	>70% viability versus time 0 at each time point	
Potency	Accumulation of molecule in cultured cells at 24 hr	SOP QC#11 and SOP#QC-12	<10% accumulation over baseline, time 0	

Prepared by/date: _______________ Approved by Quality Control/date _______________
Approved by Quality Assurance /date: _______________

we say there are 25 mg/mL of product in 1 mL per vial, then there, in fact, should be 25 mg/mL and 1 mL in each vial.

- **Potency.** The most challenging trait to measure, potency tells us that the product, in fact, does have the intended biological effect.

This is the first step in drafting a CoA. The next step, as shown in the second column of Table 7.1 and Table 7.2, is to identify analytical methods to measure each of these attributes.

Select Analytical Methods

This section provides information on selecting a test to measure each attribute; by way of example, it identifies a few assays commonly applied to recombinant protein biopharmaceuticals in BS (Table 7.1). Selection of tests for FP is discussed in a later section and a list of assays, with technical descriptions, is presented there and in Table 7.2.

Laboratory tests used in quality control, also referred to as analytical methods or just "methods," can be classified in several ways. Tests described in a compendium (e.g., a pharmacopeia) are nationally or internationally recognized and performed in a very standard manner, no matter what the product. Tests are matched to attributes in the example CoA and, in the third column of Table 7.1, reference is made to the specific manner in which each test is performed on BS.

A compendial test, such as those for sterility, microbial limits or endotoxin, is applied to a wide range of biological and pharmaceutical products. Examples of compendia are given in Box 7.1. An outline and contents of the U.S. pharmacopeia (USP), a much used compendium and quality control reference, are given in Box 7.2, along with the titles of some compendial tests

TABLE 7.2

Certificate of Analysis for Biopharmaceutical Final Product (r-Protein Example)

Attribute	Analytical Method	Reference to Method	Specification	Result
Appearance	Visual inspection of FDP in final container	SOP# QC-21	Clear, colorless liquid without particulates or aggregates	
Safety	Endotoxin Gel Clot LAL	USP <85>	≤10 EU/1 mL dose	
Safety	Sterility Compendial	USP <71> 21CFR 610.12	Sterile	
Safety	General Safety Test	21 CFR 610.11	Pass	
Safety	Osmolality by Osmometer	SOP#QC26	10.0 ± 0.1 mOs	
Safety	pH by pH meter, microprobe	SOP#QC27	7.1 ± 0.2	
Identity	N-terminal sequence	SOP# H411	Confirm known sequence	
Identity	SDS-PAGE	SOP#QC-22	Single band at 30 Kd	
Purity	SDS-PAGE	SOP#QC-23	Single band at 30 Kd; comparable to reference standard	
Purity	HPLC	SOP# QC-24	Single peak integrated >98% material in sample	
Purity	Excipient Glycerol by Atomic Absorption	SOP#11-4422C	1 ± 0.1 mcg/mL	
Purity	Excipient, Human Serum Albumin by Enzyme-linked Immunosorbent Assay	SOP#11-2244C	200 ± 20 mcg/mL	
Purity	Aggregated Protein by Light Scatter	SOP#QC-28	<1 mcg aggregated protein/mg total protein	
Strength	Total Protein by BCA	SOP#QC-25	1.0 ± 0.1 mg/1 mL dose and per vial	

Continued

TABLE 7.2 (*Continued*)

Certificate of Analysis for Biopharmaceutical Final Product (r-Protein Example)

Attribute	Analytical Method	Reference to Method	Specification	Result
Potency	Viability of cultured cells at 1, 2, 4, 6, 12 h	SOP#QC-09 SOP#QC-10	>70% receptors versus time 0 at each time point	
Potency	Accumulation of molecule in cultured cells at 24 h	SOP #QC-11 and SOP#QC-12	<10% accumulation over baseline, time 0	
Potency	Reduction of disease; transgenic mouse model	SOP#QC23 SOP#QC24 SOP#QC25	>50% reduction as compared to control	

Prepared by/date: ______________ Approved by Quality Control/date ______________
Approved by Quality Assurance /date: ______________

commonly used to measure attributes of biopharmaceuticals. Methods of compendial tests cannot be easily modified, but they sometimes are adapted for novel applications.

Another group of analytical methods, referred to here as *generic tests*, are used to measure attributes of identity, purity, strength and, sometimes, potency of biotechnology products, but are often not found in a pharmacopeia. Even though they are not compendial, there may be industry or regulatory precedence, procedures or even quality standards for their performance. This depends on the nature and history of the product. These quality control tests can be established in most laboratories or, if they require expensive instrumentation, can be performed by a contract research organization (CRO). Most are readily adaptable to a variety of products and, in many cases, their methods can be changed to suit specific purposes. General safety testing, described in FDA regulations, and visual appearance of a product would be considered generic tests. While commonly used, the tests high pressure liquid chromatography (HPLC) or polyacrylamide gel electrophoresis (PAGE) are performed in a very specific manner for each analyate.

A third group of tests includes those developed for one product or a very few closely related products. These methods (some examples are given later in this chapter) often originate in a research laboratory and are further developed, adapted and refined by the quality control laboratory for use as a quality control test to measure an attribute of a specific biological product.

Quality control assays also are classified in yet another way—by their intended use or application. These include tests for raw materials, for in-process testing, for drug substance, for drug product, or for stability testing of any material, substance or product. Hence, a given method might be applied at one or more points in the product manufacturing cycle.

BOX 7.1 EXAMPLES OF COMPENDIA AND REFERENCE TEXTS FOR QUALITY CONTROL

United States Pharmacopeia (USP). The official pharmacopeia for the United States, the USP, along with a sister publication the National Formulary or USP-NF, provides standards and references for medicines and other products. If a standard applicable to a biopharmaceutical, in fact, is published in the USP, then it is very likely that FDA will demand it meet this criterion.

European Pharmacopeia (EP). The official pharmacopeia for the European Union.

British Pharmacopeia (BP). The official pharmacopeia for Great Britain.

Merck Index. This book provides precise and comprehensive information on chemicals, drugs and biologicals written as monographs and carefully indexed. It is used in biopharmaceutical development of research materials and products that are well-characterized and may be predicate or comparators for a novel compound in early development.

Merck Manual. A medical encyclopedia, organized by disorders or diseases of various systems or organs. Explains the symptoms or diseases, their diagnosis and treatment. A leading medical reference.

Martindale's. The Complete Drug Reference. A very complete guide giving monographs, albeit brief, on thousands of drugs and biologicals with reference citations and manufacturers. It is carefully indexed.

Physician's Desk Reference (PDR). A collection of the product labeling of the most commonly used drugs and biologicals in the United States. It has information on drug indications, dosages, side effects as well as detailed instructions for use.

As noted earlier, another means of classifying quality control tests is by their intended outcome or application. Examples of test applications are appearance or description, identity, purity, impurities, potency, quality and special tests. It is not unusual for a very adaptable test method to be used to measure two attributes. For example, one test may be used to measure both identity and purity. A test is classified as well according to the method's enabling technology, such as identification of bacteria and yeast, pH measurement, HPLC, peptide mapping or receptor binding.

Now, to describe the information entered into the third column in the CoA (Table 7.1), the procedure used to complete each test. For compendial tests,

BOX 7.2 AN OUTLINE OF THE UNITED STATES PHARMACOPEIA (USP) AND EXAMPLES OF USP SECTIONS RELATED TO BIOPHARMACEUTICALS OR DRUGS

General Information. Provides guidance on a variety of product classes spanning the spectrum of drug and biopharmaceutical development. Examples are general guidelines for ophthalmic preparations and water for pharmaceutical purposes. Section <1231> identifies classes (qualities) of waters and their standards.

Reagents. All types of reagents and their quality standards are described in this section. Examples are acetic acid, diluted, used in various test procedures, and Pancreatic Digest of Casein, used in culture media.

National Formulary (NF). Here are recipes for making a host of products that are used in or as drugs, including many over-the-counter preparations. An example is the procedure for preparing the peppermint solution added to certain oral drugs.

Official Monographs. In this section of the USP are monographs on commonly used drugs, and many described here are long-standing and generic. An example is Aspirin Delayed Release Tablets. Others are products used largely in medical treatment facilities, such as Lactated Ringers and Dextrose solution.

General Tests. Many tests that are used in drug and biopharmaceutical quality control are found under this heading. Certain tests used for biotechnology products are described in great detail. Examples include:

<111> Chromatography: gas, paper and column
<85> Bacterial Endotoxin Test
<71> Sterility Test
<61> Microbial Limits Test

Also under *General Tests* is information that provides guidance and procedures for biopharmaceutical testing and assay development, in general. Examples include:

<1041> Biologics
<1045> Biotechnology Derived Articles
<1046> Cell and Gene Therapy Products
<1047> Biotechnology-Derived Articles—Tests
<1048> Quality of Biotechnological Products: Analysis of the Expression Construct in Cells Used for Production of r-DNA Derived Protein Products
<1049> Quality of Biotechnological Products: Stability Testing of Biotechnological/Biological Products
<1050> Viral Safety Evaluation of Biotechnology Products Derived from Cell Lines of Human or Animal Origin

reference is made to a section of a monograph (e.g., USP <71>). For noncompendial tests, standard operating procedures (SOP) are identified in this column, citing the number of the SOP. If a CRO performs a test, then that laboratory and the SOP used by them is identified under the test method column of the CoA.

To summarize, a key element of quality control planning and performance is matching the proper test to an attribute, and this requires knowing which tests are available to the analyst and understanding how the attribute relates to each available test. Only then can a meaningful panel of tests be selected for the product. We return to our example in an attempt to better explain the matching process of attribute to method. First is the identity test. Identity tests reveal whether or not a product, in fact, is the intended material, the fingerprint of a biopharmaceutical. For the r-protein example, we know from research it is a globular protein of known molecular weight and a defined sequence. How might we go about ensuring that the material we manufactured is that r-protein? A common approach is to sequence, by amino acid determination, from the N-terminus of the r-protein until about the 10th amino acid. It is highly unlikely that another protein would have the same 10 amino acids in that order at the N-terminus. One might also identify the isoelectric point of a protein when analyzed by electrophoresis at various conditions, e.g., various pH or ionic strengths. Another approach to demonstrating identity is to measure the molecular weight of the molecule under reducing or nonreducing conditions by gel electrophoresis. While not as definitive as N-terminus sequencing, this test differentiates the analyate from many other proteins. Gel electrophoresis is made more powerful as an identity test if, after the electrophoresis step, the protein is blotted to an inert but absorbent membrane and then probed with an antibody specific for the product. This is known as a Western blot test.

A sample of reference standard (i.e., a protein known to really be the desired protein) is always tested in parallel with a test sample. If results of the test sample match theoretical or expected values and the results obtained from testing match the reference standard, then there is a high probability the test sample is indeed the intended molecule.

The strength of a preparation is a general measure of how much of the desired active ingredient is in the product. For the r-protein, strength might be reflected in the total amount of protein as long as the vast majority of protein in the sample is, in fact, r-protein. How do we ensure this is the case? First, we perform identity testing on that sample to inform if the molecule in the sample is indeed the intended r-protein. Second, and as described below, we measure the purity of the molecule in a sample of product. Back to the concept of strength, for a recombinant protein this may be measured by a total protein assay, such as bicinchoninic acid (BCA) or by ultraviolet (UV) absorbance at a specific wavelength in a spectrophotometer. Then, this measurement, given in milligrams per milliliter, is multiplied by the percent purity and total volume in milliliters to give the total amount, in milligrams, of the desired protein. This paradigm demonstrates the interdependence of various quality control tests and the need to interpret the result any one assay may give in relation to a result from another assay.

Purity of any product is usually measured using several tests. Further, major impurities are characterized. Methods are developed to measure the level of product purity, i.e., the percent or actual amount of the stated product and the percent or amount of impurities and contaminants. Other methods are used to identify the major impurities or contaminants to include even tiny amounts of potentially toxic or otherwise undesirable substances that might have entered into the production stream and remained in the product. The development of testing schemes and the selection of tests are based on an in-depth understanding of the raw materials, equipment and processes used in manufacturing, and of the scope of possible impurities or contaminants. Impurities and contaminants are further discussed in relation to biomanufacturing (see Chapter 6).

Returning to our example of the r-protein, we consider the purity of that molecule as well as impurities that might exist with the r-protein as bulk substance. Here, r-protein purity is measured by sodium dodecyl sulfate in polyacrylamide gel electrophoresis (SDS-PAGE), an assay also used as an identity test. However, this assay is, at best, only semiquantitative in that it cannot accurately measure the amount of the r-protein or amount of impurities in the sample of BS. It may, however, give a reasonable estimate of purity. The SDS-PAGE test is often used in-process to follow progress in purification of molecules. A more exact measure is applied to measurement of r-protein purity using an analytical chromatographic method; HPLC is a common choice. Here, the example r-protein should appear on the chromatogram as an independent, major peak while impurities might appear as smaller side peaks or shoulders of the major peak. Further, we might use size exclusion chromatography (SEC) to show that the molecule in our preparations, in fact, is not aggregated. This test demonstrates both that the r-protein has maintained a native form and that it has not otherwise distorted through clumping. Yet, another purity test, peptide mapping, may reveal that molecular integrity remains. Protein aggregates are also measured by light dispersion, if their presence is suspected. Special tests, such as mass spectroscopy, nuclear magnetic resonance or capillary electrophoresis may be considered. Carbohydrate analysis also may be applied if there is a need to examine certain posttranslational modifications. Today, many newer methods are applied to molecular characterization of certain products and some are noted later in this chapter. Because purity is critically important to molecular integrity and function, and because impurities must be characterized and measured to prevent them from causing undesirable reactions in the user and from increasing in amount with later production, several purity and impurity tests are typically applied to a product, both as BS and as FP.

Detection of contaminants presents a different challenge because these can enter the stream from so many sources, especially if a failure went undetected during processing. For example, most proteins are filtered at some point during purification and filtration materials may release fibers, particles or fragments into the product stream, thus contaminating the product. Contaminants, particulate and soluble may enter the product stream from raw materials or virtually any substance that contacts the product stream. Given that one cannot test for every possible material that might contaminate a biotechnology manufacturing process, what should the quality control scientist consider as possible contaminants for any given product? Anything that might be toxic or otherwise dangerous to the user in small amounts and is part of the process comes first to mind. For example, if there is a possibility for bacterial growth, then endotoxin testing, described below, is a possibility. If there is a piece of biomanufacturing equipment that is essential, but is known to sometimes shed particulates of silicon lubricant into the product stream, then it might be wise to test for silicon lubricant. Both endotoxin and silicon lubricant are considered in the example CoA for BS, as shown in Table 7.1. Clearly, an effective yet affordable contaminant testing program involves discussion between manufacturing and control staff, and decisions based on fully understanding the production processes.

Potency assays are critical to a quality control testing scheme because they are used to predict whether the product will function as it was designed to function. It would be futile to produce any biopharmaceutical product and test it for purity, identity and safety, but not know if it could function as intended. Unfortunately, this is far too often the case with biopharmaceutical development programs. For testing BS, potency assays are often a surrogate assay, meaning they do not directly measure the biological function in a complex system, such as a whole animal, but instead measure a physiological attribute of the product in an *in vitro* or a cell-based assay. Surrogates are used in all aspects of biotechnology development, but any surrogate measure must be appropriate, well designed and, eventually, validated against the intended use. The quality control scientist developing a surrogate assay must be both knowledgeable about the product and the therapeutic indication, particularly the mechanism of action and the biology and molecular biology involved in the product's therapeutic effect. Using this knowledge, scientists become inventive, even crafty, in finding analytical methods that predict potency (or lack thereof), while keeping that test as simple, inexpensive and practical as possible. The best source of new potency tests is the research laboratory.

Returning to the example of the r-protein and examining Table 7.1, we see two potency assays were developed for this bulk substance. One measures specific binding of the r-protein to its cell-surface receptor. The rationale is that receptor binding is a critical step in the pathway to molecular activity and a biologically active r-protein must bind to that receptor. For this example, the receptor was identified and the gene was cloned in a research laboratory, so that it is now produced in small amounts (enough for testing purposes). Using analytical instruments, quality control scientists next develop an *in vitro* assay that measures the amount of r-protein that binds to a given amount of receptor. For example, between 0.6 to 1.0 μ of the recombinant product binds to 1.0 μg of the receptor. Upon repeatedly testing three batches of recombinant protein with the newly developed assay, scientists determine that between 0.75 and 0.92 μg of product, in fact, does bind to the receptor in this test. This test is chosen as one of two potency assays for r-protein in the bulk substance.

For an attribute as important as potency, one always considers two or more complementary assays. This is because a single assay measures only one aspect of a product's potency attribute. In the r-protein example, the quality control scientist chooses a second assay, also developed in the research laboratory, to measure the degree to which r-protein inhibits the buildup of the undesirable molecule within the target cell. This is a relevant surrogate to the intended biological response because it measures the ultimate activity associated with therapeutic value, at least at the molecular and cellular level. Shown in Table 7.1 as the last assay on the CoA, the measurement is an *in vitro* assay, likely rapid and inexpensive, but hopefully very precise and sensitive to product activity.

This section of Chapter 7 provided an overview of early test selection, basing each assay on a product attribute. After this is done, the next step in the planning process is consideration of specifications. Later in this chapter, we discuss in greater detail analytical tests and their application to the QC of biopharmaceutical products.

Develop Specifications

Identifying a test to measure each attribute is important, but it is also critical to know whether product passed or failed based on the results obtained when the test is applied to a particular product. The word *specification* carries great meaning to both QC scientists and biopharmaceutical development projects. A specification is a descriptor, numerical or verbal, that a product must achieve to be considered suitable for use. It also serves as a requirement or condition upon which product is accepted or rejected. Specifications may be established by regulation, by precedence and proven value and capability, by outside guidance or by a product development team. Often a specification is quantitative, stated as a range of values (e.g., ≤12.5 units or 1.0 to 3.0 mg/mL or 2.0 ± 1.0 mg/mL), but it may be qualitative, a term that compares it to a reference, such as "comparable to values of reference standard #0017," or it can be purely a descriptor, like "clear, colorless solution free of particulate matter." Examples of specifications for BS are shown in the fourth column of the CoA, Table 7.1.

Prior to market approval of a product, specifications may be considered interim or temporary. Other specifications, those codified (e.g., sterility test in 21 Code of Federal Regulations (CFR)) or established by precedence or compendium (e.g., sterility test in USP), are less flexible and must remain intact during the development life cycle. Specifications are taken quite seriously by both regulatory authorities and the sponsor, and final or ultimate specifications established through the validation process in Phase 3 often guide the release of a marketed biotechnology product for years to come. Once established, specifications may be changed, but only with scientific evidence to support the adjustment, and this follows strict change control rules. Hence, there is a great need to carefully choose and then fully develop a specification, basing interim and final decisions on experimental data generated by testing multiple lots of manufactured product.

Advances in analytical technologies for biopharmaceutical products have increased the number of tests used on any product. Regulatory authorities are quick to suggest yet another test that might ensure safety or better predict efficacy of a product. Specifications themselves have become more complex, quantitative and sensitive as well. Indeed, the role and importance of the quality control function itself to biotechnology development has grown considerably over the past 30 years.

Establishment of specifications for purity or impurities is challenging for biopharmaceutical product development teams by raising questions that have no simple answers when data are limited or do not exist at the time. What quantitative limits are acceptable for purity and for impurities and what is allowed and how much? The correct answer must be experimentally determined for each product and a final specification is not finally established until late phase development. Also, there are no guidelines that can be considered during QC planning phases of development for each product. First, the sponsor must expect that a product will not be 100% pure and that impurities will exist in both the BS and the FP. Manufacturing generates, and processing concentrates, certain impurities and contaminants; no product is expected to reach 100% purity using current purification technologies. Upstream production is a dirty process, while downstream processing, notably chromatography, does introduce and concentrate novel yet undesirable contaminants, even while removing other impurities or contaminants and concentrating the product. Contaminants enter the stream out of necessity, that is, because they are inherent to a required process, a necessary evil. Examples of contaminants are endotoxin, in some systems shed by the very bacteria making the product, chromatography gels or matrices, particles from vessels and tubing, and chemicals leached into the product stream from various contacts and surfaces. Impurities are molecules that are derived from the ingredients used to make the product, but are not wanted in the BS or FP. They include cellular debris and molecules derived from the cells in which the recombinant protein or tissue was itself produced, and it includes components of the nutrient medium that fed those cells. Impurities also are seen as breakdown products of the desirable biomolecules, and include improperly folded protein, shortened versions of the protein, posttranslational product variants and fragments or aggregates of the protein.

As noted before, a rule of thumb for establishing purity and impurity specifications in biotechnology is this: The BS or FP specification should be at least 95% pure product. However, this rule does not apply to every biopharmaceutical, and specifications for purity must be developed based on the intended use and attributes of the product and the nature of the impurities themselves. For example, a protein that will be used to enrich cattle feed might be fine at 75% purity, as long as none of the impurities were toxic to cattle or man and consisted largely of protein fragments and aggregates. A biopharmaceutical intended for injection at large doses into human patients with serious disease might need to be over 99% pure and completely free of any aggregated protein. Here again, careful planning is required to ensure that analytical methods and specifications developed for purity and impurities match exactly the intended use and other attributes, e.g., safety of a product. Impurities, such as a virus or lethal toxin, are simply not allowed in a biopharmaceutical and, if such materials could possibly have been introduced, then methods must be designed to ensure their removal and highly sensitive and specific tests introduced to ensure that product is free of such substances. Here specifications are very stringent. The guidelines for setting specifications for other impurities or contaminants are established based on prior experience, probability that they, in fact, do exist, availability of tests to identify or measure them and, mostly, on common sense and good scientific practice. As noted in Chapter 6, specifications for levels of impurities or contaminants are, in the end, often negotiated between the sponsor and the regulatory agency. This brings up a final point on the subject of setting specifications for impurities, both qualitative and quantitative. National populations and their regulatory agencies vary greatly on perceived risk to the user of certain impurities and the biotechnology firm would be well advised to consider all marketplaces, not just the United States, when establishing specifications for a product.

There are other outcomes to the processes of establishing and applying test methods and specifications in this cycle. Sometimes a well-considered analytical method fails miserably and is not predictive of the attribute or is otherwise unable to predict product quality. There is no need to consider further refinement of the test or of the specification. In other instances, the hypothetical value established by QC scientists is not at all close to experimental values. In such cases, the assay may be further studied or it may be reworked to either explain differences or to optimize a method, respectively. More often than not, however, a well-considered analytical method is meaningful to measuring product quality with only the need to adjust the specification and thus ensure consistent quality for future batches or lots of product.

Some QC tests have, in the eyes of regulatory authorities, absolute specification requirements. As noted before, sterility tests are performed by compendial methods and they must meet standards published in a compendium or by regulatory agencies. There is but one definition of sterility and adjustment of the sterility specification is simply not acceptable for a biopharmaceutical, such as the r-protein mentioned in the earlier example. Other historical tests, even certain compendial methods, may allow specification variance; however, this depends on the nature of the product and risk to the user associated with a change in specification. Such changes must be negotiated with regulatory authorities. For example, the appearance and description for a parenteral biopharmaceutical, formulated as protein in a buffer, is expected to read "Clear solution without particulates." However, for some proteins at high concentrations in a buffer solution, it may be normal for the product to be cloudy or opaque. Hence, recombinant protein in the example might be allowed to deviate from the standard specification for most protein solutions and instead be considered acceptable if it had a specification for appearance of "cloudy, colorless solution without particulates or foreign matter."

The establishment of specifications for contaminants and impurities is a more challenging task. It is impossible to know in early development how an impurity might impact the safety or efficacy of a product. How does one evaluate the impact of minute amounts of a given contaminant unless it is a known toxin? Establishing purity and impurity guidelines has led to long discussions within the international biopharmaceutical community, often based on the risk posed by certain impurities found in some products. An example relevant to biopharmaceutical protein preparations is the impact that protein aggregates might have on parental products. Years ago, it was felt that they had little impact on product safety or potency unless they were present in significant amounts. Even then, a definition of significance was elusive. Recent evidence suggests, but does not prove, that protein aggregates, even in small amounts, may be immunogenic and potentially elicit an antibody response by a patient to a recombinant protein. If this is the case, the impact is probably quite variable depending on route of delivery, amount given over the lifetime of the patient, exact nature of the protein and the aggregate, and the patient. How then does one establish a specification for maximum allowable amounts of aggregate for any given protein product? Panels of experts may be called to address such questions, but even then outcomes are fuzzy. Hence, challenges to planning product attributes, tests and specifications continue through the cycle of product development and into marketing phases as well.

As with other attributes, specifications for potency of product in BS are established early in development, even when little experimental data are available. Potency tests, identified in the planning process, are established in a laboratory and then used to analyze some early batches or lots of BS or FP, respectively. Referring back to development of the r-protein receptor binding potency assay (Table 7.1), we know that 1.0 μg of the recombinant product should, in theory, bind to 1.0 μg of the receptor, and also that laboratory testing revealed a range of binding activity, between 0.75 and 0.92 μg of pure r-protein to 1.0 μg of receptor. The quality control scientist must at this time establish a specification for this potency assay based on both theoretical and derived experimental data, but with the knowledge this data might, by chance, reflect an incorrectly high or low estimate of actual binding activity. The early or hypothetical specification, referred to as S-1 in the QC cycle drawing (Figure 7.1), is the following specification in this example: "Range of protein binding, 0.60 to 1.05 μg of product per 1.0 μg of receptor" (Table 7.1). Testing additional and subsequent batches of bulk substance, the sponsor might discover the range of values is too broad and, based on the data, narrow the range of acceptable values in the specification, perhaps to 0.80 to 0.90 μg protein per 1.0 μg of receptor (e.g., the refined specification, S-2, in Figure 7.1). Alternatively, and with additional data, the specification could be a "lucky estimate," holding up to experimental results and remaining the same through the product development life cycle. This then is the accepted process, experimentally intensive, but proved effective to establishing meaningful tests, test results and specifications (Figure 7.1).

The other potency assay for BS used in our example (Table 7.1) measures another trait important for measuring activity of the r-protein, the ability to halt buildup of a molecule inside cells. The specification was established in the same manner.

In summary, the early establishment and then later adjustment of a specification is a normal part of QC testing in the overall product cycle. In early development, the process is part scientific, part iterative and part intuitive, and later it becomes heavily scientific, driven by data from clinical, nonclinical and laboratory studies.

Enter Test Results

Results are added to the fifth and final column of a CoA (Table 7.1, Table 7.2) after testing has been completed. Results are given in the same format as specifications to allow for comparison. For example, if the specification for Range of Protein Binding is 0.60 to 1.05 mg of recombinant protein bound per 1.0 μg of receptor, then the result could only be given as milligrams of protein bound per 1.0 μg of receptor. If the result was 0.73 mg, then this batch of BS would "Pass" by this standard. However, if the result was 0.55, then it would "Fail." The subject of handling a failure to meet a specification is discussed later in this chapter.

Certificate of Analysis for Drug Product

The process of planning QC test methods and specifications is not unique to BS and this process is also applied to developing testing plans for FP. Indeed, the process of potency test development is often more challenging with the final formulated product than it is with the bulk substance.

Once it has been tested and released, BS is formulated, then filled into a final container, such as a vial or syringe, and finally labeled and packaged. Once finished, testing begins and results are included in a CoA for FP, a sample of which is shown in Table 7.2. This certificate, once signed and reviewed, becomes one of several documents supporting release of FP to the user or, if product fails, for destruction of this material. The contents of a CoA for a lot of FP are important and so great care is taken during QC planning to choose the correct analytical tools and specifications. Again, the choices are based on the nature and attributes of, and the indication for, the FP.

Quality control tests for FP may be more stringent than tests for BS and they are always focused on an attribute that is important to the intended use and the well-being of the user. In designing the tests that apply to a biopharmaceutical FP, we consider many aspects of quality. The appearance test is performed on a representative sample of FP when it is in the final container, filled and finished. The specification for appearance is designed to ensure that the quality control examiner inspects a representative number of vials for certain attributes as well as ensure the absence of undesirable and visible impurities or contaminants. For example, with the r-protein, we expect the FP to appear colorless and to not contain any aggregates or particulate matter. The appearance test illustrates an important point of quality control testing: The SOP must be written in such a way that the operator or inspector, in fact, will examine for these attributes and properly identify substandard FP. This general procedure, use of trained operators or technicians, adequate equipment and SOPs, and exact reporting, are applied to every test that is performed in a QC laboratory and to results reported to a CoA.

FP is subjected to several safety tests because this is a most important attribute of any biopharmaceutical. Most biopharmaceuticals are given parentally and, hence, they must be sterile. They also must be free of or have very low levels of endotoxin or other toxic substances. Tests for specific types of undesirable contaminants or impurities are defined in a compendium or in regulations, and described elsewhere in this chapter. In addition, the general safety test is performed in the United States on most biopharmaceutical FP to detect any highly toxic properties the product might have. The CoA in Table 7.2 identifies one test for appearance and three methods for safety testing.

The active ingredient in FP must be exactly what it purports to be on the label, and nothing else. Hence, identity testing is performed on FP. A variety of generic tests, such as SDS-PAGE, perhaps in conjunction with a Western blot using a specific monoclonal antibody, HPLC, N-terminal sequencing, tryptic digest mapping, and cell karyotyping or phenotyping, may be employed in a panel of identity tests. In Table 7.2, the chosen identity tests for an FP, our example r-protein, are N-terminal sequencing and SDS-PAGE.

Despite the fact that purity was demonstrated and impurities were identified at the BS stage, it is important to test FP for purity and impurities. This is because impurities or contaminants may have been generated or introduced during formulation and fill, the processing steps from BS to FP. In the example, r-protein, might have broken down during processing. Microscopic contaminants could enter the FP if, for example, the containers were not scrupulously clean. In the CoA for FP (Table 7.2), two purity tests, SDS-PAGE and HPLC, are used to detect and to identify (or measure) macromolecular impurities in FP.

Aggregate formation in FP is a problem with formulations of certain biopharmaceuticals. They are considered impurities, but can impact safety and potency of FP as well. A test for measuring protein aggregates is added to the CoA for the r-protein (Table 7.2).

FP is also tested for concentrations of any excipients, materials that were added during processing, and should exist in the FP. In the example (Table 7.2), both glycerol and human serum albumin were added to the formulation and the amounts of each are measured to ensure they meet specified concentrations.

Strength of FP is determined by an assay that measures the total amount of active substance. In the case of a recombinant protein, this might be a BCA assay to measure total protein (Table 7.2). A variety of analytical methods are available to measure in FP all macromolecules, or count and measure the amount of cells or tissues.

Other tests performed on FP measure attributes of the formulation that are important to product purity, potency and stability. In the example (Table 7.2), osmolality and pH were measured to ensure the salt concentration was correct in the formulation buffer. Maintaining pH is important as well to maintaining the stability of most biopharmaceutical products.

The development of relevant potency tests for FP challenges the design and subsequent execution of any quality control plan and requires considerable abstract thinking, laboratory testing and interaction with scientific colleagues. FP potency tests must be meaningful and practical. A potency test that measures a noncritical potency criterion is not very helpful and any test that takes over 60 days to complete and report is impractical. Consider also that a potency assay must stand as surrogate for the ultimate potency test: performance in many human users. There probably never was, and never will be, a single "perfect" potency assay, one that stands alone to predict the biological efficacy of an FP. Therefore, sponsors seldom rely on a single potency assay, but instead apply three or more potency assays, each of which may be "imperfect." This does not always happen from the start of product development, but it should begin in the early phase so that multiple FP potency assays are available in mid- and late-phase development.

Another objective of potency testing is to learn if the FP possesses attributes that result in optimal performance for the end user. This is difficult to achieve because we often do not understand every biological factor that leads to optimal and consistent efficacy in all users. For biopharmaceutical development, this means, in theory if not always in practice, that the potency test applied to FP mirrors exactly the potency when it is used in man. A single test seldom, if ever, achieves this objective, but application of multiple potency assays may support such conclusions.

Biological responses and biological molecules or cellular systems are complex. Application of a biopharmaceutical to a biological system is an attempt to disrupt or bring back to equilibrium a biological system. Thus, biopharmaceutical treatment may further complicate a biological system already out of control. As compared to small drug molecules, many biopharmaceuticals themselves are complex biological entities. Given this information, now consider how difficult it is to measure a product's potency in a complex system. Hence, multiple potency assays provide a greater chance of ensuring product efficacy than does a single potency test because several potency tests evaluate the impact of the product at multiple points in complex biological pathways. While the use of a complete living organism (e.g., a whole animal) for FP potency testing brings into play all biological influences on the product and allows measurement of product potency, it is often difficult to develop and validate an appropriate animal model mimicking the human situation. Often, however, it is worth considering animal models for potency testing over *in vitro* models.

Consider again the example of our r-protein, indicated to treat a disease based on buildup of an undesirable molecule inside certain cells. To reach the desired biological endpoint, the r-protein must function properly at several cellular locations and in a number of ways. Unlike the situation of testing for potency in a simple and highly defined *in vitro* laboratory model, there are other extracellular influences that impact this molecule as it exists in a human. Hence, a well-designed panel of potency assays for this biopharmaceutical takes into consideration functions at the cellular level. For the r-protein, the initial QC plan considers three potency assays: one that examines potency in a living animal, another focusing on the r-protein entering the target cell, and a third involving measurement of a specific desired activity within the cell. This plan applies a commonly used and practical approach to ensure a potent biotechnology product in every lot of FP, performance of product under multiple potency tests, each of which measures a different aspect of the potency attribute. Could an animal model possibly be used to measure potency of r-protein? Perhaps. As noted earlier, the best potency tests are developed in or adapted from the sponsor's research laboratory where the technology may already be applied for other purposes.

In-Process Testing

The concept of in-process testing during product manufacture was introduced and several examples are provided for various products in Chapter 6 and in Figure 6.9 and Figure 6.12 through Figure 6.17. Analysis of any product is important to establishing effective manufacturing processes and to maintaining quality. First, timely feedback regarding product, impurities or contaminants in the product flow allows staff to make adjustments and resolve issues that arise. Stopping a process midstream and then reworking a particular step is usually a much less costly solution than uncovering a problem at the end of manufacture and then having to repeat the entire process. In-process testing also gives a level of assurance that product will be pure and potent once it is tested at the end of production. This, too, can save time and resources.

While in-process testing is included in manufacturing protocols and samples are taken by manufacturing staff, typically the quality control scientist will develop in-process assays and test samples supplied to the laboratory by manufacturing staff. In-process test results appear on manufacturing documents like Batch Production Records (see Chapter 5, Chapter 6). Most in-process tests characterize product using chemical or simple biological measurements, examine strength, measure the amount of product, or test either for product purity or for specific impurities and contaminants. Many in-process tests are those already developed for release testing of BS or FP and others are modifications of tests found on those CoAs. Whatever the test, it must be designed to generate results in a short period of time, usually hours or a few days. This means that many tests, such as complex biological potency assays and those methods performed by CRO laboratories distant from the manufacturing facility, are unlikely candidates for in-process use. Also, in-process tests are simple to perform and do not require expensive, dedicated instrumentation or staff with special analytical skills. Despite these disclaimers, a number of tests mentioned in our discussions on release of BS and FP can be adapted and many more are available from the methods listed later in this chapter. Some excellent in-process tests are commercially available and used to rapidly measure protein concentration, count cells, or assess their viability. Most are adaptable to measure various products for two or more attributes; examples are simple analytical chromatography methods, such as HPLC, and various types of electrophoresis, notably rapid procedures like SDS-PAGE.

Analytical Methods

Having now introduced the quality control test development and specification development cycle (Figure 7.1), we now examine traits—technical aspects, attributes and limitations—of several analytical tests commonly applied to biotechnology product quality control. Analytical methods, like specifications, are adapted from many sources by quality control scientists. Some methods, such as sterility testing, are compendial, performed only using very specific recipes and reagents and with industry standard specifications. Other methods are traditional or generic in basic design, but adopted for use on a specific product or group of related biotechnology products. Generic methods provide some flexibility in method of performance and the specifications are product specific. There are also novel tests, analytical methods developed for a special measurement of one product. Methods are classified in other ways: by analytical instrument, degree of difficulty, foundation technology, or type of product or level of product manufacture or development. The quality control test matrix has, in part, been introduced earlier in this chapter. Presented below, but not classified or listed in any special manner, are certain tests commonly applied to biopharmaceutical development. Potency assays are given little attention because, as mentioned before, they are often "homemade" and relevant to one or a very few products. Potency assays for a variety of biopharmaceuticals are, however, listed in the manufacturing descriptions and figures (flow charts) of Chapter 6.

- **Sterility Test.** Sterility testing is described in great detail in pharmacopeias. The sterility test, if passed, ensures that a biopharmaceutical is at the sterility assurance level required for a parenteral product. The USP sterility test provides a 95% assurance that no more than one vial of product in one million vials will have a bacterium or fungus. Because it is impossible to test one million vials in each lot of product, the assurance level is a statistical relationship to the actual number tested, which can be surprising quite low. The test must be performed with great care, to exact procedures and with many controls. Sampling protocols must be carefully designed to ensure representative product is selected for testing. A specification for sterility test, USP <71>, might read: *"Sterile"* or *"No Growth,"* because growth of organisms is the measurement made on a sample.
- **Microbial Limits Test (MLT).** For in-process materials and often for BS, another compendial method, the MLT, is used in place of the sterility test to determine the microbial load or burden. The MLT, USP <61>, is designed to enumerate the total bioburden in a sample and to identify a few select and highly undesirable bacteria and a fungus, should they grow from that sample. The specification for the MLT is expressed as colony-forming units per sample (dose or milliliter). A result might be: *<2 colony forming units per mL sample and no pathogens detected*. Both MLT and sterility testing are very specialized and highly regulated endeavors, hence, many biopharmaceutical firms contract this work to specialty laboratories.
- **Endotoxin Test.** This test measures the amount of a toxic molecule, endotoxin, that is produced and shed by many species of Gram-negative bacteria. A gel clot, lymph amoebocyte lysis (LAL) assay, is commonly used, but there are other accepted methods. It is an important test performed on most biotechnology products because the molecule itself can cause adverse events, inflammation and even

toxic shock and death in humans. Endotoxin thus serves as a sentinel for past or present bacterial contamination of a product. Some expression vectors themselves produce endotoxin, shedding it into the product stream. Endotoxin is sticky, adhering to surfaces or other molecules, and it persists, so absence or low levels of endotoxin signals good production and purification techniques for a biopharmaceutical. As is the case for MLT and sterility testing, national regulatory agencies or international advisory groups have established specifications for endotoxin, maximum acceptable levels that cannot be exceeded in certain classes of products intended for a specific use or route of administration. A typical result might be: *4.3 EU/ml endotoxin by Gel Clot LAL.*

- **Appearance.** Appearance tests measure attributes, such as color, presence or absence of visible particulates or aggregates, and clarity. The appearance test is often performed visually by trained operators inspecting a representative number of containers, these selected at random at various times during the fill operation. Inspection of syringes or vials is typically before an indirect bright light and a dark/light background. Through training and carefully written SOPs, the examiner becomes proficient at identifying certain undesirable traits, such as coloration, opaqueness, aggregates or particulates. Instructions for reporting results are critical to success of appearance testing. Instrumentation also is used by some biopharmaceutical QC laboratories to scan vials of FP or samples of BS to measure appearance. A typical specification might be: *Clear, colorless solution, free of visible particulates or aggregates.*
- **General Safety Test.** The General Safety Test, an historical method required by CBER (Center for Biologics Evaluation and Review), U.S. FDA, for release of most biopharmaceutical FP, is exactly described in 21 CFR 610.11. But, for unusual biopharmaceuticals, and many are unique, CBER allows flexibility in performance or specification for this safety test. A dose of FP is injected into the peritoneal cavity of mice and guinea pigs and the animals are monitored for health. If the animals live and gain weight, the product passes the general safety test. As noted, creative variations of this test for unusual products are used by biopharmaceutical laboratories, proving that the intent, not the letter, of a regulation rules in many cases. A result might read: *Pass General Safety Test.*
- **Osmolality.** The ionic strength of FP, and sometimes of BS, is determined using an osmometer. Ionic strength is important to the stability of many products and reflects proper formulation. A typical result might read: *10±1 mOsM.*
- **pH.** A basic but important measurement, especially for FP, is pH. Virtually all biopharmaceuticals are formulated in buffered salt

solutions because they must be kept at a particular pH. Otherwise, they become unstable and degrade. pH measurements are made using a pH meter with a microprobe. Typical result might be: *pH 7.25.*

- **N-terminal Sequencing.** Used as an identity assay and described earlier, this inexpensive method establishes a unique identifier for a protein. It is performed by sequencing, from the N-terminus, the first 10 amino acids in a recombinant protein. Typical result: *KQENMEVRLL versus known and reference standard KQENMEVRLL.*
- **Polyacrylamide Gel Electrophoresis, Native or Reduced Molecule (PAGE or SDS-PAGE).** This assay will determine molecular weight of a molecule and it can disclose impurities in a preparation. It is typically used with proteins and glycoproteins. The sample is subjected to an electric field, "electrophoresed," in a polyacrylamide gel matrix and the gel is stained with a vital dye to disclose bands of proteins, distributed by molecular weight. It is not generally considered a quantitative test, but an estimate of purity can be calculated if comparisons are made on the same gel of test material to qualified reference standards of known purity, e.g., 100%, 95%, 90%, 80% or 60%. PAGE does rapidly provide semiquantitative information and other valuable insights into product identity, structure (secondary, tertiary quaternary) and purity. When testing unreduced or native proteins, shape or change isoforms can be found, even in small amounts, and compared to reference standard. A "high load" of sample can reveal oligomeric or aggregate species at the top of the loaded lane. A typical test result might read: *Native (nonreduced): Single dominant band at MW 56 kD and two faint bands at approximately 20 and 10 kD. Minimal amount of material at top of load lane. Reduced (SDS): Dominant bands at MW 20 kD and 36 kD and faint bands at MW 10, 15 and 26 kD.*
- **Electrophoretic Methods.** Many other electrophoretic methods are used and each method aims at identifying a unique attribute and, hence, each uses a different matrix or format to retain macromolecules while they are subjected to an electric field, two or even three dimensional. Immunoelectrophoresis (IEF) is one example. Most electrophoretic methods are commercially available and easily adaptable to suitable quality control testing protocols. Further, some electrophoretic methods can be immediately followed by immunological assays on the sample to further identify each type of molecule (e.g., PAGE and Western blot testing). The adaptive possibilities are many and varied.
- **Western Blot (of PAGE or SDS-PAGE).** Often used as an identity test, but adapted to also detect certain impurities, a profile by Western Blot Analysis is performed by transferring the molecules from a PAGE gel to a membrane and then treating that membrane with polyvalent antiserum, or with a monoclonal antibody, specifically

reactive against the test protein product or an epitope of that test protein. The reaction is developed by immunohistochemical methods to demonstrate colored band(s), which should fall at the molecular weight location of the protein and be comparable to immunoblot bands of the reference material. Oligomeric or aggregated species also may be detected at the top of the lanes. Using antisera specific for known impurities, the Western blot may identify those materials. Protein reference standards and negative control antisera or monoclonal antibodies are used. Western blots are not quantitative. A result might read: *Protein X, major band at 56 kD recognized as major band at 56 kD by polyclonal rabbit serum to Protein X and by monoclonal antibodies 3D7, 8F8 and 4D2, specific for epitopes Ala3, Leu29 and Try54pf of the Protein X molecule. No other bands detected.*

- **Host Cell Protein.** If a molecule is produced in a cell-based system then in the upstream material, that is before purification, the product contains impurities consisting of a variety of host cell substances. Several of these are proteins: enzymes or structural molecules. Host cell proteins can be identified by a variety of methods and popular ones are antibody-based assays, such as an enzyme-linked immunoassay (ELISA). This measurement is based on a polyvalent animal antiserum raised against all proteins of the host cell. A result might read: *<0.10 mg host cell protein per 100 mg recombinant protein in solution.*
- **Host Cell DNA.** DNA is usually an impurity, unless the product itself is composed of DNA. It is measured by a variety of methods, many commercially available, with great accuracy and specificity. The fluorescent polymerase chain reaction (PCR) probe method is adopted by many laboratories. A result might read: *<10 μg DNA/1.0 mg recombinant protein.*
- **Host Cell RNA.** Some host cell production systems can yield a considerable amount of undesirable host cell RNA. Commercial test kits are used to measure this molecular impurity. Results could read: *<10 μg RNA/1.0 mg recombinant protein.*
- **Carbohydrate.** Some biopharmaceuticals must be glycosylated to show activity. Indeed, gylcosylation and a unique pattern (e.g., location on protein, composition and structure or pattern of gylcosylation of each carbohydrate side chain) can be important to potency. A variety of methods, adapted from classical carbohydrate chemistry and some now semiautomated, are applied as surrogate measures of potency and to demonstrate identity of some biomolecules.
- **Light Scatter for Aggregates.** Subvisible molecular miniparticles and aggregates can be disclosed and measured using instruments that measure the scatter of light as it passes through a solution of product. A result might be: *<0.1% scatter at wavelength 300 nm.*

- **Protein Measurement.** A variety of tests, some commercial and others developed in or adopted by the quality control laboratory, are on the market to measure the amount of protein in solution. Each has advantages and limitations so the quality control scientist picks a test carefully to meet a particular need and qualifies it for use with a given protein. BCA reagent-based tests are commonly used in biotechnology. A result might read: *1.06 ±0.1 mg/mL.*
- **Peptide Mapping.** This is another identity test. A protein in solution is digested with an enzyme, e.g., trypsin, and the fragments are subjected to an electric current (electrophoresis) in a matrix and then stained. The pattern of fragments is characteristic to a given protein. The result would be obtained from the peptide map and might be given as: *Matches predicted map and reference map.*
- **Size Exclusion Chromatography.** The SEC method examines a sample of protein for purity and impurities using a chromatography gel that distributes proteins on the basis of size. Aggregates of protein are detected by SEC. The results provide a measure of purity and identify impurities based on molecular size. A result might read: *Dominant peaks at MW 20 kD and 36 kD and faint peaks at MW 10, 15 and 26 kD.*
- **Isoform Characterization.** The isoelectric focus assay, a high resolution method allowing the separation of proteins based on their isoelectric point, evaluates the charge characteristics of a protein and can demonstrate isoforms, major variants, of the protein. Isoelectric focus gels are scanned and bands can be measured and identified by pI. Results are not quantitative, but estimates may be made from scans. Example of an IEF result: *Major sample band lies between pI markers of 5.20 and 5.85 and is comparable to reference standard. Minor variants constitute under 10% of total protein.*
- **Amino Acid Composition.** This method may be used as an identity test. An analytical instrument determines the amount of each amino acid and then calculates their ratio. This ratio is compared to the expected ratio and to that measured for a reference standard. The ratio relative to a reference amino acid, say l-leucine, may also be determined and compared to theoretical and reference standard. To perform the test, a sample of protein is hydrolyzed with acid and the amino acid composition is determined by an automated method. Typical Result: *Correct amino acid composition ±10%.*
- **High Pressure Liquid Chromatography.** HPLC has been applied to measure several attributes of many biological molecules. It is relatively fast and inexpensive and quite adaptable. Sample is separated into components that appear on the output as distinct peaks or even bumps or "shoulders" on a major peak. Also useful is the ability to

measure the amount of material under each of those peaks and, for reference purposes, to "spike" known molecules, such as impurities or contaminants, into a sample of highly purified reference product. Indeed, this is how experiments are designed and initial results are seen using several other modern chromatography and spectroscopy methods. Typical result: *Major peak at 15.8" comprising 97.68%. Two minor peaks at 22" (0.88%) and at 18" (0.27%) with slight shoulders at leading and trailing edge of major peak. Matches reference standard.*

- **Electrospray Mass Spectrometry.** This instrument-based test is used to measure the mass of a molecule and the result is compared both to a reference standard and to the theoretical mass of the molecule of interest. Controls might include proteins of known mass, especially those in the molecular weight range of the test material. Results are reported in Dalton (Da) unit of protein mass. Typical result: *72,811 Da (versus Theoretical 72,806.3 Da).*
- **N-terminal and C-terminal analysis by Liquid Chromatography–Mass Spectrometry (LC-MS).** This method uses physical separation, LC, with mass analysis by spectrometry, MS. It is highly sensitive and specific and can be applied to characterize a variety of proteins. It will confirm both the N-terminal and the C-terminal sequence. Typical Result: *N-terminal, KQEN; C-terminal EIGGY; comparable to the reference standard.*
- **Protein Folding/Unfolding by Intrinsic Fluorescence.** The amino acids tryptophan and tyrosine are in proteins and they fluoresce under specific wavelengths of light. Proteins in the native or correctly folded state demonstrate a unique fluorescence signal. Denaturation of a protein results in a shift (e.g., red-shift) of the fluorescence emission barycentric mean (BCM) value, which can be derived from an analytical instrument. A specification for batch-to-batch variance can be established. An example is BCM Lambda_{nm} <358 for fluorescence measured between 300 and 400 nm. A result might read: *Lambda_{nm} 340 for fluorescence measured at 380 nm.*
- **Mass Spectrometry–Time of Flight (MS-TOF).** This method relies on an expensive piece of equipment and represents one of several new and promising analytical tools that may be used to characterize proteins and other biological molecules.

Additional Analytical Tools and Concepts

A number of tests have been identified above and most of these are used to test attributes of biotechnology products that exist as molecular entities. But, what about testing for appearance, safety, identity, purity and strength of living biopharmaceutical products, such as the attenuated organisms used in vaccines, retroviral vectors and somatic or pluripotent-derived cell and tissue products, as introduced in Chapter 6? Here we review, with very general descriptions, a few of the many methods used in the quality control of various biopharmaceuticals, notably live materials.

- **Cell Karyotyping.** A karyotype represents the appearance and number of chromosomes in the nucleus of a eukaryotic cell. A cytogenetic analysis of a cell's karyotype is used as an identity or a purity test to demonstrate quality of a cell line, especially if that line were used as a somatic or pluripotent cell-derived therapeutic. Karyotyping is performed in specialty laboratories and reference cell lines are required. The number of cells with an abnormal karyotype is measured. The species origin of the cells is also confirmed. The result would be identity of a cell line, as compared to a reference cell line, and might also attest to its purity.
- **Cell Phenotyping.** Any cell trait or characteristic distinctive of that cell line is considered phenotypic. Cell phenotyping, used when a cell line or tissue is derived from somatic or pluripotent cell sources, measures one or more molecular parameters to demonstrate that cells are identical to those intended and that the cells have not differentiated or dedifferentiated nor that they have become contaminated. The result would be presented to identify the cell line and attest to its purity.
- **Microbial Identification.** Bacteria, fungi, yeast and viruses, including retrovirus and bacteriophage, are biotechnology products. They are derived from many sources and manufactured in various ways, both with and without cell-based systems. Identity testing demonstrates in various ways that the microbe is as purported. Bacteria and yeast are identified by traditional methods, such as culture on selective media and metabolic properties. Viruses are identified by growth characteristics on selective cell lines. All microbes may be further identified using species-specific antibodies to agglutinate or label with fluorescent dyes, to kill them in the presence of complement or to neutralize their activity. Describing their morphology or ultrastructure is also an effective means. PCR is increasingly used to identify live organisms or DNA molecular products.
- **Monoxenic Nature of Microbial Product.** The purity of a microbial product may be demonstrated to show that all organisms in a product have the same trait. Cell phenotyping or karyotyping or microbial identification methods, described above, can be applied for this purpose. Other methods use growth characteristics or a panel of chemical or immunological reagents to selectively identify possible

contaminants. An example is the use of selective media that support the growth of most bacteria, but not the strain or species comprising the product. PCR, using probes against DNA from a variety of possible contaminants, is another test applied to reveal purity of a culture.

- **Attenuation of Microbial Product.** Many products, notably those intended for genetic therapy or vaccines, are attenuated so they will not produce disease. QC tests for safety focus on ensuring markers of attenuation, such as inability to grow on certain substrates or in specific cell lines. These traits, or lack thereof, also may be evident using antibody or molecular probes, such as immunofluorescence assays or PCR.
- **Expression of a Molecule by a Vector or Host Cell.** Other biotechnology products are engineered to express a molecule and, thus, to exert an immunological or therapeutic effect. Quality control tests are based in methods mentioned above, but instead of searching for absence of a trait, they focus on ensuring an attribute is present and active. PCR and other genetic probes identify a gene of interest in any host. Immunological and molecular probes are used to ensure that an expression product is both expressed and in the expected location, such as on the cell surface.
- **Adventitious Agent Testing.** Adventitious agents may be found in a variety of cell products and cell banks. Considerable effort is spent examining samples for microbial agents, e.g., mycoplasma and retrovirus. Some testing is performed to identify, through culture or immunological methods, the agents themselves. More often, indirect measures of adventitious agents, such as electron microscopy to detect viral or viral-like particles or PCR, are used to locate and, sometimes, to identify undesirable microbes or the DNA fingerprint they leave behind. If infectious agents are not obvious at the outset, they may be induced by applying a variety of chemical or biological stimulants to the cell line. For example, endogenous retrovirus can be induced with nucleotide analogues and then detected by electron microscopy as viral particles. Viruses can be detected by injecting cell lysate into newborn animals and then examining the animals months later.

Quality Control of Cell Banks

As noted in Chapter 6, the identity, purity and viability of cell banks, microbial, mammalian, or insect cells, are very important attributes to overall manufacturing quality and success. Every cell bank is tested both at the time of release and at specific intervals. The attributes considered in most test protocols are identity, sterility, purity and viability (potency). Following are tests applied to the quality control of cell banks, both master and working cell banks as well as the progeny of cell banks that are used in extended manufacturing campaigns.

- **Identity of Cells.** The identity of cells in a bank is established by applying tests to ensure they are, in fact, the species and strain intended. Particular methods are described above, under the headings of Microbial Identification, Cell Karyotyping and Cell Phenotyping.
- **Identity of Insert.** If a cell line has been genetically engineered to express a product, then the process of expression and the expression product are tested to ensure banked cells do possess these intended capacities. Methods described in earlier sections of this chapter are applied as appropriate for the nature and function of the molecule or other attribute.
- **Potency.** This is defined as the capacity for cells in a cell bank to divide following removal from storage. This function is clearly basic to the intended use and so quantitative potency testing is an annual QC "physical examination" for any cell line.
- **Purity.** To ensure that a cell line has not been contaminated with another product, and this does happen, then the monoxenic nature of the cell line must be established using methods described above.

Samples and Sampling

Consideration of sampling methods, assay controls and reference standards is an important aspect of quality control planning because it is critical for each test sample to represent the whole of that batch or lot of BS or FP. Thus, sampling must follow standard procedures, with methods tailored to the intended scope and nature of each test and heeding statistical considerations like sample size or random selection. Indeed, papers and books are written about sampling methods for quality control of various consumer products, including pharmaceuticals. Beyond the number of samples taken, there must be a plan for sampling performance. For example, if one wished to sample 100 glass vials containing a biopharmaceutical from a total lot of 10,000, it would be important to take vials periodically, perhaps 10 vials at each of 10 time points, throughout the fill, as opposed to grabbing the first or last 100 vials off the fill line.

Also, when seeking a representative sample from a single container, it is necessary to take the sample in the proper manner. For example, if one is taking a sample for aggregate testing from a small vial containing a recombinant protein, the container is first gently stirred and then the sample is taken, ensuring aggregates, that tend to settle to the bottom of a vial over time, are fairly representated in the samples. Sampling raw materials that are provided by a vendor in large containers drives the need to both ensure that containers are selected in a representative yet random manner and, for each selected container, to randomly take material from within that container. The general concepts of sampling are included in a QC Plan and specific sampling methods for each assay are written into the test protocol or SOP. If not every sample is subsequently assayed, then it is important to ensure that tested samples are indeed representative of all samples taken. Consultation with a statistician is often helpful throughout the sampling process.

Analytical Controls and Reference Standards

Every assay requires one or more controls, but a full set of control materials is too often missing from an otherwise adequate test procedure. In establishing a quality control test, a panel of controls is carefully chosen and then applied correctly within the test scheme. Each control reagent is certified for a particular purpose, and during assay qualification or validation, it is shown to produce the intended outcome. Positive and negative control materials are generated through biomanufacturing or purchase from vendors, prior to beginning analytical work. Controls are stored in a manner that retain all desired attributes. Controls for quantitative assays, such as those requiring generation of a standard curve, require particular attention.

A reference standard is test material that consistently produces, over a long period of time, the same result when used in a given assay. It also must be the same or nearly the same as the actual test material, the product or analyte. Reference standards are by definition used with every assay or panel of assays. In some instances, more than one reference standard is required for an assay, such as when a standard curve is produced by limiting dilution of the standard.

The source of each reference standard becomes part of the QC plan as well. For some products, it is best to use a material from a natural source. For example, when a recombinant protein cannot serve as a reference standard due to instability, it may become necessary to purify and store small amounts of the native molecule. This can be a very resource-intensive project and must be regarded in light of other possibilities. Another option is to use samples from the first batch or lot of product as first reference standard. Over time and as manufacture and control expertise improves, this first reference standard is replaced with material from new batches of BS or lots of FP. It is not unusual for five or more individual, product-derived reference standard lots to be used over the development cycle of a single assay. Cross-over studies are performed to compare, in great detail, an old to a new reference standard. To further complicate the matter, reference standards, like controls, are used and relied upon over a long period of time and, thus, they are kept in extended storage. Early in development, and often for years into the development cycle, there is limited information on stability profiles of controls and reference standards. This creates unknowns regarding their reliability when used to test product over that same period or into the future. A plan to meet each of these challenges becomes a part of the QC plan and this is typically done for each assay.

Test Failures, Out-of-Specification Results and Retesting

As one might expect, failures are experienced in the QC laboratory just as they are in other aspects of biotechnology operations. In some endeavors, such as biomanufacture, a failed process can be repeated, at least if an explanation is a possibility and time and money are available. Failure in testing, specifically the failure for a test to meet a specification, can be difficult to resolve. It is no simple matter to "retest" a product in light of missing specifications with a single test. Most QC operations refer to such situations as *deviations*, meaning that results did not meet established specifications. There are strict controls on managing deviations in a regulated environment and biopharmaceutical quality control follows guidelines and regulations promulgated by the FDA. Further, some highly publicized judicial actions taken by the FDA originated from improper retesting and reporting of QC test results.

Deviations in quality control testing demand, by regulation, an internal investigation. Investigations involve a complex process, having four major components that are described in greater detail in Chapter 5. Briefly, an investigation uses established methods, a root cause analysis, a corrective and preventive actions plan (CAPA) and approval of the outcomes and recommendations by supervisors, quality control and quality assurance. Further, findings of the investigation may lead to a recommendation of significant actions, such as qualifying or even replacing an assay before it can be used to further test a product. Other resolutions, such as closely monitoring the performance of the assay, replacing key components, such as a reference standard, and establishment of formal audits or documentation systems, may be recommended by the quality assurance unit. Because management is ultimately responsible, the failure may be raised to this level.

It is important to prevent test failures and this is achieved in several ways. First, full quality control planning prevents many failures and sticking to the accepted plan avoids others. Following a quality control development cycle (Figure 7.1), i.e., developing each assay one step at a time and not skipping a step or a critical experiment, is another successful approach. The most frequently cited reason for failures in quality control testing is, in this author's experience, setting unrealistic (meaning too stringent) specifications in early and middle phases of the development cycle. In effect, this means the development team establishes a specification from blind optimism and before adequate data are available to support that specification. Following the hallmarks of quality and abiding by a quality system, current Good Manufacturing Procedures (cGMPs) for most biopharmaceutical quality control operations is another way to ensure success with quality control endeavors.

Testing for Product Stability

Once FP has been made and released for further processing or use, it is stored for a period of time, then transported and stored at least once again before it is used. As this storage and shipment is repeated between final production and time of use, it is important for the product to remain pure, potent and safe. BS is also stored by most producers and it may be transported from the manufacturer to a fill operation located at some distance.

To ensure quality of product during this period, experimental stability protocols are designed by quality control and manufacturing staff to evaluate attributes and desired traits of BS and FP as might be expected under conditions of storage, handling and shipment. Most biotechnology products today are kept refrigerated or frozen, but all products are subject to fluctuations in temperature and humidity, or they face exposure to light. Environmental conditions certainly change as product is moved from manufacturer to truck, then to wholesale warehouse to truck, then to pharmacy, automobile and finally to the consumer, and stability studies consider this. Also, if a product will be shipped, stored and used in regions with a warm moist climate where it is difficult to always maintain a cold chain, or in very cold regions where it might be exposed to winter weather, then stability studies need to be especially rigorous. Stability data provide a sponsor with very useful information because they aid in developing proper product formulation and approaches to marketing, transport and storage. Because one cannot sell degraded product, stability test results impact both business and marketing plans. There are significant economic advantages for a product with a long shelf life at ambient temperature and there are marketing hurdles for the unfortunate sponsor of a product that must be kept frozen, especially one that must be kept at -80° C. Consider also the difficulties of manufacturing, stocking and rotating supply of a product with a shelf life of just one year. While experimental in nature, the design of stability protocols is driven by quality, business and market interests because the information derived from stability studies in part ensures sales of that product. Hence, stability testing plans are an important aspect of the overall quality control plan.

Stability testing, typically a quality control function, is a formal process under which the actual shelf life of a biopharmaceutical is identified through experimentation. Several items should be considered as quality control scientists and operational colleagues develop a stability plan and protocols. First, it is important to know if a product could be stored at room temperature or in the refrigerator, as opposed to in a freezer, as ambient storage reduces cost and complications of shipping and storing the product. Second, shelf life is determined empirically for each proposed storage condition. Third, we often attempt to improve shelf life or simplify storage conditions for a product by applying different formulations. For a given product, we know, in general, that certain chemicals preserve cells or proteins better than do others. Finally, for products that must be frozen or refrigerated, we seek an understanding of how long that product could withstand room temperature or even an elevated temperature before it loses potency.

Stability testing is an experimental endeavor, designed as a matrix experiment and guided by a written protocol. It applies not one but a panel of assays, each capable of measuring a stability-indicating attribute for the product. Further, to provide enough material for testing, many samples of product, both BS and FP, are stored in a variety of configurations. For example, storage at three or four temperatures and, for FP, in two positions (upright and inverted). With testing by three to five assays required at seven different time points, sample requirements are in the hundreds or thousands of vials or syringes of FP or samples of BS for every lot or batch, respectively. During stability testing, storage conditions must be carefully controlled and documented. Thus, there is a need for many samples and a laboratory infrastructure able to accommodate various environmental conditions and perform many tests. Stability testing is very labor intensive and expensive, but it is absolutely necessary both from a regulatory and business standpoint.

Stability-indicating assays, if not available as release assays, are modified from existing research or development assays or they are developed by QC scientists to fit this purpose. They are selected by first identifying those product attributes providing meaningful information about the shelf life of that product. Because any assay used to measure a product attribute is, in theory, a stability indicating assay, the CoAs for release of BS (Table 7.1) and FP (Table 7.2) provide a foundation for developing stability testing protocols, shown in Table 7.3 for FP and Table 7.4 for BS. However, a CoA intended for product release lists far more assays than are necessary or could ever be performed on the variety of stability test samples and at every time point. To down-select choices to perhaps three to six assays per stability protocol, the quality control scientist must first determine which attributes of the product are most likely to be stability indicating. Next, they choose those attributes and experimentally determine if in fact they are stability indicating.

TABLE 7.3

Tabular Synopsis of a Stability Protocol for Biopharmaceutical Final Product (Example of Formulated r-Protein in Vial)

	Postmanufacture (Months)								
Attribute/Test/Specification	**1**	**2**	**3**	**6**	**12**	**24**	**36**	**48**	**60**
Appearance. Visual. Clear, colorless liquid without particulates or aggregates	X	X	X	X	X	X	X	X	X
Safety. Microprobe. pH. 7.1 ±0.2	X	X	X	X	X	X	X	X	X
Purity. SDS PAGE. Single band at 30 Kd. Comparable to reference standard	X	X	X	X	X	X	X	X	X
Purity. HPLC. Single peak integrated >98% material in sample	X	X	X	X	X	X	X	X	X
Potency. Blocking Assay. Cultured Cells. >60% inhibition of secretion as compared to reference standard	X	X	X	X	X	X	X	X	X
Safety. Sterility. USP <71>. Sterile			X		X	X	X	X	X

Note: X = This test will be performed at this time point. Accelerated stability testing only to 12-month time point.

This protocol is performed on FP that has been stored at static conditions in the upright position at the following temperatures (one set per temperature in static conditions): 6 ± 2°C (recommended storage temperature), 21 ± 2°C (room temperature) and, during the first 12 months, 37 ± 1°C (accelerated temperature).

In addition, one set is performed following storage at 6 ± 2°C with vials kept in both upright and inverted positions.

Another set of FDP vials is tested following static storage at the 6 ± 2°C temperature with vials in upright position, but only after sample vial has been subjected to five freeze–thaw cycles prior to testing.

Yet, another set is tested at all time points and at the 6 ± 2°C temperature storage condition in an upright position, but only after sample vial has been subjected to shaking at 30 oscillations per minute for six hours immediately prior to testing.

What are reasons for choosing an attribute and assay for inclusion in a stability protocol? First, some assays are, by tradition, stability indicating for most biopharmaceuticals. Second, it is good to have a balance, choosing one or two methods for the attribute of purity or impurities, one for potency, and one or two others for safety. The nature of the product gives a clue to what constitutes a good stability-indicating assay, as does understanding the mechanism of action. Further, each test must be sensitive so that loss of product integrity or activity is, in fact, detected early and long before the product is completely unsafe, degraded, impure or impotent. For some products and tests, this selection is challenging. For the r-protein in our example, instability might be reflected in denaturation or breakdown of the molecule. This would result in an inability to perform the key biological functions and reach the desired endpoint. Referring again to the r-protein, and focusing only on FP, initial consideration is given to several key assays for early stability experiments. SDS-PAGE and HPLC are selected for purity and impurities and the potency assay is chosen for blocking accumulated

TABLE 7.4

Tabular Synopsis of a Stability Protocol for Biopharmaceutical Bulk Substance (Example of r-Protein)

	Time Postmanufacture (Months)								
Attribute/Test/Specification	**1**	**2**	**3**	**6**	**12**	**24**	**36**	**48**	**60**
Appearance. Clear, straw-colored liquid without particulates or aggregates	X	X	X	X	X	X	X	X	X
Safety. pH 7.1 ± 0.2	X	X	X	X	X	X	X	X	X
Purity. SDS-PAGE. Single band at 30 Kd; comparable to reference standard	X	X	X	X	X	X	X	X	X
Purity. HPLC. Single peak integrated >98% material in sample	X	X	X	X	X	X	X	X	X
Potency. Blocking Assay, cultured cells. >60% inhibition of secretion as compared to reference standard	X	X	X	X	X	X	X	X	X
Safety. Bioburden. USP <61>. <5 cfu/mL and no evidence of pathogenic organisms			X		X	X	X	X	X

Note: X = This test will be performed at this time point.

This protocol is performed on samples of BS that have been stored at static conditions in the upright position at the following temperatures (one set per temperature in static conditions): –70 ± 10°C (recommended storage temperature); during the first three months at 21 ± 2°C (room temperature); during the first three months at 37 ± 1°C (accelerated temperature).

Another set of BS samples is tested following static storage at –70 ± 10°C, but after having been subjected to three freeze–thaw cycles prior to testing.

molecules by cultured cells. These are chosen in part because both are meaningful to this protein, and its biological activity, and because each is simple, accurate, well controlled and can be performed on large numbers of samples relatively easily and inexpensively. Because vials are opened to test the product anyway, we might add tests to record appearance and measure pH at every time point, because changes in pH are often related to protein instability and because degradation often discolors or adds precipitate to a vial of product. Regulatory authorities ask that sterility be examined once each year because sterility is a test of container (vial) integrity. Returning to the example, SDS-PAGE provides an indication of whether or not the r-protein was breaking down under stressful conditions. HPLC, a much more sensitive test, confirms and extends any observations of protein breakdown, perhaps detecting changes earlier than seen by SDS-PAGE. Besides, HPLC analysis of r-protein might provide, on the chromatogram trace, a clue as to breakdown products, impurities that accumulate over time. A stability-indicating assay also is chosen for the attribute of potency, this based on the requirement that r-proteins inhibit buildup of intracellular carbohydrate molecules. The assay uses cultured cells, providing a well-characterized and much used method that should reflect biological activity, or loss thereof, in a relatively simple and reproducible format. Now, this stability testing concept is designed into a written FP Stability Testing Protocol and the test scheme is summarized in tabular outline, as shown in Table 7.3.

The same process is followed for developing a stability protocol for BS, but some different assays may be chosen. An example of a BS stability protocol is shown in Table 7.4.

Consider also that controls are included with each assay to demonstrate that a stability assay will detect unstable product. These controls, partially and fully degraded or inactivated product, are tailored for use with a particular assay. To develop a control, it is critical to first understand the most likely routes to degradation and then mimic this degradation in a meaningful way. For the r-protein example, it is not adequate to simply boil the r-protein or completely digest it in trypsin; this would not mimic possible environments in actual storage or transport. Instead, it is necessary to carefully devise "accelerated" degradation using conditions that mimic possible real environments. These might include using room temperature as seen in temperate and tropical environments or one or more cycles of freeze–thaw. Producing controls is itself challenging and requires time and knowledge of possible degradation pathways.

Stability testing is performed on FP and BS kept at recommended (labeled) conditions, say refrigerated or frozen, but also at several suboptimal conditions that might be experienced in a real situation. For the r-protein example, the intended storage conditions are refrigerated, but for stability testing the quality control scientist also tests samples kept at room temperature and at a temperature above room temperature. Such protocols are referred to as accelerated stability, when, in fact, they are accelerated degradation in most cases. As noted earlier, vial handling is an important variable. Initially, the vials are kept upright, but in later stages of development vials are placed in other configurations of storage, such as inverted or horizontal. Shaking and exposure to light also might be included in late-phase stability studies. Finally, several time points must be tested at each of the chosen conditions because it is not known exactly how long the product will remain stable under any given set of conditions. The end result is a set of protocols that provide a stability test matrix. While stability testing adds considerable effort and expense to a product's quality control program, it results in extremely valuable information and becomes the basis for ensuring proper handling, shipment and storage of the FP. And, of course, it ensures a high quality product—pure, potent and safe—is sold to the user. This, in turn, enhances marketing opportunities and prevents future recalls or complaints.

Quality Control Testing of Raw Materials

The quality of BS or FP reflects not only the manufacturing process and release testing, but also the quality of each raw material that goes into making that product. Therefore, raw materials are carefully selected and controlled with involvement of staff from manufacturing, quality control and quality assurance. Manufacturing staff have primary responsibility for selecting the finest materials to use in each process. Quality assurance professionals approve these selections and are further involved by later signing off on use of each lot of raw material once it has arrived at the manufacturing facility. Quality control staff considers the technical quality of raw materials, either reviewing the results of tests performed by the vendor or testing or retesting the material at the sponsor's laboratory.

Quality attributes of raw materials differ based on their intended use and integration into the product. Items that do not contact product, such as a detergent used to clean the floors in a manufacturing facility, receive the least attention. For example, quality control at the biopharmaceutical firm would review the CoA provided by the detergent's supplier to ensure that it had been tested and did meet specifications, but the biopharmaceutical firm would not retest this detergent to verify this CoA. In this manner, the CoA of many products attests to the safe, pure and potent nature of a raw material. Certain other raw materials, such as culture media or salts provided to the biopharmaceutical manufacturer by an established and reputable vendor and carrying both a USP certification on the label and a CoA, might not be routinely retested. The quality assurance department might audit these vendors periodically (see Chapter 5), but unless there are potential issues with the vendor or the raw material, laboratory testing might not be repeated. The third and greatest level of attention is given to raw materials that become part of the product or directly contact the product and do not carry a recognized certification (e.g., USP), those that are notoriously difficult to control or those having even a remote possibility of harboring adventitious agents or toxic materials. This third class of raw materials, examples of which are given below, demand retesting in the biomanufacturer's QC laboratory or they should receive other special consideration to ensure they are, in fact, safe, pure and potent. General testing requirements for a few classes of raw materials commonly used in biomanufacturing are given below.

- **Solid Containers and Process Equipment.** Containers, such as hold vessels, and process equipment, like plastic tubing and filters, have been known to shed particles or to allow chemicals to dissolve (leachate) into the product stream. Biomanufacturers take great care in choosing process materials that will not release particles or chemicals. In steps such as holds in which this might happen even with the highest quality materials, the quality control laboratory may be called on to test for those possible contaminants in the raw material, in-process samples or the product. Standard assays are available to the biopharmaceutical industry for particles and many leachable chemicals, and adventitious agent testing was described earlier in this chapter.

- **Water.** As noted in Chapter 6, large amounts of water are used in biomanufacturing and this must be of the highest purity and without any microbial contamination or undesirable dissolved chemicals. Water is purchased as "purified" or water for injection (WFI) by some operations, but biomanufacturers often purify water themselves, beginning with an excellent source: tap or well water. Because it is used in large volumes, impurities or microbes that enter upstream processing may be carried through or even concentrated during the process, and placed into the product. Hence, water of all grades must be tested for traces of chemicals, such as total dissolved carbon, particles, endotoxin as well as for yeast and bacteria. Compendia (e.g., USP <1231>) describe the various levels of water quality and the tests used by quality control laboratories to ensure this critical reagent remains pure and safe.
- **Inorganic and Organic Chemicals.** Large amounts of salts or saline solutions are purchased or prepared during biomanufacturing. Whenever possible, compendial grade or otherwise certified reagents are purchased from a reputable vendor and the Certificate of Analysis is carefully reviewed prior to acceptance. When USP-grade materials are not available, then the quality control laboratory may test a raw material to ensure that it meets established specifications.
- **Culture Media and Supplements.** These raw materials are critical to most biotechnology operations because recombinant molecules, cells and tissues must be grown in basal media, typically enriched with a variety of natural or synthetic products. While manufacturing operations attempt to use only well-characterized or chemically simple materials, this is not always possible. Also, it is sometimes necessary to use plant- or animal-derived materials to produce a biopharmaceutical. As with inorganic and organic chemicals, it is important to understand the nature and origin of each product used to grow and maintain cells or tissues. Quality control plays an important role in analyzing the nature of each raw material and ensuring it is purchased and inspected carefully. Again, vendor-supplied certificates of analysis are scrutinized and, for some items, they are again tested. While synthetic and plant-derived natural supplements, such as vitamins or growth factors, are of moderate concern, animal-derived products are of great concern to both sponsors and regulators. This is because animal products can carry animal viruses or prions and, if present in a supplement, these agents could be transferred via the product to humans. CoA are scrutinized, vendors are asked to certify the origin and microbial purity of such products and vendor audits are commonly performed. In rare instances, such as with a special animal serum, it may be necessary to further process and test the product at the sponsor's laboratory, or if a risk of disease

transmission even exists, to simply find an alternative source or another supplement.

Quality Control and the Manufacturing Environment

As noted in Chapter 6, there is a great need to ensure a consistent and high quality environment in areas of a facility dedicated to aseptic manufacture of a sterile product. One means of demonstrating compliance with environmental standards is to test swabs taken from personnel, equipment or facility surfaces, and also samples of air and water. The quality control laboratory is often responsible for this environmental sampling and testing. Samples are taken during periods without manufacturing activity (static environments), and they also are taken during actual manufacturing operations (active environments). Instruments are used to sample air or water, to count the number of nonviable particles, and to culture and count the colonies representing viable particles, i.e., bacteria and yeast. Samples are taken by swabbing the uniforms and gloves of operators as well as work surfaces, walls and floors. This information is used to better maintain a clean environment and to alert or alarm the manufacturer whenever the aseptic or low-particle nature or a work area has been compromised. Further, the quality control laboratory may be responsible as well for testing for residual product or cleaning agents, using analytical methods to identify trace amounts of product or undesirable chemicals on work areas and process equipment.

Qualification, Validation and Verification of Analytical Methods

Validation was mentioned in Chapter 6 as an important aspect of biomanufacturing. Validation or the related process of qualification also are performed on individual analytical tests. Typically, an assay is validated during the late phase of the development cycle. Assays are often qualified in mid-phase, but some critical assays may be qualified even earlier in development and these tests may be validated in early phase as well. Critical assays are those important to the safety of the product or to ensuring potency prior to use in clinical studies. In contrast to validation, verification is a word applied to ensuring proper application and use of compendial assays.

Regulatory agencies demand that analytical tests be validated prior to licensure. The USP defines validation of an analytical procedure as: "…

the process by which it is established, by laboratory studies, that the performance characteristics of the procedure meet the requirements for the intended analytical applications." And the ICH adds that validation must "... demonstrate that the procedure is suitable for its intended purpose." Validation of an assay thus ensures the established specification is appropriate for a particular use and product. While tentative or working specifications may be established long before assay validation begins, the validation process is a means of confirming or adjusting those specifications. Hence, assay validation and establishment of final specifications are highly integrated processes.

Assay Validation

Assay validation is an experimental endeavor and the process is always performed under a written and approved protocol, one designed to achieve specific purposes and perform exact experiments for each assay–product combination. The protocol states a purpose and scope, and provides pass versus fail rules for method validation outcomes. In general, the purpose is to prove that the analytical method can perform adequately as a written SOP for the intended purpose. The scope of use is also stated, along with the specification that is currently under consideration. A series of experimental procedures then challenge that assay to demonstrate suitability by a number of criteria or traits that are established prior to testing. The most commonly applied traits were briefly mentioned earlier because they are important to consider well before validation begins. These are described below as: System Suitability, Precision or Repeatability, Linearity, Accuracy and Range, Specificity, and Robustness.

- **System Suitability.** This is the ability for an analytical system to achieve the objectives of the assay and is defined in several ways. First is the need to ensure equipment is suitable for the intended purpose and is installed and operational (see Chapter 6). For some analytical tests using highly sensitive instruments, the conditions and settings are carefully defined. Reagents are shown to be suitable and reference standards and controls adequate and well characterized. Interfering substances in any reagent, including the sample matrix, are considered because any substance could, in theory, artificially enhance or inhibit an analytical system. Sampling and sample preparation studies are completed to ensure matrix compatibility with the test method and other reagents. A system suitability report is prepared to demonstrate that the complete analytical system, including instrumentation, is suitable for the intended use and for further validation of the test method.
- **Limit of Detection (LOD) and Limit of Quantitation (LOQ).** An early step in validation of quantitative assays is determination of the

lowest amount of analyte that can be detected, the LOD. The lowest amount that can be accurately measured in a quantitative assay is the LOQ and this is shown experimentally, as well.

- **Linearity.** For quantitative measurements, it is necessary to demonstrate linearity, the ability, within a given range of analyte in sample, to obtain test results that are directly proportional to the concentration of that analyte. In other words, a linear curve must be generated within the dilution work area. Standards, shown to be of the highest quality by other methods, are prepared at concentrations in a range and with a matrix that matches that of the intended test sample. For example, if one expects the r-protein to exist in a sample of BS in phosphate buffered saline at concentrations of 15 ± 5 mg/mL, then one might establish test dilutions of standard BS in phosphate-buffered saline of from 5 to 30 mg/mL and then test them using the standard assay. Duplicate assays might be performed on each of five days and the results plotted as a linear regression. Acceptance criteria for this example is based on statistical analysis and might be a coefficient of determination, $r^2 > 0.98$, and y-intercept of the linear regression. Linearity demonstrates that the assay results may be extrapolated from the linear curve, within this range of values, and that this can be achieved repeatedly.
- **Precision.** Analytical precision refers to the ability of an assay to repeatedly produce the same or very similar measurements on repeated testing when variables are held constant. This is referred to as the degree of scatter. Critical variables, such as measurement of sample volume or weight by a standard procedure, may need to be evaluated first to ensure that preanalytical procedures are precise. Then the test is performed repeatedly in the same manner. A test is precise if the results have little scatter. Acceptance criteria for precision of an assay are usually given as the percent relative standard deviation for a given number of sequential tests of the same sample. Along with robustness experiments, precision validation ensures repeatability, intermediate precision and reproducibility.
- **Accuracy and Range.** Accuracy is a measure of an assay to agree with a known true value. Put another way, it identifies the total error of the method, considering both the systemic or inherent technical and random errors for that test. Range is the interval allowed between the upper and lower concentration of analyte in the sample. In a practical sense and for many assays, the acceptable range is demonstrated as an outcome of linearity testing. To validate accuracy and range using the example of the r-protein, one might make several dilutions, within and just below and above the limits used to produce the linearity curve. Each of these dilution points is repeatedly tested several times to produce multiple results at each dilution

point. The linearity of the response is evaluated to determine if indeed the linearity testing results are confirmed and to learn of the actual range of acceptable accuracy values. The percent relative standard deviation would be calculated for each dilution point and compared to acceptance criteria, perhaps 95 to 105% of theoretical value. Typically, in such plots of multiple dilutions, the accuracy falls outside these acceptance criteria at dilutions above or below those determined for the linear curve. This further determines a range of values, calculated as baseline value and here milligrams per milliliter of total r-protein, that then are used to accept samples for the assay. It also provides a percent relative standard deviation for each dilution within this range, a value that can be applied to reference standards in the future.

- **Specificity.** Specificity is the degree to which the measurement is due to the analyte of interest and not to other substances that could interfere with the assay or confound analytical results. Such substances might include components of the matrix, such as macromolecules or buffer salts, impurities or degradation products, or similar but undesired molecules. The assay is shown to exactly identify an analyte, to differentiate analyate from impurities, and, when desired, to provide an accurate or exact result related to other product attributes. Specificity validation requires input of these substances by "spiking" a known pure sample of product. Purity of a known sample, often the reference standard, is achieved by adding in substances that are expected to interfere with the assay and might be present in a sample. Following multiple analyses of the various samples, reference and reference-plus-substance, or other substance alone, one can determine the degree to which the assay is specific for the intended analyte. Acceptance criteria might be given as a percent of the reference standard, such as "±10% of reference standard." Any values obtained for a sample outside the reference standard range or value would indicate undesirable interference.
- **Robustness.** This refers to overall reproducibility of the test results obtained when aliquots from a homogenous lot of sample are analyzed under normal, expected operational conditions, given that even the most consistent conditions introduce small variations. Hence, variations between instruments, reagents or test conditions are introduced during validation experiments. For example, a given reference standard might be tested under the same procedure by different operators using one instrument on the same day, or it might be run using three different instruments but by the same operator on the same day or it might be run using the same instrument and operator but on different days. An experimental matrix is developed and parameters are carefully chosen and then

varied to ensure meaningful and affordable robustness testing in multiple experiments.

Some assays require limited time or effort to validate while other assay validation protocols demand months of planning and experimentation and consume significant resources. For example, an HPLC assay of a well-characterized vaccine recombinant protein might be simple to validate, but an immunopotency assay of the same protein performed in rabbits to determine the immunogenicity might require 12 months of effort and 10 times the resources. Assay validation must be considered for each assay and be carefully planned well in advance. Indeed, it takes much time and requires the input of many experts to even develop a good validation protocol for one test method. Validation applies to assays used to test BDS and FDP, both for release and stability purposes. It is necessary in validation both to test multiple batches or lots of product because consistency of results is important, and to have available fully qualified controls and reference standards. The product, BS or FP, that is used in assay validation protocols is made by product manufacturing processes that are or that exactly mimic commercial procedures. Clearly, quality control assay validation and manufacturing scale up and assay validation require a tremendous effort and a significant investment of time and money. Because of this, the QC plan must be carefully devised and the assays themselves must be scientifically sound before the sponsor enters these endeavors in late stages of the product development cycle.

Selected assays may be qualified before they are validated. Qualification is in many respects a "minivalidation" as it focuses only on important aspects of assay validation and is performed under abbreviated protocols. In contrast to validation, qualification is completed earlier in development and only with assays considered critical to demonstrating purity and potency or those in which confidence is lacking due to their newness, uniqueness or complexity. Qualification also may serve to establish product release specifications for critical attributes, as multiple lots of product are tested using qualified assay procedures. Another purpose for assay qualification is to give a sponsor confidence that an assay is predictive of product quality for use in early clinical trials. Results of qualification are predictive of validation; if an assay fails qualification, then it is a bad candidate for full validation. Finally, qualification is sometimes recommended by a regulatory agency to a sponsor in early development so as to alleviate fears of using impure or subpotent product in clinical trials.

Assay verification refers to a process applied to commonly used assays, notably those published as a standard method in a pharmacopeia or other authoritative reference or regulation. Verification ensures that a method has been established correctly when adapted into a new laboratory. A compendial assay may be established in a biotechnology laboratory that has little or no experience with that test. In such cases, verification is a formal process, similar to qualification, in which the sponsor ensures proper performance and outcomes in the hands of less-experienced operators or at a new laboratory. Assays, such as sterility (USP <71>) or endotoxin test (USP <85>), are candidates for verification because they are critical to product safety yet are already well characterized and have highly detailed standard procedures.

Application of Statistics in Assay Performance and Validation

Utilizing good statistical practices throughout the assay development and validation life cycle is important to ensuring correct performance while minimizing bias. Perhaps the greatest threat to proper test performance and interpretation of results is bias, a systemic distortion of results. Bias often appears unbeknownst to the quality control scientist; indeed this is inherent in the definition. Factors generating or influencing bias must be identified and statistical analysis is an important means of detecting bias. Also, statistics is key to correctly analyzing measurements, especially those considered a quantitative measurement of an important attribute.

Statistical analysis is important to the quality control scientist because quantitative or semiquantitative assays, and most potency assays and many purity tests fall into these categories, require comparison of results to a standard curve; this in turn requires constant calibration and ensuring linearity of these tests. Demonstrating a linear response is needed to ensure results are meaningful and statistical analyses are applied to experimental results. For tests based on linearity analysis, the statistical methods chosen have a great impact on assay performance metrics, such as accuracy or reportable range of values. Statistical tests also are applied to assay qualification and validation and appropriate data analysis methods have been established for these endeavors.

Indeed, two metrics are considered for any assay, the measurements themselves and variability of those measurements. The measurement must be specific and accurate. Quantitative tests are either demonstrated to be linear or, in some assays (e.g., dose-response), they are nonlinear, but require curve fitting with a specific equation. Variability takes into consideration precision, range, LOD, LOQ and robustness. For these, statistical rules are applied to interpretation of actual results. For example, assay precision is determined by calculating the mean, the standard deviation and the variance or coefficient of variation (CV). Certain statistical rules argue for focusing on the standard deviation and its corresponding 95% confidence interval, and considering CV of lesser importance. In the case of a potency assay for a biopharmaceutical product in which there may be considerable inherent variance, such statistical rules must be considered in design, performance and validation of the assay. Acceptance criteria are established only after careful statistical analysis of data generated by extensive use of the assay. Good statistical practices are seriously considered from the outset of QC planning and then throughout the QC cycle because proper application of statistical methods to analytical endeavors leads to reduced development times, ensures that testing does indeed meet intended use and prevents bias from entering into any analytical test.

Summary of Quality Control

Quality control is a technical or laboratory function to ensure, in part, the purity, potency and safety of biopharmaceuticals. Planning for and development of quality control for a given product is based on the attributes of that product because each test focuses on a particular attribute. Attributes are appearance, safety, identity, strength, purity and potency. Certain analytical tools are available to sponsors at contract laboratories, but many tests must be developed specifically for each product. This is especially true for tests to measure purity and potency. Specifications are established for each test beginning with early manufacture and each serves as a boundary to establish whether a product passes or fails testing. However, as data are generated, specifications change during the development cycle as greater experience is gained on each test and batch or lot of product. This information—attribute, test and specification—is written into a batch- or lot-specific document referred to as a Certificate of Analysis, along with the specific test results. This certificate thus compares specification to actual result and is used to decide whether or not a batch or lot of product meets, by test results, specifications and, hence, whether it may or may not be released (pass or fail) for use.

A large number of analytical tools are available to quality control scientists and many more are being developed each year. In quality control planning, it is incumbent on the scientist to choose the correct tests to determine quality of a product. Other considerations early in quality control development include use of reference standards and test controls, samples and sampling and the need to establish in-house, special tests that are not available elsewhere. Quality control tests are used not only to release product, but also to measure stability of product after it has been transported or stored under various conditions. Hence, stability protocols are developed for each product and analytical methods included in these protocols. The quality control laboratory is responsible for monitoring the environment of a manufacturing facility and operation and for testing raw materials to ensure that whatever goes into a product is of appropriate quality. Finally, assay qualification, verification and validation are performed during the development life cycle to ensure that analytical tools perform as intended.

8

Nonclinical Studies

Nonclinical Studies and Risk Assessment

The assessment of risk versus benefit for any candidate biotechnology product is performed by experimentation in the laboratory, in animal models and in man. Specifically, this involves understanding the nature of the biological construct or molecule, its purity and potency following manufacture, and the safety and efficacy profile. Nonclinical studies, performed *in vitro* and in animals, are primary means of measuring potential product risk and much of this testing precedes clinical trials. Results of nonclinical studies serve to better ensure product benefit will indeed outweigh risk once it reaches clinical studies and the marketplace. Nonclinical study activities precede clinical research for good reason. It is the user, oftentimes the human subject enrolled in a clinical trial, who bears the burden of risks associated with product use. Thus, the sponsor of a novel biopharmaceutical provides clear experimental evidence that risks are tolerable and the product itself is unlikely to result in disease or death to the human subjects or, upon marketing approval, the public.

Information demonstrating safety and tolerability of a candidate biopharmaceutical is presented in the Investigational New Drug (IND) application, specifically in Section 8, Pharmacology and Toxicology. Here, test results, *in vitro* laboratory and animal test data demonstrate in various ways both how the biopharmaceutical behaves in biological systems (pharmacology) and whether or not it is toxic (toxicology). Some pharmacology and toxicology test systems are simple, applied to samples of the biopharmaceutical in a laboratory setting and using tests focused on answering a single question, e.g., the mutagenic potential of a compound. Other tests are performed in appropriate animal models and these are supplemented with additional laboratory testing. An adequate and well-controlled panel of nonclinical studies, an example of which is shown in a general scheme in Figure 8.1, can demonstrate beyond reasonable doubt that the biopharmaceutical, in fact, does possess desired pharmacological attributes and at what levels of exposure it is safe, not toxic and well tolerated in animals.

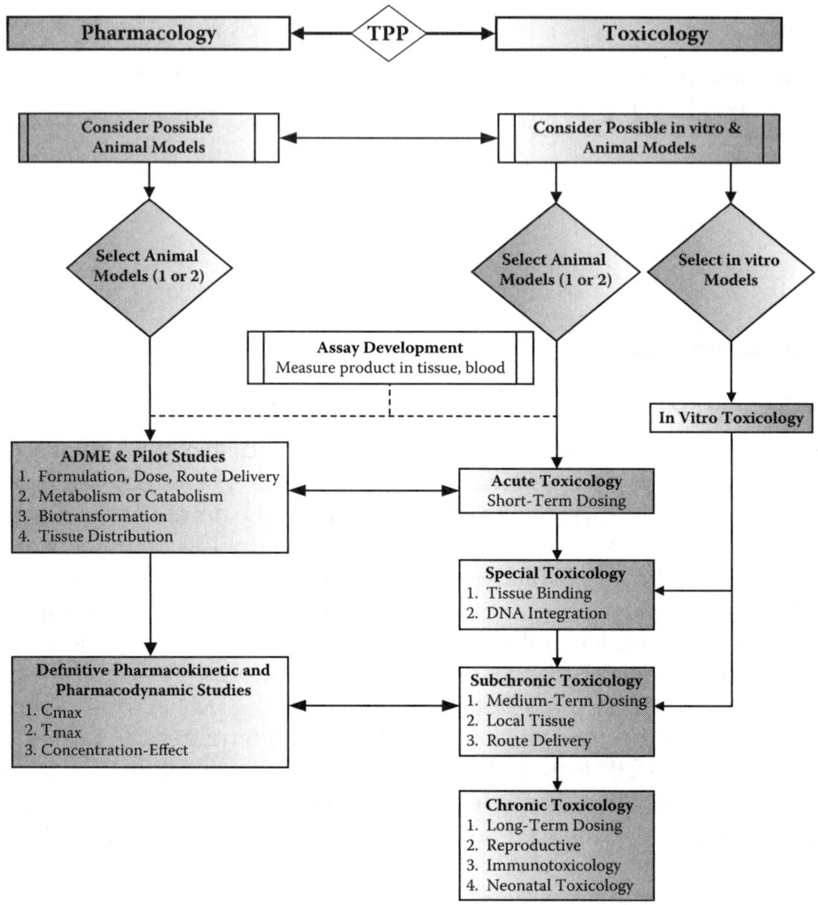

FIGURE 8.1
Scheme of nonclinical activities in biopharmaceutical development. The Targeted Product Profile (TPP) is instrumental in developing a nonclinical plan. Both pharmacology and toxicology studies are performed to identify a safe and effective dose. Assays are developed to measure product and metabolites in blood and tissue and appropriate animal models are used in various types of studies.

Many biotechnology products that are not biopharmaceuticals are exposed to humans. Products used for environmental, industrial or agricultural (including as food) purposes are studied in formal toxicology tests, both laboratory and animal. Yet biopharmaceuticals are given the greatest safety scrutiny and testing because they are directly given to large numbers of humans, sometimes over long periods, and many are injected into the body.

Nonclinical testing of biopharmaceuticals has its foundation in the drug industry where a general understanding and appreciation for the value of pharmacology and toxicology has led to successful development and marketing of small molecule drugs. Studies of dose-response, pharmacokinetics and pharmacodynamic relationships and toxicology, as well as to the development and application of *in vitro* laboratory tests and animal models, are the scientific tools routinely used by scientists and described in this chapter. However, because drugs and biopharmaceuticals do differ in many respects, the panel of tests required for biopharmaceuticals is often unique, even if the basic principles are the same. Consider that biotechnology products are typically large molecules, living cells or microbial products, these having unique patterns of biodistribution and distinctive toxicities. Drugs are usually small organic molecules. Biopharmaceuticals do not always lend themselves to testing in "traditional" *in vitro* tests or animal models that have been developed to assess safety of drugs. Nonclinical biopharmaceutical scientists do apply knowledge and experience borrowed from the small molecule drug industry, but they have also developed unique methods to effectively study the pharmacology and toxicology of biological molecules and cells. Further, as compared to testing drugs, pharmacology and toxicology testing of biopharmaceuticals often requires unique and expensive tests and development of applicable animal models.

Biopharmaceutical Delivery, Pharmacokinetics and Pharmacodynamics

Product Delivery to the Body

Most biopharmaceuticals and drugs are transferred from the final container, such as a vial or syringe, to an initial target tissue; only then is it distributed to the target organ or tissue where it has the intended therapeutic effect. There are many ways to achieve this objective and some are listed in Box 8.1. Many drugs are given orally because they are taken up in the digestive tract without first being metabolized. This is rarely the case with biopharmaceuticals and most are given parenterally. Products given intravenously are designed to be distributed throughout the body as rapidly as possible. Other parental routes of delivery are intravenous, subcutaneous or intramuscular. Monoclonal antibodies and therapeutic proteins are often given by one of these routes. Vaccines are usually given subcutaneously or intramuscularly, but some are given intranasally and others intradermally. Certain cellular therapies are delivered parenterally, often by direct injection into a target organ or tissue. Oral ingestion is by mouth, but, in this case, the biopharmaceutical must be specially formulated so that gastric and intestinal acids and enzymes do not degrade the product before it crosses the intestinal or gastric mucosa. Also, special consideration is given to the size of a molecule taken orally, as large molecules, such as an antibody, would not be readily absorbed in appreciable amounts. Pulmonary delivery for lung absorption is sometimes applied to

BOX 8.1 ROUTES OF ADMINISTRATION TO ANIMALS OR MAN FOR A BIOPHARMACEUTICAL

1. **Parenteral (injected)**
   Intravenous
   Intraarterial
   Intramuscular
   Subcutaneous
   Intradermal
   Intracardial
   Intraocular
   Intraperitoneal
   Epidural
2. **Oral**
3. **Inhalation**
4. **Body Cavity**
   Intranasal
   Sublingual
   Rectal
   Intravaginal
   Intrauterine
   Intraurethral
   Intraauricular
5. **Topical**
   Percutaneous (Transdermal)
   Cutaneous
   Ophthalmic

smaller biomolecules, such as insulin. And topical application of biopharmaceuticals is another delivery route, exemplified by transcutaneous delivery of vaccines or therapeutic peptides. In development are a host of special delivery methods for biopharmaceuticals, such as patches, microneedles, controlled release preparations and special injection devices.

Absorption, Distribution, Elimination and Metabolism (ADME)

Once it has crossed all barriers and is in the blood, a biopharmaceutical must reach a target organ or tissue, an exact location to produce its therapeutic effect. For some biopharmaceuticals, this step can be challenging. It must be absorbed, usually into the bloodstream, if it is to be distributed, and a method of delivery may fail to achieve this objective. For example, many therapeutic monoclonal antibodies are injected into the subcutaneous tissue or muscle even though the target organ, rheumatic joints, for example, is some distance away. The biopharmaceutical must be distributed and

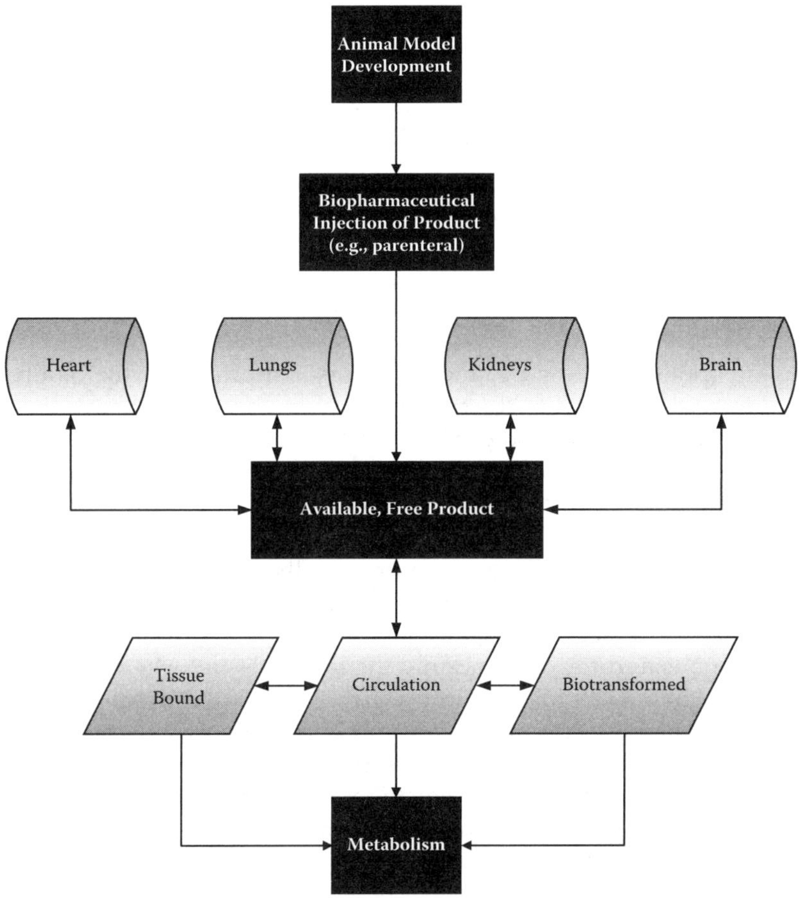

FIGURE 8.2
Outline of absorption, distribution, metabolism, and excretion (ADME) studies. Using animal models and sensitive and specific assays, product is measured in various tissues and blood. Additional assays are used to measure metabolites.

absorbed in adequate amounts and before it results in local reactivity or is metabolized or otherwise eliminated. Biopharmaceutical products, therefore, are designed to be absorbed, distributed and then metabolized and eliminated (but not too rapidly). These functions, known as ADME, are studied and reported for each biopharmaceutical product because they are of critical importance to success in clinical trials. Figure 8.2 outlines possible tissue relationships between each of these functions.

Absorption

How is absorption defined for a biopharmaceutical? Oral absorption is an unlikely route because most biopharmaceuticals, composed of protein, cells,

RNA or DNA, are recognized as just another food substance; the gastrointestinal juices and enzymes digest them. Few biopharmaceuticals might be absorbed across the skin (topical) or mucosal surfaces (transmucosal) because they are simply too large to diffuse intact across such barriers. Hence, most biopharmaceuticals are given by the parenteral route, meaning they are directly injected into either the bloodstream or a tissue. From here, biopharmaceuticals either have a local effect or they are absorbed into the blood or lymph systems and then distributed to other tissues.

Distribution

Following application of a product, it is very important for the biopharmaceutical to be distributed to the tissue or organ where it will have the greatest therapeutic effect. It should not reach tissues where it might be toxic. It also is critical to sustain a certain level of biopharmaceutical in blood and tissue. Many biopharmaceuticals reach a state of equilibrium upon reaching the blood stream, while for other products there is a rapid drop in blood levels following injection into or absorption by blood. The pharmacological and toxicological implications of parenteral delivery may be considerable because a biopharmaceutical in the blood is rapidly distributed and exposed to many organs and tissues, not only the target tissue. Alternatively, a biopharmaceutical may be injected into a firm or semisolid tissue, such as subcutaneous, where it resides in a depot and is slowly released and distributed by the bloodstream. For some products, this can result in sustained levels of biopharmaceutical with blood absorption and then distribution over a longer period, as compared to intravenous injection.

However, for other biopharmaceuticals, it is not desirable to distribute product to certain tissues because the molecule might be toxic to certain tissues or organs. Some molecules accumulate in one or another tissue or cellular compartment, a reservoir, and this may be desirable or it might lead to toxicities. For example, a product could accumulate in the liver where it may be hundreds of times more concentrated than in other tissues. If the product is therapeutic in the liver, then it may be best to have that biopharmaceutical largely concentrated right there and unequally distributed in the body. But, if it is toxic to liver cells at high concentrations, then it is not good for the biopharmaceutical to accumulate there. Sometimes it is best to avoid the bloodstream when possible and deliver the biopharmaceutical directly to the target organ. However, this can be challenging with certain types of products that target hard to reach organs, such as pancreas or brain. Biodistribution studies to determine these pharmacological parameters are critical to understanding pharmacology of a product. Also, information gleaned from distribution studies is used to support development of new formulations and delivery methods aimed at improving the therapeutic value and reducing toxicity of a biopharmaceutical.

Metabolism and Biotransformation

Biopharmaceuticals are eventually changed in the body to another form, becoming metabolically inactive through normal processes, such as enzymatic degradation. A few biomolecules, such as the DNA in a genetic therapy or a pluripotent cell-derived product, may not follow this rule because they are developed for the purpose of longevity in the body. Yet, biotransformation, a term used to describe any biological process that converts the original product to another molecular format, is the rule with most biopharmaceuticals. In some cases, biotransformation can enhance therapeutic activity, while in others it decreases, limits or terminates biological activity. Physiological, genetic and environmental factors may be, and often are, involved in biotransformation. While we can establish the average time of biological activity in a given population, it has been nearly impossible to reliably predict, for a single individual or animal, how long a particular biopharmaceutical will remain active. Living organisms are quite diverse when it comes to processing biopharmaceuticals. Further complicating the picture, the co-administration of two compounds can have unexpected effects because metabolic drug interactions are possible. Drug interactions can impact absorption, distribution, pharmacokinetics, metabolism or excretion, and many patients take two or more drugs or biopharmaceuticals. Metabolism and biotransformation studies can assist in understanding the overall pharmacological profile of any product.

Excretion

Clearance is a process in which a biopharmaceutical is eliminated from fluid phases, tissues or organs. With most biopharmaceutical products, clearance is expected through the processes of metabolism and excretion, but the molecule must remain in the target tissue or organ long enough for it to have a therapeutic effect. Excretion cannot be too rapid. With many small molecule drugs, the absolute rate of clearance is a linear function of the concentration in blood. But, with biological molecules, this is not always the case and the rate of clearance is not simply the rate of elimination divided by blood concentration. And, while small molecule drugs are often cleared by liver and kidney, larger biological molecules are not often metabolized in the liver and are retained, not excreted, as they pass through the kidney. For many biopharmaceuticals, the sites of metabolism and excretion are unknown and it is assumed that components of degraded biopharmaceuticals, such as polypeptides, amino acids and nucleic acids, are simply catabolized to a certain degree and then used by the body to produce energy and to build other macromolecules.

Pharmacokinetics and Pharmacodynamics

The science of pharmacokinetics attempts, for a given biopharmaceutical, to understand ADME and to explain the outcomes following dosage of that

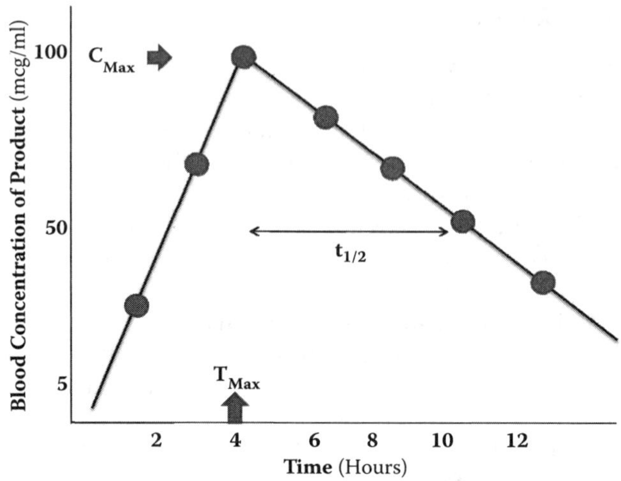

FIGURE 8.3
Example of biopharmaceutical concentrations in blood over time. Following injection of a biopharmaceutical at Time "0," the concentration rises in blood until it reaches a maximum concentration (C_{MAX}) at the fourth hour (T_{MAX}). The concentration then falls due to metabolism and excretion until one-half the maximum concentration is reached at the 10th hour, resulting in value of $t_{1/2}$ for the period.

product. Pharmacokinetics is the study of complex interactions between an active compound and the cells, tissues and organs upon which it has an impact. Only certain aspects of this science are discussed here. Pharmacokinetic studies for pharmaceuticals and drugs differ in many respects because small molecules and the large molecule biopharmaceuticals are sometimes quite different in biological properties and mechanisms of action.

However, some rules of pharmacokinetics do explain the behavior of many biopharmaceuticals. For example, with specific and sensitive analytical tools we can measure the maximum concentration of a biopharmaceutical that is reached after a certain dose is given by a route of injection. This value is called the maximum concentration or C_{max} and it is shown in Figure 8.3. From pharmacokinetic studies, we also determine the amount of time it takes from injection of a biopharmaceutical until C_{max} has been reached. This is referred to as T_{max}, also shown in Figure 8.3. From the same experiment, it is possible to measure the half-life of the biopharmaceutical in the blood and this is referred to as $t_{1/2}$. $t_{1/2}$ is a derived value based on both clearance and volume of distribution. The period from injection to C_{max} is called the absorption phase. While this may be a very short time for biopharmaceuticals given intravenously, it is an important parameter for products that are injected into subcutaneous or other tissues. The period beginning at C_{max} and lasting until all product has been eliminated from the blood is called the elimination phase. If a product is given in multiple injections, then the blood level rises until it reaches a plateau, referred to as the steady state for that dose and dosing regimen. These rules do not apply to some biopharmaceuticals; for example, vaccines seldom reach the blood in appreciable quantities and it is difficult or impossible to measure small amounts of recombinant proteins or live cells in a solid tissue.

Clearance, discussed above, is an important aspect of any pharmacokinetic profile. The apparent volume of distribution is another parameter and it is abbreviated as V (volume of distribution) and is equal to the amount of biopharmaceutical administered, divided by C, the concentration of product in drug or plasma. This value varies widely depending on the amount of tissue binding, the degree to which the product is hidden by or binds to other materials. Using real-time models in animal or human studies, and assuming assays are available to measure a biopharmaceutical in blood, it is possible to measure plasma concentration time curves for a product.

Bioavailability measures the amount of biopharmaceutical that is available for use by a tissue at any given time. For most products, bioavailability is maintained through multiple doses; this keeps biopharmaceutical levels at a reasonably constant, albeit fluctuating (within a range of values), level in blood or tissue. Controlled bioavailability means the product is consistently available to the patient's tissue and ensures therapeutic effect at all times. Each of the factors—absorption, distribution, metabolism and excretion—has a great impact on bioavailability as do calculated values, such as T_{max}, C_{max} and $t_{1/2}$. For successful therapy with many biopharmaceuticals, pharmacokinetic experiments develop information that, in turn, allows one to design and optimize dosage regimens based on the desired effect and the amount of product available to produce that effect.

This might sound logical, even simple, but, in fact, deriving these values is a very complicated process, made more difficult by the lack of adequate animal models in which to study most biopharmaceutical products. Nonetheless, the product development program must take this information, generated in animals, into consideration as the target level, maintenance dose, loading dose and individualized dose is calculated for man. Oftentimes, the human target dose calculated from pharmacokinetic studies in animals differs significantly from the target dose estimates made, in the absence of experimental data, in the Target Product Profile (TPP). For large differences, it is wise to ask why this happened and perhaps do further pharmacokinetic experimentation. Each biopharmaceutical molecule is unique, in molecular characteristics, how the product is dosed and the indication. For example, the loading dose of a gene therapy might be high and without maintenance dose while a monoclonal antibody to treat a chronic disease might require a specific dose through years of treatment without need for a higher loading dose at the onset of therapy. The possibilities are endless and must be experimentally determined for each product.

In summary, the experimentally derived target level is simply the amount of biopharmaceutical that hypothetically will produce the desired effect when given in a particular formulation and route of delivery. The loading dose is the amount of product that will be given at the initiation of therapy to rapidly achieve the target level. The maintenance dose is the amount that must be given at set intervals following the loading dose and to maintain the target level.

And speaking to the future, many traditional methods used to measure pharmacokinetic parameters in small molecule drug studies have not always been successfully applied to experimental pharmacokinetics of biopharmaceuticals. Given that necessity is the mother of invention, there are new methods that seem quite relevant for measuring various parameters, such as biodistribution and clearance, of biological products. Optical imaging using bioluminescence and dyes, and variations of computer-assisted tomography, are promising in this respect. Future studies of biopharmaceuticals will certainly be more informative, although the measured endpoints might seem quite different from those of small molecule drugs.

The science of pharmacodynamics studies biochemical and physiological effects and the mechanism of action of biopharmaceuticals. The concept of drug-to-drug receptor interactions underlies most pharmacodynamics for small molecules. Because most biological products exert their effects by interactions with molecular or cellular components of an organism, pharmacodynamics also is important for development of biopharmaceuticals. To properly progress a product to clinical development, a scientist should have some idea of how a therapeutic effect is generated within a complex organism. For some biopharmaceuticals, such as a monoclonal antibody targeted to the receptor of a particular cell type, therapeutic effect might be well known from discovery research. Indeed, "designer molecule" biopharmaceuticals are developed for a specific purpose, such as binding to a receptor having a known physiological function. For other products, such as most recombinant protein vaccines, the effectors' mechanism remains unknown at the time it is first tested in man. Some classes of molecules, oftentimes those given in one or a few doses, are never completely understood because their exact mechanism remains in a mysterious and seemingly complex "black box," even through market approval. So, many biopharmaceuticals are an exception to the recommendation of clearly understanding the mechanism of action prior to using the product in man. However, attempts are still made to understand pharmacodynamics of a biopharmaceutical and the information derived from these experiments is applied to designing pharmacokinetic and nonclinical safety studies and estimating clinical doses and dose regimens.

What types of pharmacodynamic information can be derived for a biopharmaceutical during nonclinical development? It is important to understand the relationship between concentration of product and magnitude of the response to that biopharmaceutical. However, the response may be complex and even unpredictable in some individuals, animals or man. A concentration–effect curve can be constructed. As shown in Figure 8.4, this experimentally derived information provides a wealth of information

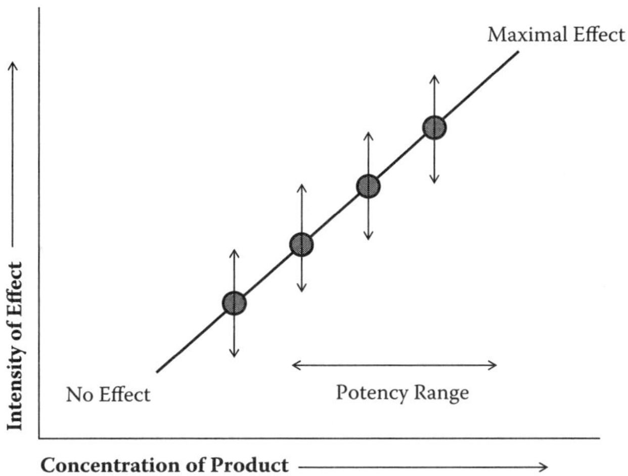

FIGURE 8.4
Concentration–effect curve for a biopharmaceutical. The intensity of effect is proportional to the blood (or tissue) concentration of biopharmaceutical. By measuring physiological effect, the potency range is determined as are concentrations with no effect or with maximal effect.

regarding pharmacodynamic properties of a biopharmaceutical. Potency is that part of the curve where an effect can be measured. Potency is clearly based on the concentration of the biopharmaceutical, but can be quite variable within a given population. Maximal efficacy is that amount of biopharmaceutical that produces maximal effect in that individual or in a population. There is also a slope to the concentration–effect relationship and this reflects the mechanism of action of the biopharmaceutical and is seen in data from a population of animals or humans. Finally, there is biological variability, seen as standard deviation from the line traced by the population value. Much individual variability is due to genetic and other factors. At any given point on the curve, this can be significant for some products and is referred to as the individual effective concentration.

However, there are caveats to pharmacodynamic studies using biopharmaceuticals. The standard concentration–effect relationship for a given biopharmaceutical considers a normal population, matched by age, gender, disease status, etc. As we organisms (humans) age, our response to a given biopharmaceutical changes and the concentration–effect of an aged population may look quite different from that of a young adult population. The same could be said for populations composed of infants, children, adolescents and those with certain underlying diseases. Pharmacodynamic variability demands that a biopharmaceutical development program understand the kinetics and toxicity of each product and carefully consider every population it might intend to treat. This is no easy task. Pharmacodynamic variability, the individual variation in response to a biopharmaceutical based on the mechanism of action, is an important issue in product development.

This variability is a factor even after extensive pharmacokinetic and pharmacodynamic information is experimentally derived from a population of animals or human volunteers.

Bioequivalence is a term that, in its purest definition, suggests two different biopharmaceuticals have an equivalent effect. Stated another way, the concentration–effect relationships of two molecules are very much alike. Consider two super-imposed concentration–effect lines as pictured for the single line in Figure 8.4. But, bioequivalence has other, sometimes more practical, meanings. It can mean an active ingredient has the same effect even when formulated in two different ways or that one biopharmaceutical has the same effect when given by two different routes of injection. If a reliable model with minimal variability can be developed and applied, bioequivalence testing can be an important aspect of nonclinical development as optimal formulations, routes of delivery and other variables maybe tested, first in animals and then in man.

Bioavailability, first introduced under the pharmacokinetic discussion, above, is also relevant to pharmacodynamics. Defined in pharmacology as the fraction of biopharmaceutical, of total amount given, available to systemic circulation, bioavailability has additional meaning for certain biological products that are given to produce a local effect and rely little on blood concentration. For example, a monoclonal antibody given intravenously has, upon injection, 100% bioavailability to the circulatory system. However, if that same product must reach a tumor mass to produce an effect, then it most likely has a much lower percentage of bioavailability where it counts, within the tumor. Experimental measurements of pharmacodynamic properties of this monoclonal antibody are not meaningful unless pharmacokinetic information is available and considered. Thus, the impact of bioavailability on bioequivalence is carefully considered with every new biopharmaceutical and for each new indication for existing products.

For this and other reasons, biodynamic experimentation is challenging for many biopharmaceuticals. Consider a few examples. A gene therapy should replace a receptor, missing from birth, on a particular type of cell. This might be achieved by inserting into host cells the gene for a molecular analog or by adding a (pluripotent-derived) cell that expresses the receptor. In either case the therapy replaces the missing activity. But, how does one measure, in a manner that is meaningful to the human situation, the pharmacodynamics of either therapeutic approach? How does the biopharmaceutical scientist determine which of the two approaches might be most successful at replacing the receptor and how is this tested in a whole body situation? Some would take an experimental approach in an animal model, while others would argue that animal studies are not meaningful and the product should skip animal studies and instead pharmacodynamics should be considered first in human subjects (and, of course, after safety studies had been completed).

In a second example, the pharmacodynamics of a monoclonal antibody are unknown. The antibody could directly bind and neutralize a molecule excreted by a cell or it could bind to a cell receptor and reduce the excretion of the same molecule. How does one approach pharmacokinetic studies in this case? Are the studies best done in animals or should they be performed in Phase 2 clinical studies?

Application of Pharmacokinetics and Pharmacodynamics in Biopharmaceutical Development

The pharmacokinetic and pharmacodynamic properties of each biopharmaceutical are important information because it is applied to various decisions, including the selection of an efficacious dose and route of delivery and to ensure that the new product is tested for safety at a correct dosage. Indeed, nonclinical studies should demonstrate that new products have little chance of causing unexpected and undesirable effect when given to humans in subsequent clinical studies. If this is so, then it is necessary to understand the pharmacokinetics and pharmacodynamics of the product before embarking on an extensive program of safety testing in animals and certainly prior to introducing the biopharmaceutical into man.

As suggested earlier, certain tools are required for pharmacokinetic and pharmacodynamic development. One need is to have an applicable animal model in which to perform this testing. A model may be available from studies of other products. However, no matter how well proven the animal model is for another particular disease or class of product, it is impossible to know if that model will be applicable to a particular biopharmaceutical or its indication. For some biopharmaceuticals, it is difficult to even begin the process, this due to inherent biological complexity of the product, animal physiology and the disease. For example, some of the most common biopharmaceuticals, such as vaccines and genetic therapies, are themselves very difficult to classify in part due to the "black box" or great unknowns concerning exact mechanism of action. And, we often lack a full understanding of pathogenesis of the disease that is to be prevented or treated. In the example of a recombinant protein used as a vaccine to prevent an infectious disease, we commonly lack knowledge on how the infectious organism is pathogenic and we do not know exactly how the immune system of our body fights that disease.

Still, it is incumbent upon the product developer to use available scientific information and attempt to understand the pharmacokinetics and pharmacodynamics of each new product. With proper methods to detect active biopharmaceutical cells or cell receptors in blood and tissues, pharmacokinetic and pharmacodynamic measurements can be made in animals. Using more than one species, single dose experiments can yield information on ADME, preferred route of delivery, optimal dose, dose-linearity (Figure 8.3), concentration–effect (Figure 8.4), interspecies differences, metabolism and excretion. Multiple dose pharmacokinetic and pharmacodynamic studies can then be initiated or they can be performed in conjunction with toxicology studies. In addition to studies in animals, it is often helpful to perform *in vitro* studies. For example, it is informative to determine if a candidate therapeutic monoclonal antibody binds specific cell types and this might be achieved by using human cells, derived from various tissues, in culture. Also, there are tests to study intracellular metabolism of compounds using cell and organ cultures and there are even *in vitro* methods to screen for induction of immune responses, again using cell or organ cultures.

To begin nonclinical experimentation, the biotechnology firm must have a formulation (see Chapter 6) for each candidate product and there must be enough of the material to allow extensive testing in the laboratory and in animals. This is sometimes referred to as the optimized clinical formulation, meaning it is the formulation of product intended for Phase 1 clinical studies. There also must be a decision on how the product will eventually be delivered to humans. This again brings to mind the need to have animal models. A pharmacokinetic and pharmacodynamic program should be designed to meet current regulatory requirements. Finally, there must be a precise means of measuring the product as it exists in a matrix, such as animal blood or tissue. This means extensive analytical support (see Chapter 7) to detect and measure exactly the biopharmaceutical of interest.

Results of pharmacokinetic and pharmacodynamic studies are carefully examined. Of particular importance are findings with safety implications, such as undesirable localization, notably vital organs or tissues like the central nervous system, the heart or kidneys, of therapeutic molecules or cells. Such information is then applied to the design of toxicology studies and to monitor safety of subjects in human clinical trials. Nonclinical data also provides a foundation on which to establish rational toxicological studies to support clinical trials. It gives clues to development or application of the best animal models or future pharmacokinetic, pharmacodynamic or toxicology testing. Specifically, study data may point toward an animal model that mimics the situation expected in man. For example, if a monoclonal antibody was expected to have a therapeutic effect only if it remained in the human body for at least two weeks, it would be unwise to use, for toxicology testing, an animal in which the same molecule was undetectable one day after it was injected. The same can be said for route of delivery. If a vaccine would be given intramuscularly to man as 200 mcg in a volume of 1.0 mL then one would not choose a mouse model because it is impossible to put this volume into a single mouse muscle.

Pharmacodynamic and pharmacokinetic studies not infrequently lead a developer to change formulations from the anticipated or most readily available format. If a therapeutic DNA molecule does not remain as required at the subcutaneous site of injection for 24 hours in an animal model, then it might need a new formulation, one that results in a depot effect to enhance longevity in that tissue. Improvements in dosing, based on data from well-designed pharmacokinetic studies, can be another positive outcome. Hence, well-designed nonclinical studies of biopharmaceutical absorption, distribution, metabolism and excretion typically provide valuable information that allows for improved and more efficient safety assessment studies and also provides a basis for mechanism of action.

In conclusion, pharmacokinetic and pharmacodynamic studies are performed with careful planning, as a series of experiments and with close coordination with other functional area experts, notably individuals from manufacturing, regulatory affairs, quality assurance and clinical studies. Primary pharmacokinetic and pharmacodynamic effects are studied in animal models and using a variety of *in vitro* laboratory methods. Experimental design focuses on the specificity of biopharmaceutical activity. If at all possible, levels of biopharmaceutical should be tested in humans based upon information derived from these studies. For many types of cells and molecules, it is important to determine where the molecule goes within the animal, to define tissue or cellular interactions and identify how long it remains in any given location, and how and where it is metabolized or excreted.

Safety Assessment of Biopharmaceuticals

Toxicology

Toxicology is a science, specifically the study of adverse effects of agents—physical, chemical and biological—on living organisms. Because any molecule could produce adverse effects, toxicology is important to all biotechnology products, not just biopharmaceuticals. This science covers acute, chronic and long-term risks using a variety of established methods, many of them biological. Toxicology assesses risks, the probability of adverse events, caused by such effects. Toxicological studies go beyond measuring risks and provide data that determine possible causal relationships, help to establish limits of safety and design rationale and safe clinical studies. This discussion covers general approaches to toxicology while using examples derived from the safety testing of biopharmaceutical products.

The term toxicology immediately brings to mind chemical toxins, acids, bases or organic solvents created by man for the purpose of producing other chemicals or life-style products. It also conjures images of physical agents, such as ionizing and nonionizing radiation and ultraviolet light. Further, we consider the target of these agents to be a biological system—plants, animals and, most notably, humans like ourselves. Chemical and physical agents can cause damage to living organisms. Drugs and medical devices are considered chemical and physical agents, respectively, and they have for decades been studied for toxic effects. Toxicology studies are routinely performed on drugs and medical devices using both laboratory, *in vitro,* and animal, *in vivo,* methods.

In the recent past, a third type of agent, biological, has been added to the list of potential toxic agents. We have been aware of natural toxins, snake venoms and poisons from plants, but had not until recently had the ability to manipulate the structure and function of biological molecules or cells and then use them to prevent or treat disease. With the advent of biotechnology, scientists began to develop biopharmaceuticals. These compounds, manmade (or at least man-designed) and unique in nature, were intended for human exposure, sometimes repeatedly and over long dosing periods. Further, biopharmaceuticals are designed to change the physiology or biological status quo of the user. However, changing the physiological balance also can have undesirable effects. A recombinant molecule or cell that produces a desirable effect in one tissue might well cause an undesirable, or even toxic, effect in another tissue or organ of the same species. Hence, with the advent of biopharmaceuticals, it became clear that each molecule would be subjected to toxicology testing in the same manner as drugs and medical devices. Clearly, biological substances could cause undesirable effects by interacting with cells or tissues every bit as much as chemical or physical agents. Biopharmaceuticals could, if used properly or improperly or in too great a dose, harm body structures or processes and some might even pass these effects on to subsequent offspring. Thus, each biopharmaceutical must be studied to evaluate toxicity, or the potential to be toxic, when used in a particular manner for a given indication.

Design of a Safety Assessment Program

An effective safety assessment program must be carefully planned and elements of planning for biopharmaceutical safety studies are outlined in Chapter 1. Here, we delve into some factors influencing biopharmaceutical toxicity, discuss the tools used in these studies and consider common study designs. The toxicologist has, or should have, four assets at his or her disposal. These include:

1. Scientific and design precedence established, over decades, for a host of biological, chemical and physical agents.
2. *In vitro* methods to serve as rapid screening tests.
3. Animal models and the ability to test agents in these complex organisms.
4. Established testing protocols, such as acute, subchronic, chronic, reproductive, carcinogenicity, local tissue, immunological and respiratory toxicology.

The key to completing a meaningful safety program for a new biopharmaceutical is to use these tools wisely under a product-specific and indication-driven experimental strategy.

The nonclinical plan is based on elements of TPP, notably the intended indication or disease and intended product safety profile, as been developed for the candidate biopharmaceutical. Consider that some biopharmaceutical products, such as a therapeutic for a terminal illness like metastatic cancer, have a very different profile from a product, such as a vaccine, intended to prevent a non-life-threatening disease in infants. The plan also is based on an estimated clinical dose and dosing schedule, as provided in the draft clinical plan. From this information, and relying upon experience with similar products, regulatory guidelines for the class of biopharmaceutical, and any available research data, the nonclinical professional can outline the intended approach for safety testing. Using a recombinant therapeutic protein as an example, Figure 8.5 presents a general scheme for nonclinical testing of a biopharmaceutical by phase of development. It demonstrates how the flow of events in a safety testing program structures a tiered testing approach. With each tier or ascending phase of clinical development, the product is used in both greater numbers and a more diverse population of individuals. This, in turn, can demand more detailed and stringent safety testing prior to each clinical phase.

The earliest nonclinical testing focuses on understanding the pharmacokinetic and pharmacodynamic properties of the biopharmaceutical, as noted above. With this information, the intended human dose or doses can be estimated for the Phase 1 clinical study. Toxicity testing applies the intended clinical doses and dosing regimen to design of nonclinical studies that are completed and reported prior to filing an IND application or initiating the first clinical study. These early nonclinical studies include acute, subchronic or other types of investigations that may be considered by a sponsor or required by regulatory agencies prior to initiating a Phase 1 study with this class of product.

Subchronic and even some chronic testing may be required prior to entering Phase 2. The route of delivery or the formulation may be adjusted, based on Phase 1 study results, and so additional acute testing may be necessary or local tolerance testing may be advised. And, because midstage clinical trials may expand into previously untested human populations (e.g., women of childbearing potential or individuals with tumors or an underdeveloped immune response), it is wise to consider specialized toxicity testing prior to initiating these clinical studies. Because Phase 3 means testing in a more diverse and much larger population and involves doses given over longer periods of time, it may be advisable to complete chronic toxicity studies at this stage of development. Besides, at Phase 3, the dose and dosing regimen will have been established, reducing the risk of having to repeat long and costly chronic toxicology studies. Based on the intended population and use, it is also wise to consider and plan for applicable specialty studies, such as those directed toward an organ system, e.g., reproductive toxicology, neurological toxicology or immunotoxicology studies, and those focused on a product-related issue, e.g., tissue binding or DNA integration studies.

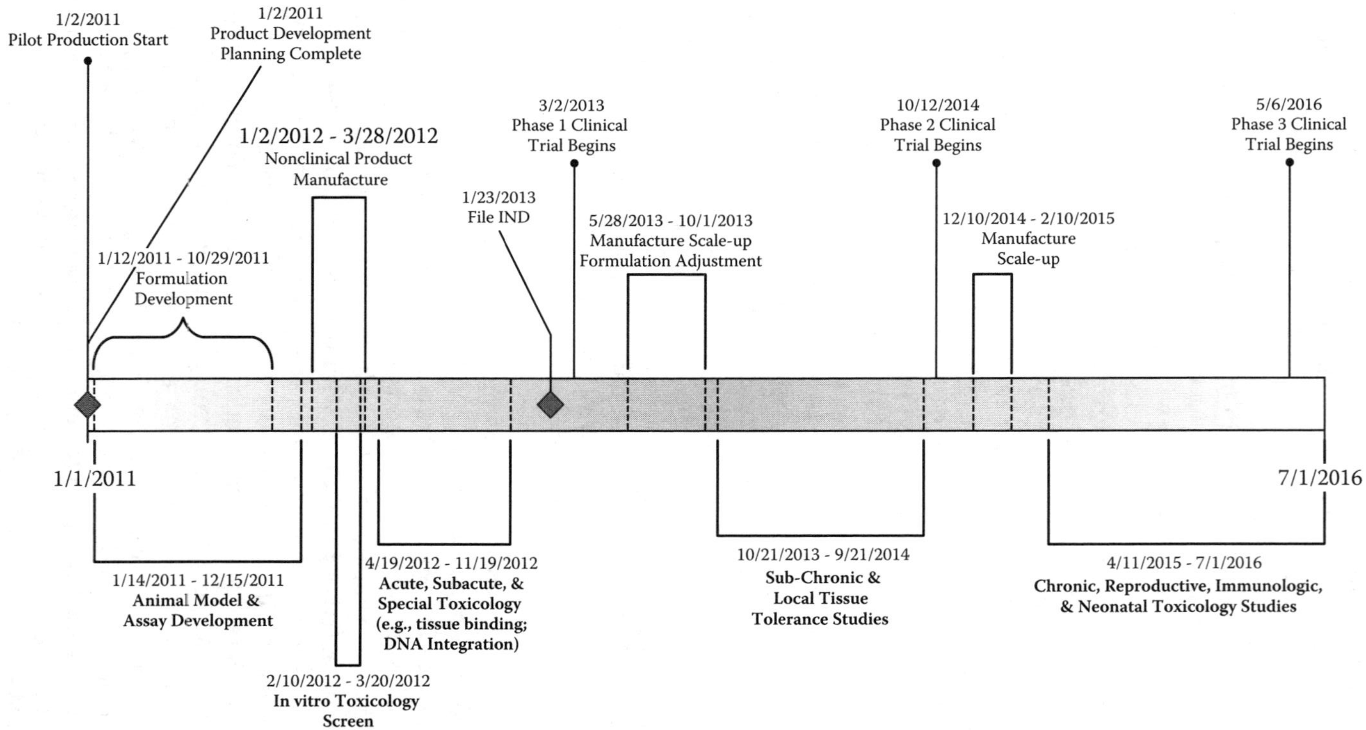

FIGURE 8.5
Schedule of nonclinical studies in a biopharmaceutical development project. Animal model and assay development is completed early in development and while formulation and product manufacture are underway. The initial nonclinical studies, acute or subacute and special and product specific (e.g., tissue binding or DNA integration), are performed prior to Phase 1 clinical trials. Subchronic studies are conducted in midstage development. Chronic, reproductive, immunologic, neonatal and other specialized studies are conducted in late-stage development.

The planning process considers current regulatory guidelines, both national (FDA) and international. It is important to consider that regulatory agencies have responsibilities to the safety and welfare of human subjects voluntarily taking investigational products and for public health regarding marketed products. Nonclinical safety testing plays a major role in meeting these responsibilities. The International Committee for Harmonization (ICH) (see Chapter 4) provides excellent guidance in regards to nonclinical testing strategies and this has the advantage of worldwide harmonization. Regulatory agencies also ensure that adequate and well-controlled, nonclinical studies have been completed according to performance standards (e.g., current Good Laboratory Practices (cGLPs)). Studies always meet generally accepted scientific guidelines as well. This means, by way of example, using experimental designs that test well-considered hypotheses, testing adequate numbers of the correct animal species, dosing only with material that matches the quality of product to be used in clinical studies and applying proper statistical tests when analyzing results. Conclusions drawn in nonclinical study reports must be supported by data.

In Vitro Screens: Surrogate Measures of Toxicity

A relatively simple and inexpensive means of beginning a series of nonclinical studies is to rely on screening tests. Unlike toxicology testing in whole animals, *in vitro* screens are individual tests, each with a specific purpose and activity criterion. And, in some cases, *in vitro* screening tests are performed in a series to provide information on a single subject from a variety of tests. Many screens are available for chemical and drug compounds, due to the large number of candidate products tested and a long history of nonclinical development. Some *in vitro* tests are performed by contract testing laboratories while others are amenable to use in a sponsor's laboratory. While plentiful and popular for drug and chemical development, these screens may not be useful for measuring activity criterion on biopharmaceuticals because test criteria might not match assay requirements and because biological molecules are often incompatible with the test matrix or design. Also, due to limited experience, there may be questions regarding relevancy of test results when applied to biotechnology products. Thus, specific concerns can focus on the sensitivity, specificity, accuracy and reproducibility of *in vitro* screening tests when used with a biopharmaceutical.

Nonetheless, when properly applied to a nonclinical safety testing program, *in vitro* screening tests can provide valuable information used to design more complex *in vivo* studies. Examples of *in vitro* tests are discussed here and a longer list can be found in Box 8.2. Mutagenicity testing, exemplified by the Ames tests, screens compounds for mutagenic potential. The Ames test relies on salmonella bacteria as a substrate and measures alteration in structure of a gene following application of a test compound. Other

BOX 8.2 EXAMPLES OF *IN VITRO* SAFETY TESTS USED FOR DRUG OR BIOPHARMACEUTICAL SCREENING

In Vitro Safety Test	Purpose of Test
Ames Screening Test	Genetic toxicology for mutagenicity in bacteria
Chromosome Aberration Test	Chromosome aberrations and mitotic indices in cells
Micronucleus Tests	Potential to induce genetic damage measured as induced micronuclei in a variety of cell types
Aneuploidy Tests	Chemically induced anuploidy in cells
Cytotoxicity Tests	Cytotoxic activity using a variety of cell types
Metabolism Tests	Evaluation of metabolic stability of various cells
Mammalian Cell Mutation Tests	Genetic toxicology by mutagenicity of various mammalian cell types
Sister Chromatid Exchange Test	Genetic damage as manifest by sister chromatid exchange in various cell types
DNA Repair Tests	Potential to damage DNA
Cell Transformation Tests	Potential to cause genetic damage as manifest by induced morphological cell transformation
Human Skin Permeation Test	Prediction of dermal and ungula permeation
Mitochondrial Toxicity Test	Damage to mitochondrial function Damage to red blood cells
Drug Interaction Tests	Ability for product to be metabolized in presence of other drugs
Hepatotoxicity tests	Measures drug-induced liver injury
Cytokine/Chemokine Secretion Test	Potential for triggering release of cytokines or chemokines
Apoptosis Assay	Induction of cell death
Cellular Anabolism	Ability to affect protein anabolism in cells
Cellular respiration	Ability to affect cellular respiration measured as ATP/ADP
Cellular proliferation	Induction of cellular division and proliferation
In Vitro Drug Metabolism	Metabolism of products by cells

eukaryotic or prokaryotic cells may be used in the same manner, as long as there is a reliable readout for demonstrating mutagenicity. Carcinogenicity testing takes mutagenicity one step further by asking if the mutagenic or genotoxic potential of a compound also results in development of carcinogenic potential. Since not all mutagens are carcinogens and because not all carcinogens are mutagens, the mutagenicity and carcinogenicity tests can give distinct answers about product. While screening tests can be helpful in making early decisions, they are not definitive and carcinogenicity testing is considered in animals for compounds that might have this potential.

In its simplest format, lethality testing places test material in contact with living, cultured cells and determines if the material kills the cells. Various cell types can be used in these studies. Variations of the *in vitro* test measure biochemical or physiological parameters of cell health that would indicate a cell might die, or at least not thrive, in the presence of test product. Special types of cells—cardiomyocytes, neural, epidermal or gastrointestinal epithelial—can be used in such studies giving rise to test protocols with specific purposes, such as neurotoxicity testing. A number of screening assays, each providing a specific outcome, are currently in development or validation with focused application. However, due to the unique nature of most biopharmaceuticals, their utility as a toxicology screen is somewhat limited.

Developmental toxicology measures toxicity of a compound as it relates to development of a fetus and, for biopharmaceuticals, is especially important for molecules that might be used in women of childbearing age. *In vitro* developmental toxicity screens would be considered insufficient by themselves for risk assessment in this area. However, some *in vivo* models using pregnant rodents also can be applied to screening compounds and are more rapid, but perhaps less sensitive, as compared to chronic developmental toxicology studies.

In Vivo Safety Testing of Biopharmaceuticals

While it is possible to perform some safety testing in various nonanimal models, it also is necessary to test most biopharmaceuticals in live animals. Toxicologists classify safety tests in three manners: (1) by the length of time an animal is dosed, that is acute, subchronic or chronic studies; (2) by an organ system of interest, such as neurotoxicology or immunotoxicology; or (3) by a particular outcome, e.g., carcinogenicity testing. Because consumers demand thorough safety testing of each biopharmaceutical they use and because there are ethical questions regarding the use of vertebrates in such tests, toxicology testing in animals is a serious scientific endeavor, professionally performed and regulated by government agencies.

Animal Model Development

An animal model is a nonhuman, living vertebrate used in nonclinical research to ensure that a product is reasonably safe before use in man. Consider that every model is imperfect in some way, certainly as compared to the human situation; this is why it is referred to as a *model*. Nonetheless, an animal model allows scientists to gain an understanding of a broad range of toxicological processes and outcomes and this means the sponsor can make informed decisions regarding the intended use of a biopharmaceutical in man. Thus, animal study results help the sponsor to avoid the risk of causing harm to humans, while at the same time allowing the biopharmaceutical to provide intended benefit at a particular dosing regimen. For example, if a biopharmaceutical dose of 1 mg/kg per week is toxic, but a dose of 0.5 mg/kg per week is not toxic in an animal model, then the sponsor can design a clinical trial to test the lower dose and avoid the higher dose. Normally, animal models are chosen or developed for a specific study design, not visa versa. As discussed in more detail later, the sequence of planning events in safety assessment is first a hypothesis or question and then a well-considered concept study design to drive the selection of the proper model. Selection of the correct animal model or models is challenging and, in the end, compromises are made and the best model is identified. Because a toxicology study is designed to answer a specific question or set of questions regarding the safety of a particular biopharmaceutical, it is often necessary to use a different species to answer each question posed.

For most drugs, safety testing requires the use of two or more animal species over the development life cycle, while for other products, notably many biopharmaceuticals, one animal will meet regulatory requirements. The first animal is normal or healthy. However, it may be necessary to use a second model as well, an animal with a similar or identical disease, because the disease process itself may greatly modify the toxicology profile of the product. General guidelines for selection of animal model species are listed in Box 8.3. It can be very challenging to identify two excellent animal models for acute and chronic safety testing of a biopharmaceutical. Typically, there are many trade-offs in the selection process and the perfect model may never be found. Identifying an animal model with the disease is especially challenging because the animal species must be correct and also the disease must be relevant to the human condition, having a similar etiology, pathogenesis and clinical outcome. For example, to examine safety of a therapeutic vaccine to treat Alzheimer's disease, one must identify or develop a model of the disease that arises and progresses in the same manner as the human disease and the animal must exhibit the same immune response as would be expected of a human treated with that vaccine and also having this condition. Advances in development, notably the use of transgenic or knock-out animals, offers appropriate models, but these animals can be very expensive. Having chosen an animal model, it is then important to carefully design the nonclinical study to take advantage of any attributes the animal may possess while at the same time using the correct number of animals for each question posed.

Animal models should be validated, or at least qualified, for a particular application. Unless an animal model has been widely used for a particular class of biopharmaceutical or disease, it is advisable to perform pilot studies and determine whether or not the chosen species and strain, in fact, is suitable for the intended purpose. Studies performed in the research laboratory ensure that when later used in an expensive and lengthy toxicology study there will be no technical issues and the results will be meaningful. Indeed, small pilot studies of a compound in a given animal model often lead to

BOX 8.3 CONSIDERATIONS FOR SELECTION OF ANIMAL MODELS FOR TOXICOLOGY STUDIES

- Toxonomic, anatomical and physiological similarities to humans
  - Consider overall anatomy and physiology of animal
  - Anatomy and physiology of target organ, tissue and cells
- Demonstrates similar pharmacology, pharmacokinetics and pharmacodynamics to the intended human population
  - Metabolize drug in similar manner
  - Has same receptors or mechanism of action
  - Expresses same target organ, tissue or cell responses
  - When possible, provides a model of the human disease
- Dosing parameters match human condition
  - Ability to give full human dose and dose regimen to animal
  - Consider route of delivery
- Economics
  - Use enough animals of this species to fully answer the question
  - Cost of maintaining the animals over period of study design
- Ethics
  - Is it necessary to an animal to answer the question(s) or would an *in vitro* system suffice?
  - Is it possible to use a species lower on the taxonomic chart?

improvements in both the application of the model and in the ultimate toxicology study design.

Test Product Formulations, Routes of Delivery and Dosing Designs

Development and selection of animal models is not the only consideration given prior to performing a nonclinical toxicology study. There is also the need to have final product (FP) (see Chapter 5 and Chapter 6) that matches the formulation and quality of the biopharmaceutical intended for use in humans. Far too often nonclinical studies are considered invalid because they apply a product that differs in strength or quality from the intended clinical material. Formulation, strength and quality can greatly impact biological activity, including parameters related to safety, of a molecule or cell. Hence, it makes sense to use the same or a very similar formulation in nonclinical studies as will be used in the clinical studies. This can have a significant impact on scheduling a nonclinical study because formulation, manufacturing and quality control timelines impact the nonclinical plan. However, there may be trade-offs. For example, if it is absolutely necessary to use a mouse to test a full human dose of a therapeutic protein, it may not be possible to give a full human dose due to volume constraints of animal size. The result could be to split doses or to concentrate the formulation. These issues arise constantly in design of safety studies and they must be addressed scientifically in consultation with other development scientists.

Route of delivery presents another hurdle in design of nonclinical studies. For the example given above, it might be necessary to give the product to a mouse by the intraperitoneal route because it is impossible to give a full dose to this animal by the intended human route, intravenous. Indeed, sometimes an animal species is chosen based on matching the intended route of delivery for man. Because pig skin is very similar in microanatomy and function to human skin, the pig is used in many nonclinical studies of biopharmaceuticals that will be delivered to human epidermis.

Studies for nonclinical animal studies reflect a vocabulary that is unique to this endeavor and are explained here because they are commonly used in study protocols. First are the trade terms. *Test article* is product or final product, as described in Chapter 6 and Chapter 7. It is the final formulation of the material that is being tested. *Neat* means an article is used full strength or undiluted. *Placebo* represents inactive ingredient, and may be referred to as *control* or *control article*. A *diluent* is a defined solution or buffer or formulation, such as physiological saline, used to titrate the test article or control. An *excipient* is a nonactive ingredient included in a formulation. Common excipients for biopharmaceuticals are Tween 80, a detergent that prevents protein precipitation or sugars that preserve the integrity of cells in solution. A particular type of excipient, *the vehicle*, is a chemical that serves to enhance transfer, absorption or distribution of the biopharmaceutical. Tween 80, a detergent-like molecule used to prevent aggregation, is also considered a vehicle in formulations of certain protein biopharmaceuticals.

A second consideration is the route by which a biopharmaceutical is given to an animal or human; the most commonly used routes are listed in Box 8.1. Attention also is given to physiological variables that can have an impact on dosing of a product to an animal or to man. Local effects are physiological responses of the recipient to the test article when it first reaches the recipient tissue or organ. Absorption and distribution is the process by which the test material moves away from the site of delivery and establishes itself in various tissues and organs. Formulation can have a major impact on how efficiently this occurs and where a biopharmaceutical goes in the body. Next there is the issue of metabolism, the process by which the biopharmaceutical is chemically changed, broken down or otherwise used by the body. Absorption and distribution, and, hence, also local effects, influence metabolism for many products and were discussed earlier.

Calculating dose is an underappreciated skill, but one that can have a major impact on the outcome of a nonclinical study. At the outset, the design of a nonclinical study has, from the TPP, a target dose or range of doses that clinical experts suggest might be used in human volunteers. For proposed nonclinical studies, the designer brackets the target human dose based on an understanding of the product and the chosen animal model and with an understanding of body weight or surface area of the animal versus a human. If body weight is chosen as a comparator, then dosing calculations are made as milligram of biopharmaceutical per kilogram of body mass. Alternatively, dose estimates based on body surface area are becoming more common, perhaps because of obesity in society, disease being treated, or factors related to pharmacokinetics or pharmacodynamics of biopharmaceutical products. The chosen dose can impact the calculated relative dose for some biopharmaceuticals and where small animals, e.g., mice, are employed. Whichever method is chosen, both calculations should be at least considered and compared during study design. Finally, it is important to calculate from a study design the total test article, control article and vehicle requirements as well as specifications and tests for quality, and to then discuss the outcomes and decisions on study design with colleagues responsible for biomanufacture and quality control, focusing on the total quantity and number of individual lots of product, diluents and placebo.

Protocols and Performance of Biopharmaceutical Safety Studies in Animals

A concept nonclinical study design, once found acceptable to a product development team, is then written further into a nonclinical study design document, the protocol. The purpose of the nonclinical study protocol is to guide the investigative team in performance of a study. Elements of a nonclinical study protocol are given in Box 8.4. Responsibility for preparing the protocol is given to an individual, the study director, the individual responsible for ensuring performance of the study according to his/her protocol. That individual also is responsible for completing a study report. Nonclinical protocols are reviewed by other scientists as well who serve on the nonclinical study team and they are approved by the quality assurance unit. Further, any use of animals in research requires review by an Institutional Animal Care and Use Committee (IACUC; see Chapter 4) responsible for the ethical and proper use of laboratory animals. A nonclinical study always results in a report, a carefully written scientific document that identifies the design of the study, as mandated under the protocol, reports the results and any conclusions made by the study director, and provides, in appendices, all raw data derived from the study. Nonclinical study reports are typically large documents, hundreds of pages, even for a small and simple study and are reported in full to the FDA under a filing to the IND. Finally, it is worth noting that a nonclinical study in animals costs a significant amount of money and takes considerable time (6 to 12 months for acute studies) from concept protocol design to final study report, all the more reason for careful study design and execution.

BOX 8.4 ELEMENTS OF A NONCLINICAL STUDY PROTOCOL FOR SAFETY TESTING OF BIOPHARMACEUTICALS

Title, purpose, sponsor and testing facility
Detailed identification of test and control articles and animals
Methods of identification
Description of materials used in the study, including the animal diet
Dosing levels of test material
Experimental design and methods used to control bias
Type and frequency of tests and measurements
Records to be maintained
Statistical methods
Dated approval of protocol by sponsor, study director, and quality assurance
Any changes made
Approvals

Elements of a Nonclinical Study Design

Adequate and well-controlled nonclinical study designs, no matter how the study is classified (e.g., acute, subchronic or chronic), have common elements. Content of the protocol is, to some extent, mandated by cGLP regulations and this alone harmonizes format (Box 8.4). To begin, there is a facility and it is staffed with qualified individuals capable of performing functions called for under the protocol. The facility is properly designed and equipped for the types of studies performed. As noted above, each study is defined under a protocol, and has a responsible study director. A quality assurance unit is also part of the study for audits, reviews, approvals and other quality assurance (QA) functions (see Chapter 5). If the study calls for laboratory analysis, and most studies do, then there are adequate laboratory facilities either at the study site or, for specialized tasks, at qualified contractor sites. Animals are involved in most studies and all their support is met; this involves housing, feeding, treating and inspecting or analyzing each animal as well as ensuring quality care and well-being. In addition to a protocol, routine procedures—everything from handling animals to archiving records—are fully described in instructional documents, such as standard operating procedures (SOPs). And every bit of information and data are captured on source documents and may be transferred to study data capture forms, most designed uniquely for that study. All of this information is reviewed by qualified professionals, condensed into summaries, written into a final study report and properly approved. Quality assurance involvement is essential.

However, each protocol is scientifically unique in purpose, scope and design and is written by the study director with much scientific and technical input and based on a hypothesis. One reason is that each nonclinical study is related to a unique biopharmaceutical. To test the hypothesis and answer every question posed in the study objectives, a study design includes a unique mixture of animal and laboratory treatments, tests and procedures. The design is given at three or more levels of variables. One level defines the dose, another the dosing scheme and schedule, and the length of time animals are on study. This leads to groups of animals, each group receiving unique treatment.

The second level of design provides further instruction for carrying out the essential elements of the study. An example is animal care and treatment, where observation, shaving or clipping, anesthesia, food or physical environment are written into the protocol in an effort to support the scientific design. Another example is laboratory testing. Clinical laboratory testing—hematology, clinical chemistry, urinalysis and immunology—measures the health of the animals. Other laboratory testing may measure the level of product or metabolites from that product.

The third level of design enters into this example because the analytical laboratory must develop or have on hand an assay or set of assays that are robust and both selective and sensitive for the compound being studied when it exists in animal blood or tissues. Assay qualification (see Chapter 7) is common practice to support nonclinical studies. Thus, laboratories must be capable of performing a variety of analytical methods. Very special laboratory studies may be performed at another site, either at a contract laboratory or in a sponsor's laboratory. For example, it might be necessary to send animal blood or tissue to a contract laboratory to measure levels of the excipient Tween 80 if there is a concern that it might be concentrated in the body of an animal. A special assay to measure an effect of a product on animals might be performed by the sponsor of a study if this test requires special expertise or laboratory equipment. For example, a cellular assay to measure proliferation of lymphocytes in response to a recombinant protein might have to be performed by the sponsor's laboratory on specimens taken from animals immunized with a new vaccine.

Returning to the second level of design, the planner must consider other procedures in the in-life phase of the study. The in-life phase of an animal study is defined as the time beginning with initial dosing of the animal until the time it is euthanized or released from the protocol. Most toxicology studies of animals require frequent examination of animals during the in-life phase of the study and this involves close observation and measurements of general animal health, such as weight or amount of food consumed per day. Some commonly used measures of animal health are given in Box 8.5. Following the in-life phase, and this is defined by the protocol, most animals, perhaps with the exception of some large species and nonhuman primates, are euthanized. For most safety studies, a formal necropsy is performed with rigorous gross inspection followed, after tissue preparation, by histopathological examination of every organ. Any abnormality in each of these parameters is

BOX 8.5 COMMON MEASURES OF IN-LIFE ANIMAL HEALTH AND POSTMORTEM CHANGES IN NONCLINICAL TOXICITY TESTING

In-Life Measures of Animal Health and Well-Being

- Clinical Signs
  - Weight loss–gain
  - Appearance or behavior (ruffled fur, hunching)
  - Agonal events
  - Neuromuscular, ophthalmic and other special tests
- Clinical Laboratory Pathology
  - Hematological abnormalities
  - Clinical chemistry abnormalities
  - Abnormalities in urine, secretions or other samples

Postmortem Measures of Animal Health and Well-Being

- Death (mortality)
  - LD_{50}
  - Time to death
  - Cause of death
- Gross Observations at Necropsy
  - Edema—fluid in tissues and body cavities
  - Color, size and appearance of organs
  - Obvious signs of systemic disease
  - Tumors
  - Developmental abnormalities
- Histopatholgical Examination of Tissues
  - Tissue or cellular changes
  - Signs of infection or inflammation
  - Malignancies or nonmalignant tumors
  - Developmental abnormalities
  - Signs of systemic disease or stress to a system or the body as a whole

In-Life or Postmortem Observations and Tests

- Infection demonstrated by microbiological examination or culture
- Immune response to various stimuli
- Pharmacokinetic or pharmacodynamic measurements

recorded and the results analyzed by an experienced scientist. Plans also are included for data handling, analysis by a statistician and reporting. Records are exact and complete and all procedures fall under cGLP guidelines. Data and reports follow stringent criteria for analysis and review by the laboratory and sponsor and for approval and reporting to regulatory authorities.

Hence, the design and procedures provided in nonclinical and clinical studies are carefully considered, rigorous and provided in some detail. Nonclinical designs and protocols are instructive and directive because both types of protocols are important to human subjects and users of the biopharmaceutical. There is little room and no justification for error.

Acute Toxicity Testing

Acute toxicity testing has been defined as the short-term evaluation of toxicity in animals following a single dose of a biopharmaceutical. Today the definition would, by most accounts, include study designs with multiple doses of the product, but over a brief period and again with short-term evaluation. If a devil is in the study details, then the devils here are definitions for the terms "brief period" and "short-term evaluation." For a therapeutic protein, an acute study might involve three doses over three days with completion of in-life studies on the sixth day. In contrast, for a vaccine an acute study might be three doses over 10 days with in-life studies completed on the 18th day. Hence, the definition may be adjusted depending on the type of compound, expected dose and dosing schedule, possible toxic effects, indication and animal model. However, there are established common elements and guidelines for acute toxicity tests and these are: (1) findings are suggestive and never definitive as to the overall toxicity of the biopharmaceutical; (2) it is screening toxicity and may be useful to rank toxicity as variables are adjusted (e.g., dose, timing of dose, route of delivery); and (3) it represents only an assessment of potential toxicity. To further confuse the definition, acute toxicity is defined in one manner for drugs, in another manner for compounds intended for environmental release, and, it would seem in yet another manner for certain types of biopharmaceuticals.

An acute toxicity study design does, however, contain elements deserving mention. First, some are range-finding studies, meaning they are designed to define a dose level that is suitable for the proposed clinical dose and upon which to base more definitive range-finding studies. Second, endpoints and measurements are as important in acute toxicity testing as they are to any other study. One endpoint may be finding the dose level that results in significant or measurable harm or disease (requires careful testing and definition) or it might be the lethal dose (easy to measure and define). Here, the term LD_{50}, the dose of a product that causes the death of 50% of animals receiving the product, might be used. While this endpoint is seldom measured for biopharmaceuticals, a similar concept, here referred to as XD_{50}, can be applied where X is a particular measurement of declining health, such as 50 g of weight loss in a rat. Toxicologists have identified and used numerous acute toxicology range-finding study designs, endpoints and measurements and now some of these are applied to nonclinical studies for biopharmaceuticals, notably to therapeutic products.

How are the results of acute toxicity testing of a biopharmaceutical considered and acted on in regards to further product development of the biopharmaceutical? Some outcomes are easily interpreted and other times they confound more than clarify the interpretation of results. A lethal dose (e.g., 100 mg/kg) of any biopharmaceutical in a relevant animal species would not be considered for a Phase 1 clinical study. And, if that lethal dose were near or in the range that had been designated for clinical therapy, then this product would probably not be progressed to clinical trials unless it were first reformulated or otherwise changed to significantly reduce the toxicity while retaining the therapeutic effect. Yet, there are caveats even with this example. Should the product be indicated for a life-threatening disease for which there is no other possible therapeutic, then it might be progressed to more definitive toxicology studies. Every result must be considered in context.

While not definitive, the results of acute toxicity testing certainly aid the selection of dosages and perhaps dose regimens, or at least they should if the study was designed properly, the animal model carefully chosen and the dosages bracketed the proposed human dose. Acute studies that give multiple doses also are instructive on additive effects of the biopharmaceutical on an animal, information that can be applied to later studies. Acute studies also may provide information on proper timing of doses and, if properly designed, yield meaningful pharmacokinetic and pharmacodynamic data. Again, these objectives must be considered in the acute study design and completed in the in-life and laboratory phases of the study. It is worth noting that acute toxicity studies of biopharmaceuticals are occasionally performed in research laboratories and without regard for GLP. They are considered pilot studies on which to base dose selection for an adequate and well-controlled (cGLP) subchronic and definitive toxicology study and may not be acceptable to regulatory authorities as definitive acute toxicity studies. In such cases, results that demonstrate lack of toxicity are not compelling, but when toxicity is noted, it must be reported to regulatory authorities and considered in the overall picture of product safety. Significant benefit is derived in savings of time and money, but there is some regulatory and scientific risk involved in performing such pilot studies.

Subchronic and Chronic Toxicity Testing

The definitions for study terms—chronic, subchronic and subacute—are wide ranging and cover any number of nonclinical study designs when applied to biopharmaceuticals. The terms *subacute* and *subchronic* are subjective but progression from subacute through chronic studies reflects an ever increasing exposure of animals to the biopharmaceutical beyond those applied in acute studies. The subacute study, a term used more often with drugs than biopharmaceuticals, refers to studies that are done as repeat dose and at dose levels between those of acute and subchronic studies with durations of one to three months. Subchronic studies are, perhaps, three to six months in duration and typically involve multiple doses, if indicated. They look for cumulative biological or health changes in animals. They can, however, be broad explorations, examining animals for a wide range of symptoms or diseases. They should, in the end, define toxicity as well as add to the pharmacologic body of information. Subchronic study results are qualitative and quantitative and the studies attempt to, and should in fact meet, statistical endpoints to clearly demonstrate toxicity if it exists. The hoped for result of a subchronic study, one supported by unequivocal data, is safety of the product at a dose and dosage regimen that is desirable and feasible in man. This is referred to as a "clean" dose level of the biopharmaceutical. Finally, a well-designed subchronic study forms the foundation for designing required follow-on chronic and specialty toxicity studies.

Subacute or subchronic study designs are driven by so many variables, not to mention the indication and nature of the product, that no single design should be considered authoritative. The animal model is very important to success of a subchronic study and here data from acute studies and the research laboratory are most helpful. Regulatory guidelines for drugs specify two species and often pharmaceuticals are tested in a rodent (rat) and a larger animal (dog or nonhuman primate). Today, therapeutic biopharmaceuticals reviewed by CDER (Chapter 3) often follow this guideline. But other biopharmaceuticals, such as vaccines and genetic therapies, have followed precedent established at CBER and used a single species, one shown to be suitable for the product, notably for the intended indication, dose, and dosing regimen. For subchronic studies with these biopharmaceuticals it is important to perform pilot studies to test variables and the animal model before the definitive subchronic study is performed.

Design may be single dose or, more commonly, repeat dose. It is important to ensure the chosen design is statistically valid so that enough animals of each sex are assigned to each dosing group. Different doses are tested for most biopharmaceuticals to ensure that the dose and dose regimen taken to human studies is safe and well tolerated and that the next higher dose is not unsafe. Controls are always included in subchronic studies and, for many biopharmaceuticals, this means at least two additional groups of animals, one dosed with the formulation minus the active ingredient and another, the null control, dosed with normal saline or nothing at all. Details regarding performance of a subchronic study, such as dosing animals in all groups at the same time, can be important. Another critical detail is study termination and animal sacrifice. It is important to leave enough time after dosing to allow toxicity to occur, but then again this is not a chronic toxicity design. For some biopharmaceuticals, guidelines recommend sacrificing one group of animals after the last dose is given and sacrificing another group, treated and controlled, in a similar manner and weeks after the last dose is given. These nuances in study design demand that the sponsor is very familiar with precedent within a class of biopharmaceuticals, has read all the regulatory guidelines and intends to meet with the FDA (see Chapter 3) soon after the concept design has been drafted and well before the subchronic study beings.

One definition of a chronic study is "long study," taking months to even years and examining animals repeatedly and closely for changes in health and signs of chronic disease. For many biopharmaceuticals, chronic studies are not worthwhile because the product will not be given over a long period to patients. A recombinant protein vaccine, intended to be dosed three times, on Days 0, 30 and 60, is a fine example of a product that might not require a long chronic study. In contrast, a therapeutic monoclonal antibody intended to treat patients with chronic inflammation by dosing biweekly and for many years deserves chronic toxicity testing. As does a genetic therapy that is designed to incorporate foreign DNA into the human genome and produce a lasting effect, even though it is given one time as a single dose. The challenge in designing chronic studies is not so much when they are done in the product development life cycle as in how they are designed and performed.

In chronic toxicity testing, the biopharmaceutical is administered over much of the animal's lifetime. Animals are kept on protocol, housed, fed and observed daily, for a substantial portion of the animals expected life. It is no surprise then that small animals, especially mice with a lifespan of under two years, are selected for chronic studies. Chronic toxicology studies are often confounded by findings that are a normal part of aging. For example, sudden and unexplained deaths of individual animals are a reality in all species and cancers are not uncommon findings as inbred animals age. Well-controlled studies are the key to distinguishing product-related adverse events and disease from those simply associated with aging or a common environment. Statistically valid designs thus require large numbers of animals in each group to rule out an incidence of disease or events that occur by chance alone from incidents related to product toxicity. This concern drives the design of large studies with many animals and significant numbers of tests, both in-life and after sacrifice. Hence, chronic studies tend to be large and expensive, all the more reason to ensure proper design, focus on the toxicity that really matters and ensure study protocol reviews by regulatory agencies.

Reproductive, Developmental and Teratogenicity Toxicity Testing

Biopharmaceuticals intended for use in individuals of childbearing age or in children with developing reproductive systems are further tested to ensure the product will have no undesirable effect on reproductive tissues or a developing fetus. Consumers are extremely sensitive about developmental and reproductive toxicology for good reason. Guidelines suggest a sponsor consider at least three types of studies if the product might reach the gonads or fetus. The first type of study, Segment I, is toxicity to male and female fertility and of early embryonic development to implantation. Endpoints measured in such animal studies are maturation of sperm or eggs, gonadal integrity and, in females, normalcy of gestation until the time of implantation. This calls for using an animal model of both sexes and recently mated female animals of an appropriate model species. Measurements focus on the reproductive cells and tissue. The second type of toxicology test, Segment II, is for embryo–fetal development. In these studies and following treatment with the biopharmaceutical, organogenesis is studied in pregnant animals from the time of implantation through the second gestational period. Here, the study is designed to measure abnormalities that might develop in the fetus and associated organs, such as placenta. The third type of study, Segment III, focuses on pre- and postnatal development. Here dosing of animals begins in the earliest phase of gestation and continues through birth of the animal. Examinations are performed at various times during development of the fetus and include examination of neonates. Other study designs may be used, especially if the class of biopharmaceutical is suspect of causing reproductive or development abnormalities. The need for product to actually reach the reproductive system or the developing fetus is a consideration when selecting an animal model.

Because there is no "typical" biopharmaceutical, each product must be considered on a case-by-case basis. It is worthwhile examining precedence and regulatory guidelines prior to designing a study and to consider a study laboratory and animal facility with experience in reproductive and developmental toxicology. Two examples on how and when to perform a study are instructive. For the first example, the biopharmaceutical is a therapeutic protein, intended for long-term, monthly, intravenous dosing at 100 mg/dose. Because the molecule binds receptors of white blood cells, is able to cross the placenta and is indicated for use by women during childbearing years, reproductive and developmental toxicology testing is considered advisable, probably even necessary, prior to Phase 1 clinical studies and certainly before Phase 2. The second example is a recombinant protein of a virus, a vaccine intended for the general population to include women of childbearing age and children above the age of two years. It is given intramuscularly at 5 μg/dose and in three total doses. The sponsor intends to add a label warning stating that the vaccine not be taken if a woman might conceive in the near future or is already pregnant. This vaccine might be tested for developmental toxicology in young animals prior to clinical studies in children; but for the adult population it might never need reproductive or developmental toxicology. For either product, ADME studies would be helpful in making a decision because they would demonstrate distribution of the product after injection. These examples point out the need to consider all aspects of a product, notably pharmacology and the intended treatment population, prior to designing a toxicology protocol.

Carcinogenicity Testing

Biopharmaceuticals are not commonly thought of as carcinogens but, in theory and sometimes in practice, one is found to be associated with cancer. Carcinogenicity testing is a long and expensive process, much like chronic toxicity testing. Guidelines concerning when and how to perform carcinogenicity testing on a particular biopharmaceutical product are available and there is precedence for most classes of product. Mice and rats are used almost exclusively and strains of each species must be selected on the basis of many factors: longevity, spontaneous tumors, capacity to develop tumors in response to known carcinogens and tolerability of the biopharmaceutical to be tested. Design issues, such as route of administration, doses, dosing regimens and termination, are complex and the sponsor considering carcinogenicity testing is well advised to seek expert advice and regulatory guidance before embarking on a study design. The field is rife with pitfalls, complications, uncertainties, controversies and changes in recommended practices. Interpretation of results presents another opportunity to seek expert opinion, especially if controls were limited in scope or number and if criteria for carcinogenicity were not well considered in the design and protocol. While complex and difficult, carcinogenicity studies are simply necessary for some products, sometimes prior to mid- or late-phase clinical studies.

Immunotoxicology

This relatively new field evolved out of observations and studies on the toxic effects of chemicals on the immune system. With a large number of biopharmaceuticals targeted, directly or indirectly, to the immune system and with other biologicals likely to interface with immune cells and tissues at some point in their distribution throughout the body, immunotoxicological studies deserve consideration for many classes of product. Adding to the situation is the complexity of the immune system, notably the fact that scientists do not yet understand the intricacies and control mechanisms of this system. Major immunotoxicological concerns include:

1. Adverse allergic responses to the biopharmaceutical itself because the product itself or an excipient is perceived as "foreign." This is manifest as immediate or delayed hypersensitivity reactions, some of which can be immediately life threatening.
2. Immune responses to the biopharmaceutical that neutralize the molecule and make it ineffective. This is not uncommon with products

that are taken over a long period of time. Further, when biopharmaceuticals in solution change format, such as going from soluble to microparticulate while in storage, the propensity to elicit both allergic and neutralizing immune responses may increase significantly.

3. Up-regulation of the immune response or an inappropriate immune response resulting in immunity to "self" and thus leading to autoimmune disease. Immediate up-regulation caused by biopharmaceuticals that act to release, immediately and in large amounts, cytokines or other mediators of inflammation are of special concern. Use of recombinant cytokines is especially suspect in this regard.
4. Down-regulation or suppression of the recipient's immune response. This is not uncommon in patients with preexisting conditions, such as cancer or immunodeficiency.

Immunotoxicology testing is highly recommended for certain biopharmaceuticals or for any product derived from and possibly containing molecules from certain sources. To name just a few, cytokines or cytokine-like molecules, vaccines and vaccine adjuvants, monoclonal antibodies or immunoglobulin-like molecules, allergens, products mimicking or derived from microbes that themselves stimulate untoward responses, products derived from certain plants or those mimicking plant allergens (latex, peanut), or molecules that bind to cells or receptors on cells comprising the reticuloendothelial system. Given the complete list, it becomes clear that immunotoxicology is increasingly considered in design of a safety testing program, and there are no simple templates for routine testing of a molecule as there are for acute toxicology studies of certain products. Studies are best designed based on an understanding of the immunological properties, or potential, of the molecule. From this knowledge, it is possible to consider how relevant tests might be performed, *in vitro* or *in vivo*. An efficient approach is to piggy-back, whenever possible, immunotoxicology studies with acute and subchronic toxicology studies, adding immunological measurements to the protocol. However, these measurements always require the application of immunological or cytochemical assays, some of which are expensive, time-consuming and technically challenging. Also, given the species specificity of cells and molecules involved in the immune response, it may be difficult to draw valid conclusions no matter how well designed the study or how compelling the data. For example, a humanized monoclonal antibody tested in an otherwise appropriate rat model might be, indeed is, expected to be highly immunogenic in that species. And, a recombinant vaccine antigen that is immunogenic and could lead to hypersensitivity reactions in man might not be immunogenic or allergenic in rabbit, rat or mouse. The possibilities are endless and suggest that immunotoxicology testing must be carefully considered in concept and experimental design, that pilot studies are desirable once a model has been selected and that results be expertly interpreted before conclusions are made.

Genetic Toxicology

Another relatively new subspecialty, genetic toxicology, studies the effects of chemical, biological or physical agents on nucleic acids, genes and chromosomes. Biopharmaceuticals can profoundly affect genetic material, but the mode of action is usually quite different from that of small molecule drugs or ionizing radiation, insults that result in chemical changes to nucleic acids, producing mutations, chromosomal breakage or abnormalities in controls. Cancer chemotherapeutic agents are examples. In contrast, certain biological products, notably genetic therapies and products containing DNA as the active ingredient, are designed to alter the genome through biological processes. They may deliver therapeutic DNA to the nucleus and even insert that DNA into the genome making them suspect of causing genetic toxicity, but by unique mechanisms. One example of a biopharmaceutical with the potential to cause genetic toxicity is a DNA molecule intended to repair or replace a gene within a target tissue, such as the bone marrow. The molecular delivery system enhances the chance that foreign DNA in the product will enter a host cell nucleus and integrate into the host's genome, thus enhancing its therapeutic potential. With this product, it is possible the DNA will be delivered to the wrong tissue or cell, perhaps a gonad, enter the nucleus and insert into cellular DNA of gonadal cells, perhaps even sperm or eggs. It might then be expressed in an uncontrolled or inappropriate manner or it might even be passed to the next generation. Much genetic toxicity testing required of biopharmaceuticals focuses on such possibilities, really just errors in well-intended gene delivery.

Genetic toxicity of a biopharmaceutical is typically studied both *in vitro* and *in vivo*, if a suitable animal model is available. An *in vitro* protocol might use mammalian cells in culture to determine the frequency at which inappropriate insertion or expression occurs. More definitive animal studies are designed to inject biopharmaceutical into an animal and, using sensitive nucleic acid probes, measure in various types of tissues or cells the nucleic acid that has been introduced. Additional studies of cells or tissues can determine if the therapeutic nucleic acid is actually inserted into the genome of living cells. Returning to the example, a therapeutic plasmid DNA, intended to deliver a missing gene to myeloid cells in the bone marrow, could be injected into mouse bone marrow and then located and identified with nucleic acid probes. Locating the gene in bone marrow or blood cells three days postinjection would be an expected finding and considered a desirable event. However, discovery of this DNA in testicular or ovarian tissue of the same mouse would be cause for concern because the injected product might have entered the nucleus and even inserted into the genomic

DNA of germ cells. While most biopharmaceuticals have little potential for influencing genome or altering DNA or RNA, some products have significant potential. Here again it is advisable for the sponsor to review regulatory guidance, identify and study precedence for its class of product and seek expert advice because genetic toxicology studies are long, arduous and expensive and failure to conduct them when required can lead to significant delays in development.

Tissue Binding or Local Tissue Tolerance

Biopharmaceuticals are sometimes given in large amounts to a single site on the body; most are injected. For example, a monoclonal antibody may be periodically injected subcutaneously in doses, each over 100 mg. A number of untoward reactions can result and nonclinical study designs consider how local reactions are detected at the site of injection or deposition in an animal model. The mechanism of action can be quite different for each biopharmaceutical and tissue. Local immune and inflammatory reactions can result, especially after multiple doses, and these may be chronic or acute. Cells or vaccine antigens can, by product design, remain at the site of injection and cause problems that are not anticipated and not immunologic, such as proliferation of fibrous or adipose tissue. A common method for studying tissue binding or local tissue tolerance is to add measurements of local reactivity to already designated acute, subchronic and chronic toxicology protocols. In one example, tissue samples are taken periodically by biopsy and again at the time of sacrifice and studied for signs of local toxicity.

Inappropriate cell or tissue binding by a product could result in damage to tissue or even lead to problematic reactions or disease. What can be done to ensure that a biopharmaceutical, designed to bind a particular receptor or cellular molecule, will bind only to the intended target and not to innocent bystander cells or tissues? Because many therapeutic biopharmaceuticals are developed for the purpose of binding to a certain cell surface molecule, and monoclonal antibodies directed against a number of proteins are the perfect example, this is a common concern. Toxicology studies to determine tissue binding patterns are called for whenever a product could inappropriately bind to normal cells. These studies are performed using immunohistochemistry and other methods that clearly demonstrate tissue or cell binding or the lack thereof. Substrate is various human tissues (cadaver material). Other approaches may be applied for certain types of molecules. Because these studies are somewhat artificial, i.e., performed *in vitro* and not in a living organism, the significance of cross-reactive binding study results may be unclear. Then, it can become necessary to perform additional experimentation in live animals. As noted earlier, methods are now available to track biopharmaceuticals in a living animal.

Quality of Nonclinical Studies: Current Good Laboratory Practices

Consumers and the government recognized in the 1970s a need for quality in preclinical testing of drugs. The response was the institution of a quality system by the FDA known as current Good Laboratory Practices (cGLPs). This set of regulations, outlined in Chapter 4, is applied to all safety testing of biopharmaceuticals. For non-FDA-regulated products, such testing may be required by other government agencies responsible for licensing (see Chapter 4) products released into the environment or contacting humans. It is important to consider the scope of the GLP regulation, stated in 21 CFR 58 "for conducting nonclinical laboratory studies that support or are intended to support applications for research (clinical investigation) or marketing permits for products regulated by FDA." It then defines nonclinical laboratory studies as "*in vitro* or *in vivo* experiments in which test articles are studied prospectively in test systems under laboratory conditions to determine their safety." The definition goes on to exclude clinical studies and laboratory studies that are designed as exploratory or to determine potential utility or product characteristics. The scope of GLPs is significant in three important respects. First, it does not apply to research, early development or to clinical trials; second, it excludes quality control of product (which falls under cGMP); and, third, it does apply to all laboratory studies, including animal studies, in which a claim is made for product safety. Thus, GLPs apply to most of the work we discuss in this chapter with the possible exception (if safety claims are not made on the results) of pharmacology studies. At the heart of GLP regulations are requirements for (1) a quality assurance unit (see Chapter 5) with broad authority to review and approve or disapprove just about anything and everything; (2) the requirement for a study protocol; and (3) the need for a study director. Also, cGLPs are comprehensive, with every other aspect of laboratory operations and procedures, from facility standards to elements of animal feed, found in these regulations. Elements of cGLP are listed in Chapter 4. Finally, many of the terms used throughout this chapter to guide the scientific, management and administrative aspects of a nonclinical study, including this term itself, were either introduced or institutionalized by cGLPs on their introduction in 1976.

Summary of Nonclinical Studies

This chapter reviewed methods used to assess the risk and benefit for a candidate biotechnology product, with emphasis on biopharmaceuticals, as performed in nonclinical laboratory and animal studies. These studies begin once the nature of the biological construct or molecule, its purity and potency, following early production have been clarified. Absorption, distribution, elimination and metabolism studies explain the pharmacokinetics of a product and pharmacodynamic studies describe how the product interacts with cells and tissues. A number of safety tests measure the toxicity of the product at predetermined doses and dosing regimens intended to match or exceed those to be used in man. These measure acute, subchronic or chronic effects of the product as well as screening for specific types of toxicities. Nonclinical safety testing is performed under a quality system: cGLP.

9

Clinical Trials

Introduction to Clinical Trials

A clinical trial, also referred to as clinical study or clinical research, is the overall process of evaluating the safety and efficacy of a medical product or intervention in humans. Importantly, it is investigational and the purpose of a clinical trial is to learn about a product and how it impacts humans. The intention is to treat a select group, not the general population of patients. Successful completion of clinical trials is required for U.S. Food and Drug Administration (FDA) market approval of drugs, biologics and some medical devices and, hence, clinical studies are used in the biotechnology industry to support market approval of biopharmaceuticals. However, the concepts developed for clinical studies incorporate scientific and design elements shared with field trials of other biotechnology products for which there is no testing in humans, such as field studies for environmental or agricultural products.

In the scheme of biopharmaceutical development, clinical research follows animal studies and other preclinical endeavors because a product is always tested for safety and performance in laboratory studies before it is used in man. Research in human volunteers is divided into several phases of clinical development (Figure 9.1), with each subsequent phase becoming increasingly large or complex and showing a continual shift in focus from measuring product safety to measuring both efficacy and safety. It is a long, complex and expensive process based on the scientific method.

Clinical trials are an important aspect of biotechnology development because a large number of firms develop medical products—drugs, biologicals or medical devices—and because, prior to commercialization, each medical product must be extensively tested in humans. Some biotechnology firms plan to take their product through all phases of clinical development to market approval. Others have a different business strategy and plan only to evaluate the product in early phases of clinical development until there is added value through proof of product safety in man. Only a few biopharmaceutical firms choose to forego clinical trials altogether and exit from product development prior to Phase 1. Hence, a biotechnology firm with a candidate

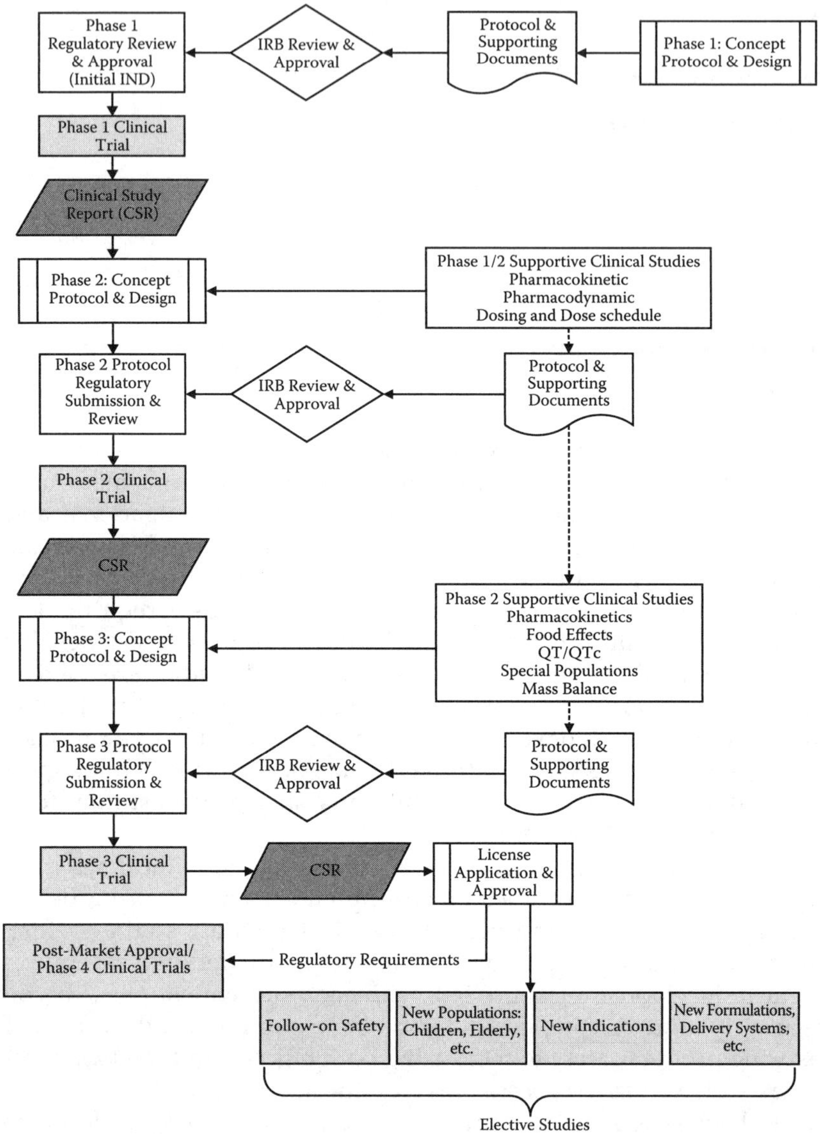

FIGURE 9.1
A typical scheme of clinical trials in biopharmaceutical development.

biopharmaceutical product may wish to add value to that product through the process of clinical research. Thus, many biotechnology firms have or will sponsor or conduct a clinical study. Despite this desire to increase product value by performing a clinical study, small biotechnology firms often have little or no experience in this endeavor. In addition to lacking experience, many biotechnology firms have a culture of intellectual liveliness that stands in contrast to the rather serious tone of clinical development, where there is an emphasis on scientific proof beyond reasonable doubt and thorough and sometimes detailed or rigid quality procedures based in government regulations and guidelines. The result is that, to be successful in clinical research, many biotechnology firms find that they must change, to some degree, their culture and means of doing business.

This chapter attempts to explain the basic elements of clinical research as they apply to development of medical products. It begins with a brief history of the field, provides an overview to introduce concepts and terms used in clinical research, and progresses to a section on clinical planning. Further information is provided on the design and conduct of clinical studies with extensive discussions concerning the people and institutions involved in a typical trial. A section on clinical trial operations discusses each phase of clinical development. Another section covers quality systems for clinical trials, current Good Clinical Practices (cGCP). The chapter ends with a discussion of ethical behavior and the importance of ensuring the well-being of human subjects enrolled in clinical research.

Background of Clinical Research

Introduction

Historically, clinical research has been considered as either observational or experimental. In observational studies, the investigator has no control over the study conditions, and investigational drugs and placebo are not given to the subjects. These studies may be referred to as epidemiological because they measure current conditions absent from novel medical interventions. While observational studies may form a foundation for conducting an experimental clinical study or for following safety of a biopharmaceutical postlicensure, they are not used to study safety and efficacy of a particular biopharmaceutical during the development process. For experimental clinical studies, also referred to as *controlled studies*, the investigator designs the conditions for the study and follows that design in an exact manner. Clinical studies of biotechnology products performed for the purpose of market application, i.e., premarketing studies, are always controlled clinical studies. Most controlled clinical trials compare two or more treatment modalities. The specific treatment schedule with the test product is predetermined and all other treatments to or medical conditions of the subjects are managed as similarly as possible. The design of a clinical study for a biopharmaceutical may compare the test product with a marketed product used to treat the same indication (a "comparator") or, if no appropriate comparator is available, to a placebo (a "sugar pill").

Historical Information on Clinical Trials

Comparative drug studies were first reported in the eighteenth century, but used infrequently until the twentieth century when, as noted in Chapter 3, laws and regulations were developed to ensure that drugs and medical devices were adequately tested for safety and effectiveness. Indeed, these regulations, in part, have influenced clinical trial designs and are based on good scientific research practices that had been used in laboratories. Notable is the need to test an hypothesis in a formal manner. Another process, randomization or the "blinding processes," i.e., assigning a patient to receive one or another drug, was first used in laboratory and agricultural field research and then was adopted as good scientific practice for clinical trials. Monitoring and auditing, now standard practice for clinical trials, was earlier used in nonclinical studies and found to be an effective means of ensuring the quality of data. Indeed, as drug development became more complex and clinical studies ever larger, many scientific and quality practices were applied to clinical trials and these are now the norm for biopharmaceutical clinical research.

In the 1970s, the quality and ethical aspects of clinical trials came to be known as current Good Clinical Practices (cGCPs) and, by the 1990s, cGCPs had gained international acceptance, worldwide, as the quality system for clinical trials. During this period, the number of clinical trials grew dramatically because more drugs, medical devices and biotechnology products entered clinical development because more countries required studies be performed on their own soil, and because regulatory agencies demanded greater numbers of ever larger studies for each product. Further, clinical investigators and statisticians devised better clinical study designs, more effective means of operating a study and improved means of analyzing data. Today, many thousands of clinical studies are underway each day in dozens of nations and throughout the world.

Clinical research plays an increasingly important role in development of biopharmaceutical products and, additionally, in their postmarket approval evaluation. Each clinical trial is also becoming increasingly complicated, expensive and, often, publically announced. Indeed, results of one or another clinical trial are related almost daily in newspapers and on television. The increased complexity and demands of clinical research have led to more regulatory requirements and an increased need for documentation. Indeed, some would argue that expansion of clinical research could someday outgrow the ability of the clinical research community to provide infrastructure for all ongoing trials. Others disagree and feel that careful planning and exact execution of clinical development make for a successful clinical study, allowing a sponsor to determine whether or not a product is safe and efficacious in man and, therefore, fit to be marketed for the intended purpose. Generalizations in either case are not correct. Well-designed and executed clinical studies are and always will be required to demonstrate the safety and effectiveness profile of any biopharmaceutical. Clinical research requires careful planning to ensure that a valid study is completed and to fully meet the rights and interests of each volunteer enrolled in any study.

To provide an overview of clinical trials, this chapter examines the structure of clinical research principles and activities as they are applied to development of medical products in the biotechnology industry. We focus on key issues one must consider for successful clinical development of any product: organization, planning, personnel, operations and processes, documentation, quality, ethics and resources.

Organization of Clinical Research

Phases of Clinical Trials

Clinical development for biopharmaceutical products is divided into four distinct and sequential phases, these proceeding from the most controlled conditions and environment (Phase I and Phase II) to a scenario that is closer to the "real world" (Phases III and IV) in which the product is used to treat patients. Other terms, including the International Conference for Harmonization (ICH) nomenclature, also are given.

Phase I. *Early Phase. Clinical Safety and Toxicology. Clinical Pharmacology. Human Pharmacology.* The first administration of a new product to man at a fixed route and schedule is considered a Phase 1 trial. Typically, the four goals of a Phase 1 study are (1) to estimate the maximum tolerated or safe dose; (2) to determine if any organ systems are affected by the product; (3) to identify any toxicity related to the product and, if there is toxicity, to measure the extent, duration and reversibility; and (4) to observe any desirable activity of the product.

Phase II. *Expanded Dosing. Pharmacokinetics. Therapeutic Exploratory.* The concept and definition applied to Phase II has been somewhat misused and broadened by the biotechnology community. A textbook definition, however, suggests that this phase of clinical development explores different dosages of the product, compares the effects of doses to those of a placebo and, ultimately, determines which dose has the best safety and efficacy profile. Results of Phase II set the stage for design of the Phase III clinical trial.

Phase III. *Pivotal Clinical Trial. Therapeutic Confirmatory.* In this phase of clinical development, the selected dose is given to a much larger number of individuals, each representing the target patient population that was established in the product labeling. A Phase III trial is referred to by regulatory agencies as a "pivotal trial" because it is *the* clinical basis for marketing approval or disapproval decisions.

Phase IV. *Therapeutic Use.* Postmarketing studies aim to further study safety and efficacy or to extend the indication or use of the product.

The Science of Clinical Research

A clinical program is designed, just as is any scientific research project, and the design must be scientifically sound, testing a hypothesis or series of hypotheses, having clear objectives and identifying measurable outcomes. Quality in performance of clinical research also is important and proper collection of data is essential. Using other scientific tools, such as blinding of procedures and application of placebo or comparator, statistical analysis of data or objective interpretation of results are essential elements of clinical research. Unfortunately, upper management in biotechnology firms may forget that good clinical research is based on the scientific method and ethical guidelines and is not simply a business or entrepreneurial endeavor.

Clinical research is the definitive step in evaluating new biotechnology products for safety and for efficacy, i.e., the prevention, diagnosis or treatment of disease. Clinical trials, and the systems that support them, are complex endeavors and require collaboration among investigators, industry, academic institutions and government agencies. Adding to their complexity, clinical trials have undergone a number of changes in the past decade. Progress in biomedical sciences and biotechnology itself has accelerated and increased the need for clinical studies and created new opportunities to improve clinical trial processes. Advances in informatics, laboratory and clinical diagnosis, and data management have led to new ways to evaluate human subjects and to report information and data. However, a new biopharmaceutical product can only be brought to market by a firm if the product is first shown to be safe and effective. Stated another way, clinical development is outcome-driven and the outcomes are issue-focused and, ultimately, based in meaningful scientific data, this based on a relevant hypothesis and clinical study design.

Quality in Clinical Research and Current Good Clinical Practices

The current Good Clinical Practice (cGCP) is an ethical and scientific quality standard and a quality system applied to designing, conducting, recording and reporting clinical trials. While cGCP has taken on a definition of regulatory compliance, it is actually more than that, representing an established means of conducting a clinical trial. There are national and international standards for cGCP, but the ICH standard is the most commonly accepted version and the one followed today by most biotechnology firms and in most nations. Quality and cGCP in clinical trials are discussed later in this chapter and elements are included throughout the text.

Clinical Development Planning

The key to successful clinical research programs in this ever-changing and fast-paced environment is effective clinical development planning. Planning must focus on the clinical claim or claims and, as noted in Chapter 1, this is best stated in the Targeted Product Profile (TPP). The primary purpose of the label claim is to inform prescribers and patients about the documented benefits and risks of a product. Clinical outcomes, derived from clinical trials, provide the basis for label claims. Further requirements for clinical planning are based in FDA regulations and guidelines. For example, a major requirement for an Investigational New Drug (IND) application is the Investigational Plan, a section of the document that outlines the sponsor's intended clinical research program. The TPP for a biopharmaceutical states medical objectives for the product, including the indication and therapeutic and safety profiles.

Using a well-conceived and written TPP with a list of desirable outcomes, a clinical development plan can be written as a predecessor to the Product Development Plan (PDP), both described in Chapter 1. Note that the clinical development plan is not a stand-alone document, it is instead an important, but integral part of and is woven into the overall PDP. Also, the clinical development plan is a living document and, like the overall PDP, can be changed at any time as long as the changes are coordinated with other aspects of the PDP and with members of the product development team.

The clinical development plan is a written document that describes how a new biotechnology product can be progressed, in an orderly and timely manner, from first administration in man through postlicensure studies. It explains to colleagues, management and regulatory agencies the proposed product's clinical development scheme and also provides critical information to the development team, such as the rationale, timeframe of Go/No Go decision points, costs (both internal and external) and an outline of proposed clinical studies. A well-constructed clinical development plan also addresses, in a "concept protocol," important scientific, medical and operational issues or factors, as listed in Chapter 1. Thus, the clinical development plan brings together all elements, scientific, management and operational, into a cohesive document.

Infrastructure for a Clinical Trial: Individuals, Documents and Investigational Product

Earlier we noted that experimental clinical studies test a hypothesis, have a written design (or protocol), use scientific research methods and are carefully controlled endeavors employing human volunteers and professional

BOX 9.1 CLINICAL TRIAL: INDIVIDUALS AND RESPONSIBILITIES

Volunteers, Patients, Human Subjects
- Sponsor and Staff
  - Medical Director
  - Safety Monitor
  - Auditor
  - Medical Writer
  - Clinical Project Manager
  - Regulatory Staff
  - Manufacturing and Clinical Trial Materials
- Principal Investigator and Staff
  - Subinvestigator(s)
  - Nursing Staff
  - Recruiter
- Statistician
- Institutional Review Board
  - Board Chair and Board Members
  - Administrative Staff
  - Quality Assurance Unit

staff. The planning aspect is referred to as clinical study design. The elements of a study design and the individuals involved in a clinical study and documents that support and control a study are discussed in this section. The positions of clinical study staff and their responsibilities are given in Box 9.1. A list of clinical study documents and the primary purpose of each are given in Box 9.2.

Design of Clinical Trials and the Clinical Protocol

Once a clinical plan has been completed, the clinical professionals at a biotechnology firm now focus on each clinical study identified in that plan. The basic elements of a study are outlined in a document, usually not exceeding 10 pages, referred to as the Concept Protocol. This is really the scientific basis for a study design and represents the initial proposal of experienced professionals, such as a medical director, a clinical project manager and various investigators. Elements of a concept clinical protocol are listed in Box 9.3. The hypothesis being tested is paramount, but other matters are also critical. Objectives are keys to success. Study size, patient population and indication also are key to experimental design. As noted below in greater detail, Phase 1 studies are smaller and focus on safety, while Phase III trials are typically large, multicenter endeavors that evaluate both the efficacy and the safety

BOX 9.2 CLINICAL TRIAL: DOCUMENTS

Concept Protocol. A brief design of a clinical study used as the basis for discussions between sponsor, investigator and regulatory authorities and is the foundation for preparing the full protocol.

Clinical Protocol or Protocol. An instructive document that identifies exactly why and how a clinical study will be performed and provides schedules of events.

Informed Consent Form (IC, CF). This document explains to a volunteer the potential risks and benefits of a clinical study. To enroll in a study, a volunteer must understand and sign the CF.

Investigator's Brochure (IB). An informative document that identifies for each member of the investigative staff information on the clinical study, the product being tested in the study and possible risks and benefits to volunteers enrolled in the study.

FDA Form 1572. Statement of investigator captures information on the investigative teams and is an agreement by the investigator to follow the protocol and regulations regarding clinical studies.

Curricula vitae. Résumés of the principal investigator and key investigational staff.

Clinical Trials Agreements. This agreement between a sponsor and an investigator or clinical CRO outlines responsibilities of each party to perform a clinical study.

FDA Form 3674. This represents certification by a sponsor to disclose clinical trials information to www.clinicaltrials.gov, according to U.S. law.

FDA Forms 3454 and 5455. Financial disclosure. In these documents, the sponsor discloses to the FDA financial arrangements that exist between clinical investigators and the sponsor.

Case Report Forms. Paper or electronic forms upon which the investigator enters medical information gathered during a clinical trial.

Patient Diary. Forms completed by subjects during the outpatient phase of a clinical trial to capture data on possible adverse events and general medical condition of the individual.

Operations or Administrative Manual. This collection of administrative and management information and instructions guides performance of the clinical trial and supplements the protocol.

BOX 9.3 ELEMENTS OF A CONCEPT CLINICAL PROTOCOL

Description of the biopharmaceutical investigational product
Previous use in man or animals
Stated indication
Protocol title
Study phase
Intervention regimens
Study objectives
Study hypothesis
Subject population, general characteristics
Major inclusion and exclusion criteria
Study design and schedule or duration; number of subjects and groups.
Study site(s)
Study schema
Study endpoints: safety and direct or surrogate efficacy
Study procedures and methods (primary, in general)
Assessments
Stopping rules
Unique scientific, ethical or medical aspects

of a product. Once the design has been drafted, the firm must then consider the management and operational elements that will support the study. This evaluation sometimes reveals the design to be overly ambitious and demands that the clinical protocol be revised to make it more feasible, from an operational point of view. Or it might suggest that the number of subjects in each group is too low if a meaningful conclusion is to be drawn. It is especially important to ensure that the nonclinical safety study plans cover the recommended dose and dosing schedule. Normally, the biotechnology firm asks experts, such as statisticians, toxicology scientists, experienced clinical investigators and, often, the FDA, to review and comment on the design contained in a Concept Protocol. Thus, the overall clinical objective and the objectives for each phase of development, both provided in the clinical plan, drive the full clinical study design, as outlined in the protocol.

The document that ultimately describes in detail the clinical study design is referred to as the Clinical Protocol and this is written by clinical study staff once a Concept Protocol is acceptable. The most important step in any clinical study is to prepare a complete, well organized and scientifically sound protocol. Protocols can be changed, or amended, but amendments take time and cost money. Therefore, to avoid delays, the protocol and other clinical documents consider and provide for every eventuality likely to occur once the study begins. Responsibility for writing a protocol may rest with the sponsor or the principal investigator. The sponsor is a representative of the biotechnology firm, while the (physician) investigator is the person responsible for conducting the study in accordance with the protocol. Thus, an investigator is retained by the sponsor. Today, national and international guidelines provide a standard organization, a template, for the Clinical Protocol.

The elements of a protocol are listed in Box 9.4. A heading sheet gives a fully descriptive title, names the investigator and their institute or employer (affiliation) and identifies the sponsor. Most institutions give a unique number to each protocol. The second sheet, a signature page, provides the names and contact information for everyone responsible for the protocol and, under a statement of compliance, prompts for the signatures of both the principal investigator and the sponsor. A summary of the protocol is typically provided in the next section, and this begins with a statement of the objectives

BOX 9.4 ELEMENTS OF A CLINICAL PROTOCOL

General Information (Title, numbers, names of investigator and sponsor)
Background Information. Description of the product and how and when to administer
Trial Objectives, with Purpose and Stated Hypothesis
Trial Design. Scientific design and factors that ensure or enhance the design
Selection and Withdrawal of Subjects. Inclusion and exclusion criteria; withdrawal of subjects
Treatment of Subjects. Administering medications, monitoring subjects
Assessment of Efficacy. Efficacy measurements and endpoints
Assessment of Safety. Safety measurements and endpoints
Statistics. Data sets and statistical analyses
Access to Source Documents or Data
Quality Control or Quality Assurance. All aspects of compliance and quality
Ethics. Ethical standards for the study
Data Handling and Record Keeping. Management of data during and after the trial
Finance and Insurance. Responsibilities for payments and liability insurance
Publication Policy. Anticipated publication and authorship policies
Appendices. For example, treatment and test charts and schedules; standard medical guidelines; publication policy; references

and, in most protocols, the formal hypothesis to be tested, and is followed by a brief summary of information on the product under investigation.

The study design is then described in some detail because it is the scientific heart of clinical research. Because most studies compare the treatment under investigation to another treatment (comparator) or to no active treatment at all (placebo), the design describes how the comparison will be made in the study design. It includes a description of the dose or doses of product, placebo or comparator; duration and intervals for giving doses (dosing); and a description of dosage forms. Typically, patients are divided into groups, or *cohorts*, and members of each group are given one or another treatments or doses. Further design criteria may include use of "randomization," whereby patients are randomly assigned to one or another group or "blinding," the process of keeping the exact treatment for each patient hidden from the subjects, principal investigator, study staff and sponsor. These elements of an experimental design prevent bias from entering into a study and are absolute requirements for late stage studies performed in the United States. Bias, a predisposition to a particular outcome or a prejudice, and any element of design in a protocol that might lead to bias, is carefully avoided in all clinical studies. Other methods may be applied to a clinical study design to avoid bias, improve study performance and validity or ensure safety and well-being of human volunteers. Stopping rules are descriptions of how a study will be halted, temporarily or permanently, should a certain type or series of adverse events be noted during the study. They provide a means of enhancing study safety and well-being of the human participants. Stopping rules are explicit and fully described in the protocol and, once approved by the FDA, they serve as an important agreement between the principal investigator, the sponsor and their regulatory agency.

Other important aspects of a clinical study design are rules for selection and enrollment of patients, i.e., "inclusion" or "exclusion criteria," and for withdrawal of the subjects. It is important to carefully select subjects, enrolling only those who meet stringent criteria. In many Phase 1 studies, the investigator wishes to enroll only healthy or "normal subjects," those best suited for studying a product upon first introduction to man. In many other studies, the subjects represent the actual patient population, as described in the TPP, that the biopharmaceutical is intended to treat. In either situation, it is important to include a specific type of individual and to exclude from the study those who have other medical conditions or disease that could put them at risk of undesirable reactions to the product. Hence, inclusion and exclusion criteria are written into a protocol. Inclusion criteria identify attributes that the patients must have to enter, or "enroll" in, the study while exclusion criteria identify issues that make a potential subject ineligible to enroll or participate in the study. For example, if one were to study a biotechnology product that was intended to lower blood pressure in otherwise healthy individuals, hypertension (high blood pressure) would be an inclusion criterion, while severe or advanced cardiac disease might be an exclusion criterion. A list of inclusion and exclusion criteria that might be applied to a Phase 1 clinical study in which normal, healthy individuals would be enrolled is given in Box 9.5.

A number of other rules, e.g., how to replace with a new subject those who withdraw from a study, might be given in the design section of a protocol. Sometimes clinical studies must be terminated or particular subjects must withdraw, either voluntarily or at the principal investigator's request. The protocol describes in the design section how this is to be decided and carried out.

The treatment of subjects with the test product is normally described in great detail by the protocol. To prevent bias, product is administered to each subject in exactly the same manner, at a prescribed amount, and on an established schedule.

Assessment of safety and efficacy are essential to the success of a clinical study and a protocol explicitly describes how each is measured. Outcomes, broad results or visible effects that form the basis for the study hypothesis, are described in medical terms. An example of an outcome in the case of a biopharmaceutical product intended for treatment of lung cancer might be to remain free of tumor for one year. Second, one or more "endpoints" are clearly stated. Endpoint is the term used to identify a measurable parameter, again exactly medically defined. Endpoints must reflect the objectives and the disease that is being treated, prevented or diagnosed. In the lung cancer example, an endpoint might be tumor mass found in the lung. Third, to adequately evaluate endpoints, measurement, the act of determining an amount or quantifiable dimension for that endpoint, is necessary. In the lung cancer example, the measurement of tumor mass, number, location and size of each, by a radiological method, once each month, might constitute a valid measurement. Safety endpoints also are measured. For example, to determine if patients became allergic to a biopharmaceutical product, the protocol might direct the investigator to carefully search for rashes following each treatment. The success of a clinical study rests upon establishing in the protocol meaningful and exact outcome, endpoints and measurements. Hence, medical experts are often used to advise this phase of protocol development and, in later phases of clinical development, these issues lead to important discussions with regulatory authorities.

Safety, as well as efficacy, endpoints are described in the protocol. Should they occur in a subject, they are recorded as adverse events (AEs) if they are mild or limited in scope and severity. If they are severe, such as anaphylactic shock, then they are considered serious adverse events (SAEs). When faced with an AE or SAE, the investigator or another physician must determine if the reaction is, or might be, related to the investigational product. Indeed, each protocol, no matter the phase, states a large number of safety measurements, such as clinical laboratory tests and physical examination, that must be taken for each subject to determine the safety and tolerability of the test product. Other measurements focus on the efficacy of the product being tested.

BOX 9.5 EXAMPLES OF INCLUSION AND EXCLUSION CRITERIA

Inclusion Criteria:

Age 18 to 50 years

Gender: Male or nonpregnant female

Good general health as demonstrated by medical history, baseline laboratory tests (urinalysis, clinical chemistry and hematology) and physical examination

Laboratory values within 1.25 times institutional stated normal values

Negative test results for HIV-1, Hepatitis-A, -B and –C

Low risk of coronary heart disease based on NHANES-1 cardiovascular risk assessment and screening electrocardiogram

Negative tests for autoimmune diseases, rheumatoid arthritis and ANA

Reliable access to the clinical test facility and availability to participate for the duration of the study

Assessment of Understanding questionnaire competed prior to enrollment and demonstration of understanding of risks and benefits associated with study participation

Ability and willingness to provide informed consent

If the participant is female and of reproductive potential:

- Have a negative serum or urine beta human chorionic gonadotropin pregnancy test performed within three days prior to study initiation
- Agree to consistently use effective contraception from 21 days prior to study initiation and for the duration of the study

Exclusion Criteria:

Planned travel outside of state during study period

Prior receipt of similar biopharmaceutical

History of confirmed diagnosis of (disease or condition) within the past two years

Use of (specific drugs) within five months of enrollment or use of (specific drugs) within two months of enrollment

Recent (within two weeks) use of (specific drugs) with (specific effects or drug indications)

Anticipated use of medications known to interact with (investigational class of biopharmaceutical)

Use of any investigational or nonregistered drug or vaccine or whole blood or blood product within 90 days of enrollment

Systemic immunosuppressive medications or cancer chemotherapeutic compounds use within past 90 days
Current or past diagnosis of Type I or Type II diabetes.
History of severe allergic reactions
Screening laboratory abnormalities beyond the limits defined in the inclusion criteria
Clinically significant medical condition, physical examination findings, other clinically significant abnormal laboratory results or past medical history that may have clinically significant implications for current health status in the opinion of the investigator
Any contraindication to phlebotomy
Body Mass Index <19% or >30%
Acute illness at time of enrollment
Pregnant or lactating female or female who intends to become pregnant during the study period
Serologic positivity for Hepatitis B or C or HIV-1
Psychiatric condition that precludes compliance with the protocol, including ongoing risk for suicide or psychosis
Suspected or known current alcohol abuse or recreational intravenous drug use within the past 12 months
Acute illness at the time of enrollment
Any other condition that, in the judgment of the investigator, would interfere with or serve as a contraindication to protocol adherence, assessment of safety or reactogenicity or a participant's inability to give informed consent or increase the risk of having an adverse experience to the study drug

The statistical section of the protocol describes analyses that will be used to make comparisons of endpoints in the overall subject population and must be described in detail. It also applies statistical principles to support the design of the study. For example, the statistical discussion provides rationale for the number of subjects enrolled in each group, treatment or placebo, of a Phase 3 clinical trial. Another section of the protocol deals with administrative issues, such as control of the test product, review and approval of study documents, methods for collecting and recording raw data, and details such as insurance or publication policy.

Most clinical trials enroll total subjects numbering in the dozens (Phase 1), hundreds (Phase 2 or Phase 3) or low thousands (Phase 3). Yet, other clinical studies of biotechnology products, notably pivotal Phase III trials, are quite large and conducted simultaneously by many investigators at several sites. Today, international trials may enroll in excess of 60,000 subjects at over 100 sites in more than 20 countries. Such big trials demand much administrative support and instruction. For these purposes, a *Clinical Operations Manual* is used in addition to the protocol. The "ops manual" is an extension of the protocol, describing in greater detail all administrative aspects of the study and providing detailed medial and managerial instruction to the many staff involved in the clinical study. Operations manuals are good business practice because they further ensure success of a scientifically well-designed study by providing consistent procedures at each trial site.

Human Subjects, Patients and Volunteers

A clinical trial includes humans willing to participate and receive either product or, perhaps, placebo. Later in this chapter, we describe the rights of those individuals who volunteer to receive investigational products on behalf of the sponsor and the principal investigator and, hopefully, for the betterment of the health and well-being of all mankind. For this, we in biotechnology greatly appreciate their participation. Because every volunteer enrolls in a clinical study of their own free will, we refer to these individuals with the general term *volunteer.* For products that could provide some benefit and are used in individuals with a disease or medical condition for which the product is indicated, the volunteers are referred to as *patients.* In some studies, such as with preventive biopharmaceuticals (e.g., a vaccine) or where the studies enroll healthy individuals (e.g., Phase 1), the volunteers are referred to as *subjects.* We use the terms volunteer, subject or patient without further definition in this chapter.

The Sponsor

The sponsor is the ultimate backer of a clinical study and, as such, takes ultimate responsibility. Responsibilities of sponsors before, during and after a clinical study are clearly defined in regulatory guidelines. First and foremost is the responsibility for ensuring the rights and well-being of every human subject or patient and for maintaining quality, through quality assurance and quality control, of the trial. To meet this obligation, the sponsor of a clinical study has policies and procedures that demonstrate exact intentions. A sponsor may delegate responsibilities to another party, but this must be specific and in writing. Such is the case when an individual principal investigator is retained to perform the study or when a contract research organization (CRO) assumes various clinical trial functions and responsibilities. Smaller biotechnology firms often delegate most or all clinical trial functions to others, but they can never transfer the ultimate responsibilities of ensuring that a study is conducted, recorded and reported properly or that patients are always treated according to medical and ethical standards. A biotechnology firm engaged in clinical studies always has, on staff or retainer, clinical trial experts. Indeed, most sponsoring biotechnology firms retain internally the functions of clinical trial and data management, monitoring and auditing, ensuring the integrity of data and attending to financial and general administrative duties.

An important and yet often overlooked responsibility of the sponsor is selection of a qualified principal investigator and, along with the investigator, the institution or CRO at which the study is to be performed. This is often a difficult task for the biotechnology firm because, with both an exciting technology and adequate investment, the firm's management may be faced with several qualified investigators, each of whom wishes to perform the study. Some may be inexperienced or otherwise unqualified to head an important clinical study, but appear knowledgeable about the product. Others may have years of experience as investigators, but be inexperienced with this type of product. The sponsor must find a person who is both qualified scientifically and has the proper experience with the product type; finding or selecting the right principal investigator can be a challenge for the sponsor and is further complicated with multicenter trials where several qualified investigators must be identified.

Sponsors ensure that all the paperwork is completed during the trial. And, as one might expect, a great amount of paperwork, in fact, is generated before, during and after each clinical study, no matter how small the trial. The sponsor, not the principal investigator, communicates directly with regulatory agencies, notifying the investigator when patients may be enrolled. The sponsor confirms that review was completed and approval was given by the Institutional Review Board (IRB). Through the investigator's brochure (IB), the sponsor informs the investigator and his or her staff about the product to be tested. (Both the IB and IRB are described later.)

The sponsor also plays important roles in relating safety information in a timely manner. First, there must be a system in place to receive, from investigators, review and, if necessary, report any safety information that is generated during the study. AEs and SAEs are collected by the sponsor in a timely manner and SAEs are immediately investigated and promptly reported to those whose job it is to influence, make or review medical decisions: the investigator, the medial monitor, the IRB and regulatory authorities. The sponsor has in place a system of expert review for AEs and SAEs; this is typically the job of the medical monitor, a physician who examines each event and reports his/her opinion regarding the significance and the relationship between SAEs and the product to the sponsor. If, in the eyes of the investigator or the medical or safety monitor, AEs or SAEs are related to the test product, and certainly if the safety of the patients is at risk, then the sponsor is responsible for reporting the events to all investigative staff and regulatory authorities and, in some cases, for stopping the study. Termination of a study, meaning that product can no longer be given, is driven by detailed rules or study "stop criteria," these also are provided for in the protocol.

To ensure that the clinical study is being conducted properly, all aspects of a study must be audited or monitored, on behalf of the sponsor, by an experienced and knowledgeable individual. Clinical trial monitoring is not the same as the role ascribed to the medical (safety) monitor, and described above. In contrast to the medical safety monitor, who reviews AEs or SAEs provided to them, trial or study monitoring is a process involving visits, by a professional auditor, to the clinical study site at regular intervals to inspect the clinical study documents and medical records. This auditing or monitoring process further ensures, among other things, that the study is being conducted according to the protocol and within guidelines established to protect the rights and safety of the subjects or patients. The trial monitor reviews study records for completeness and accuracy and auditors interview the study staff to verify that everyone is qualified to perform their assigned roles in the clinical study. Deviations, variance and deficiencies are noted by a monitor to the sponsor who, in turn, is responsible for immediately correcting the issues or, alternatively, for stopping the study until corrections are made.

Another major responsibility of a sponsor is to prepare, update and disseminate, to the principal investigator and his/her staff, a summary of clinical, nonclinical and other pertinent product information. This is done with a document called the Investigator's Brochure, written by the sponsor prior to the first clinical trial and updated as new information becomes available. The elements of an IB are provided in Box 9.6.

An important issue that can arise with a clinical study is conflict of interest, real or perceived and usually financial in nature, on the part of either or both principal investigator and sponsor. To perform an unbiased study,

BOX 9.6 ELEMENTS OF AN INVESTIGATOR'S BROCHURE

- Summary
- Introduction
- Physical, chemical and pharmaceutical properties and formulation of the product
- Results of nonclinical (i.e., safety, pharmacology and toxicology) studies
- Effects (to include safety, pharmacology, pharmacokinetics) in humans known from previous clinical studies
- Marketing experience, if any
- Summary of data and guidance for the investigator
    - Anticipated risks and adverse reactions
    - Summary of clinical data
    - Assessment and treatment
    - Toxicity management
    - Additional risks associated with this or similar products
- References

it is important that the principal investigator not be beholden to the sponsor and it is critical that any investigator assigned to a study have no significant financial interest in the sponsoring biotechnology firm. Financial interest could result in bias on the part of investigational staff and even a perception of conflict of interest or potential bias in the sponsor–investigator relationship, particularly where it involves substantial sums of money, can doom a clinical trial. Of course, investigators are remunerated for their time, expenses and professional expertise. However, compensation must be fair and open and a clinical investigator should not have a substantial interest in corporate stock options or, if they do, such interest must be reported.

The Principal Investigator and His/Her Study Staff

The principal investigator is the individual responsible for conduct of a clinical study at each clinical trial site. She or he is retained by the sponsor who delegates specific clinical responsibilities to that principal investigator. In return, the principal investigator receives reimbursement for expenses, including salary for time spent executing the study under the agreed protocol. There may be other benefits to a principal investigator, such as publication of scientific articles and ability to work at the cutting edge of his/her profession. A principal investigator may be employed at an academic institution or in private practice or he/she may be at a CRO. Whichever, the agreement between sponsor and investigator typically includes funding for additional study staff, such as nurses, administrators, clerical assistants and individuals to recruit volunteers. In some cases, the principal investigator will be asked to prepare the protocol and other clinical documents, but larger sponsoring firms frequently provide these documents and ask the principal investigator to follow the instructional documents. A physician may, in the United States, be both the sponsor and investigator, or, as happens with some biopharmaceutical firms, the sponsor may directly employ the investigator.

In effect, a principal investigator is responsible for everything that happens at the clinical trial site, to include activities by his/her staff. In the United States, a principal investigator formally accepts this responsibility in one of two ways: under a contract with the sponsor or, in an abbreviated manner, by signing an agreement with the FDA, Form 1572. Among varied investigator responsibilities, the most important is to exercise clinical oversight and medical judgment at the site. The principal investigator ensures that everyone on his/her investigational team conforms to the protocol and any other instructions (e.g., operations manual) concerning the study and that subjects' rights are fully met. Principal investigators are responsible for submitting the protocol to the IRB and then beginning the study only after receiving approval from that board. The investigator also is responsible for enrolling and then treating patients in the proper manner, for patient compliance in taking the investigational product throughout the study, for ensuring that clinical documents (such as case report forms) are correctly completed and for accountability of the investigational product. Certain administrative functions are required of the investigator as well: maintaining professional credentials, managing the research staff, communicating with the IRB, participating in study meetings or conference calls and maintaining good relations with the sponsor and other parties involved in the study. Last but not least, the principal investigator is the principal scientist in a clinical study. In the end, each of these responsibilities focuses on maintaining the safety of patients and the integrity of the data.

Clearly, a busy physician–investigator cannot complete a clinical trial without help and, so, investigational staff is employed at each study site. Principal investigators often enlist other physicians to work on a study and these are referred to as subinvestigators. Subinvestigators are qualified to serve in this capacity by education and training and, in many cases, they are professionals working closely with the principal investigator, such as medical residents or junior staff. While the principal investigator may delegate certain medical responsibilities to subinvestigators, they still accept full responsibilities in this regard. Study nurses, referred to as such because they are typically registered or licensed nurses, are used because of their medical training and experience and because a clinical trial involves medical procedures and measurements. Study nurses originate many study records; these documents are then reviewed and co-signed by the principal investigator. They educate the volunteers, ensure that informed consent is always properly administered, and take patient histories and consider AEs and SAEs. Patients or subjects do not just appear magically at the study site, but they must be recruited by someone adept at identifying potential volunteers and coordinating their initial visit. This team member is the volunteer recruiter. Administrative staff manage and organize records and files and assist recruiters and nurses. Investigational product is usually maintained and distributed by a clinical study pharmacist and, for some investigational products, this individual prepares medication according to the sponsor's instructions. Larger studies also employ data specialists, individuals dedicated to transferring data from paper to electronic databases and ensuring data integrity and accuracy.

Institutional Review Boards (IRB): Process of Informed Consent (IC) and IC Form

We as a society have, appropriately, given significant rights to individuals who volunteer for and participate in clinical trials. These rights derive from a very important document, the 1964 Declaration of Helsinki. The declaration itself is based on the Nuremburg Code of 1947. The code was drafted in response to horrific situations that occurred during the World War II, specifically when Nazi investigators conducted biomedical experiments on prisoners of Germany and without the consent of those individuals. The heart of the code is the requirement for full understanding of the risks by written consent from any human volunteer. This means that the person who is

BOX 9.7 ELEMENTS OF INFORMED CONSENT

Statements that the study involves an investigation and purposes for the research.
Description of risks or discomforts.
Description of possible benefits.
Disclosure of possible alternative treatments available to the subject.
Description of processes used to maintain confidentiality.
Explanation of potential compensation or medical treatments.
Individual to contact for answers to pertinent questions about the research or risks and benefits.
Statement that participation is voluntary and refusal or withdrawal will result in no penalty.

receiving any investigational treatment, no matter how minor, must have the legal capacity to give consent (or, in the case of children, have a responsible adult give consent), be so situated to exercise free power of choice without coercion and to have a clear understanding of the investigation, to include possible risks and benefits.

Under the 1962 Declaration of Helsinki, this guide to physicians and others for biomedical research involving human subjects, regulations for clinical research now state that legally effective Informed Consent (IC) (Box 9.7) must be obtained by an investigator before involving that subject in any clinical research. Conditions for IC have been established in most countries for usual conditions as well as for unusual situations, such as children, those with dementia and those in emergency situations, i.e., when the individual receiving the investigational product may not be capable of being fully informed. In the United States, Protection of Human Subjects is mandated by the Code of Federal Regulations, 21 CFR Part 50, a regulation with broad application.

In practice, IC is requested by the principal investigator from each subject immediately prior to enrolling that person into a clinical trial. Human subjects are asked to review a description of the clinical study, to include the design, potential benefits and possible risks. In some cases, such as novel investigational products, subjects are queried or quizzed by written examination, to demonstrate that they clearly understand the study and any risks to which they may be exposed during the course of the clinical trial. Subjects are always given the opportunity to ask questions of the principal investigator, even if his/her staff is administering IC. Once satisfied, the subject then signs an IC Form (ICF or CF) in the presence of a witness. However, the consent is always reversible and, should the subject change his/her mind, it may be negated at any time in the study. In effect, this means a subject may leave a clinical research study at any time and for any reason or for no stated reason.

The ICF is written after the protocol has been drafted and reviewed by the principal investigator and sponsor and once nonclinical toxicology information or data from previous clinical studies are available. Consent forms may be written by the principal investigator or the sponsor. Because the CF must be approved by an IRB, this board's preferred institutional format should be considered for each clinical study site. IRBs often request changes to a CF and so it is not unusual to have several slightly different versions, one for each site, in a multisite clinical trial. For a variety of reasons, it is sometimes necessary to obtain the approval of two or even three IRBs for some investigational sites. It is sometimes difficult for both sponsor and investigator to ensure each form has correct content and is acceptable under current regulations.

The Institutional Review Board (Independent Ethics Committee in some countries) is a committee, usually 5 to 10 medical professionals, clerics or lay persons, responsible for ensuring and protecting the rights and welfare of human subjects who participate in biomedical research. The IRB reviews protocols, the IC, the IB and related materials, such as recruiting advertisements. In doing so, the committee ensures the rights of the subjects. The IRB must judge whether or not possible risks to the subject outweigh potential benefits or knowledge gained through the study. IRB responsibilities do not end with approval of the study and study documents, as the IRB continues to review the program as clinical research progresses and always considers reports or changes, such as SAEs and study termination. Once the study begins, the IRB must review SAEs and other significant issues that arise. Annual reviews of each study are mandatory, whether or not there are issues related to the product, the subjects or the study itself. Of course, no member of the committee may have a conflict of interest with any study he or she reviews.

Most institutions that conduct clinical research—universities, hospitals, research centers and CROs—have IRBs. Independent IRBs also are available and are used by sponsors when the investigational site has no institutional affiliation. While no accreditation is required for IRBs, their records are reviewed by national regulatory agencies and, in recent years, IRBs at some notable institutions have been suspended for failure to follow regulations. The Department of Health and Human Services is ultimately responsible for ensuring compliance with human use regulations, but agencies such as the FDA and the National Institutes of Health also become involved. In a practical sense, each IRB is composed of individuals from different walks of life—ethicists, clerics, scientists and lay persons—so that the review is balanced in nature and considers various professional and social aspects of the proposal. The committee meets periodically and then only after each member has had an opportunity to review the clinical documents noted earlier. Following review, these documents are discussed in an IRB meeting and it is not unusual for the committee to ask for additional information or to recommend changes to a document. By working together, IRBs, sponsors and principal investigators support each other and ensure the integrity of a clinical study and protect the right of human subjects enrolled in that study.

Investigational Product

Clinical trial supplies or materials include the biopharmaceutical, the investigational product, placebo or comparator or diluents, and any device used to apply or deliver the product or otherwise ensure correct use and safety of the product as it is given to the volunteer. The product used meets specifications in terms of identity, purity, strength and quality, as discussed in Chapter 6 and Chapter 7. It is very important that the investigational product be consistent for all clinical trial sites and at all times throughout the study at any given site. Clinical trial supplies are delivered by the sponsor, i.e., the biotechnology firm manufacturing the biopharmaceutical product, in a timely manner and is kept at the proper environmental conditions (e.g., temperature) throughout the study and it is properly labeled. In a blinded study, steps are taken to identify product and placebo or comparator correctly and yet maintain the blind. All these clinical supply operations are documented accurately and the disposition of all product is fully accounted for. As noted earlier, a pharmacist with experience in clinical trials often manages these tasks and ensures that all clinical supplies are of the highest quality, fully accounted for in records, are properly stored and distributed to study staff and that any unused clinical supplies are returned to the sponsor.

Collection of Clinical Data: Case Report Forms and the Patient Diary

Accurate and timely collection of all clinical data is an absolute requirement for any clinical study. The process can be divided into four major stages:

1. Preparation of format, forms and media to collect the data
2. Collection of data during the clinical trial
3. Review or audit of data to ensure completeness, accuracy and integrity
4. Analysis of data

It takes considerable planning and effort to fully and properly collect clinical trial data. The initial raw data is referred to as source information or a "source document," the original document upon which an observation is recorded, and includes records such as laboratory reports, clinical or patient charts, memoranda, patient's diaries and pharmacy dispensing records. The raw data from a source document may be initially recorded on a Case

Report Form (CRF) or it may be transferred from a source document to a CRF by study personnel. The CRF is a printed, optical or electronic document designed to record all of the information, no matter what the source, required by instructions provided in the protocol. CRFs, and there are many for each clinical study, are drafted after the protocol has been completed and it is known what data will be collected, in what format, by whom and how frequently. Once finalized, CRFs are printed in final format and distributed to each clinical trial site.

For some studies, patients may be hospitalized throughout the course of treatment and, therefore, data are easily collected in an environment conducive to keeping complete and accurate medical records. More often than not, the investigational product is given during brief clinic visits and, if they are feeling well, the patient is sent home after a few hours (or days) at the treatment site. For other products, the patient takes the product at home and only visits the clinic initially and then periodically for follow-up physical examinations or tests. When the patient is away from the clinic, the subject diary may be used to collect data. In a diary, each subject records any symptoms he or she has noted during the study. While this is not a highly reliable means of collecting data, it does sometimes reveal drug-associated adverse events that occur between clinic visits.

Patient diaries are reviewed and CRFs are completed by study staff, particularly by study nurses and physician investigators. Then they must be reviewed and approved by the principal investigator. During the course of a study, and after CRFs have been completed by the investigational staff, they are audited by an outside representative, the clinical auditor or monitor, a representative of the sponsor. Whether paper or electronic, CRFs are carefully reviewed against source documents to ensure accuracy of the data. Today, data on paper records are usually entered into an electronic database to facilitate statistical analysis. Unfortunately, this transfer of data from source documents to CRFs or to electronic databases is prone to human error and, therefore, electronic data collection, i.e., directly recording information from a source document, e.g., a blood pressure, into an electronic database, is a common procedure. While this reduces errors of transcription, it requires a validated electronic system, hardware and software, and well-trained clinical staff.

No matter what the format, data are analyzed according to an analytical or statistical plan that was prepared by a statistician prior to beginning the trial. Several computer programs are commonly used to analyze data and to prepare tabular and graphic presentations of the information. Statisticians, experienced in clinical trial data management and analysis, are employed by the sponsor for these tasks.

Clinical Testing Laboratories

Clinical laboratory data are important to all clinical trials. Body tissues and fluids, notably blood, are collected and sent to a laboratory where they are tested for various parameters. For most of these tests, the laboratory is in a hospital or other medical center and, therefore, is certified by an accreditation agency, such as the American College of Pathologists, and regulations like the Clinical Laboratory Improvement Act. However, many clinical investigations also necessitate the performance of unique laboratory tests. These may be performed in a specialty laboratory or, in many cases, in an academic laboratory or the sponsor's laboratory. In such cases, tests must be initially qualified for accuracy and specificity and, in later stages of clinical development, they must be fully validated. Indeed, most of the quality criteria applied to product quality control tests (see Chapter 7) are applicable to tests used to measure clinical endpoints.

The sponsor is responsible for ensuring that each clinical testing laboratory meets all requirements and that laboratory testing is fully and accurately documented. The principal investigator and staff ensure that samples are taken exactly as mandated by the protocol and then properly stored and shipped to the clinical laboratory. The principal investigator also reviews the results, takes proper medical action and ensures test results reach the patient's records as a source document.

Reporting Results of Clinical Trials: Clinical Summary Reports

Once clinical trial data has been audited, analyzed and tabulated, it is included in a clinical summary report (CSR). This document, normally prepared by a medical writer with the help of biostatisticians, describes the clinical trial and reports all important aspects of the study. Because each report is reviewed by the sponsor and by regulatory agencies, it must be clear, complete and well written. Data are tabulated and presented in an unbiased yet clear and concise manner. A report relates essential elements of the protocol, clearly describing the design, treatments with investigational product and the population of human subjects. It discusses results of the study, presenting data in tables and figures, and drawing conclusions made by the principal investigator and statistician with concurrence of the sponsor. Safety issues are discussed in detail and statistically significant differences between treatment groups, as regards safety and efficacy data, are analyzed, often by several statistical tests. Conclusions and discussion of the data are written in a clinical summary report and, in many cases, a manuscript describing the study, its results and the conclusions is sent to a scientific journal.

Clinical Trial Operations

Study resources and people involved in clinical research of a biopharmaceutical product were reviewed in earlier sections. We now integrate this information by describing the planning and performance of clinical studies at each phase of development. They include tasks that are often performed in-house, by the biotechnology firm and those that are performed for the sponsor by CROs. Taken together, they are referred to as clinical operations.

Activities Leading to a Clinical Trial

Early in the development life cycle of a biopharmaceutical, the clinical plan is written and it then becomes part of the overall PDP (see Chapter 1). The decision of when to enter and exactly how to design the first, or Phase 1, clinical trial may not be established by the sponsor until a later date, perhaps after several preclinical and very early development milestones have been achieved. For example, before the Phase 1 trial is designed and scheduled, dates for the nonclinical studies, the manufacture and control of clinical trial product and filing of the IND are established. Once a tentative schedule has been set and there is some surety that investigational product will be available, it is possible to prepare a detailed Phase 1 clinical trial plan and the Phase 1 protocol.

Even the simplest clinical study requires quite a lot of coordination. Even if they do not have a formal medical department, biotechnology firms often have someone on staff with experience in managing clinical trials and this individual has responsibility for clinical planning. Alternatively, a highly qualified and recommended consultant may be retained to provide early clinical guidance. As soon as the Phase 1 planning process begins, the biotechnology firm decides if all elements of the study will be performed by CROs or if some aspects will be kept in-house. Seldom does a biotechnology firm have the resources to hire enough professionals to directly do all aspects of clinical work themselves. Thus, early decisions in clinical planning are usually to identify what, if anything, would be done in-house and if clinical support is to be performed by CROs, how this will be established. If all clinical work is to be contracted, the efforts should be divided, with functional areas, such as trial performance and quality efforts, i.e., auditing, going to a second contractor. This ensures the checks and balances so important to a successful clinical study.

Now, an experienced clinician must design the study and write a concept protocol. This may be done by a consultant or in-house staff or it may wait until the investigator and investigative site are chosen. For clinical research of many biotechnology products, the sponsor can choose from dozens of academic sites, usually medical schools, and CROs. For other products, e.g., a treatment for cancer, the sponsor might only consider sites that specialize in treating those patients. It is important to find a site that has access to the right patients, an experienced staff and the infrastructure to completely perform the Phase 1 clinical trial. It is not uncommon to find an excellent investigator who works at an unqualified site or the opposite, the ideal site but with mediocre investigators. Once potential sites are chosen, site visits are conducted by the sponsor. A site is chosen, the principal investigator is designated and agrees to do the study and, following negotiation over scope of work, budget and schedules, a contract is signed. This is followed by selection of another CRO to perform monitoring and perhaps a third and fourth group to provide other services (e.g., central laboratory). Now the clinical trial team has been established for that study and site.

Laboratory support is a hallmark of any clinical trial and it comes in two types: (1) standard or routine clinical laboratory support and (2) specialty laboratory analytics. Routine clinical laboratory support is offered by almost any hospital laboratory and includes analysis, such as hematology, clinical chemistry and basic immunodiagnostics. Small or early-phase trials, in fact, do often use hospital laboratories. However, in large clinical trials, a "central laboratory," represented by a single contract laboratory organization, is used to process, in the same technical manner, samples provided by multiple clinical study sites. Most studies also require specialty diagnostics or analytical techniques. For example, it is often necessary to measure the biopharmaceutical in samples of blood during a pharmacokinetic study. And, for vaccine studies, the immune response to the product must be measured with a variety of immunological assays, most unique and some even difficult to perform. Specialty laboratories may offer these unique testing services, but more often, these assays are adapted to or developed for clinical studies of specific products. Biotechnology firms may either do specialized assays in-house, at the firm's internal laboratory or they may identify a contract laboratory capable of developing the tests. There is no standard solution and the sponsor must carefully plan exactly how it is best achieved.

Once the clinical site has been identified, the sponsor's representative, working on behalf of the product development team, drafts the full clinical protocol. The investigator also identifies staff, e.g., subinvestigators, recruiters, study nurses and statistician, to assist in the study. Once the protocol has been written, the CF and CRFs are drafted and, along with the protocol, submitted by the investigator to the IRB. IRBs typically meet once or twice each month and it is normal for an IRB to request that changes be made in one or more documents before they are approved. Hence, the protocol approval process can take weeks or even months to complete.

While the principal investigator is leading these study and protocol development and review activities, the sponsor is actively recruiting a medical safety monitor and a clinical monitor or auditor. These individuals review the clinical trial documents before submission to the IRB and prior to ensuring quality and compliance at each study site through a prestudy site visit. At the same time, the sponsor is completing manufacture, labeling and control of the investigational product and making arrangements to have it delivered to the clinical site. Also, the sponsor is actively preparing and then submitting the regulatory documents, such as the IND application, to the FDA.

Once the regulatory agency accepts the IND and gives permission to begin the clinical study, the sponsor delivers product to the clinical site and the investigator begins the sequential processes of screening, accepting, "consenting" and enrolling subjects. The dosing phase of the clinical trial may now begin. Volunteers are given the number of doses specified in the protocol and efficacy endpoints are measured per the protocol. The volunteers are closely followed throughout the study for any sign of reaction to the product. If an SAE or a series of suspicious AEs are noted by the investigative staff, then dosing and further enrollment may be halted. In such cases, the medical safety monitor and sponsor and, subsequently, the regulatory authorities are notified. This leads to investigation and discussions; if the safety of the subjects can be ensured, the study may begin once again. However, if it appears that subjects may be at undue risk or that the product is unsafe, then the study may be terminated. Fortunately, studies of most biotechnology products are not halted due to safety concerns and most studies progress to completion, as specified in the protocol. Yet, patient follow-up is often a long process and subjects may be asked to return for physical examinations or lab tests for months or even years after the last dose of investigational product has been given. Extensive examination of subjects further ensures the safe and tolerable nature of a new product. Throughout the study, the clinical monitor visits the site to ensure that the study is being performed according to the protocol.

Once all data have been entered into CRFs, each form is screened for accuracy and completeness and information is transferred to an electronic database. Statisticians are normally responsible for these steps and the statistical data analysis that follows. In the case of biotechnology firms, the sponsor often retains a consultant statistician to perform analyses and to prepare tables and figures reflecting the data. A clinical summary report is then written by the investigator or a medical writer. This clinical summary report is first provided to the sponsor, for review, and then to regulatory authorities as definitive results of the clinical study.

The above description only lists the most important tasks, and their integration, involved in a typical clinical study. A host of other issues, some financial, others medical, and many administrative, must be considered in the design and execution of every clinical trial.

Phase 1 Clinical Trial: First-Time-in-Man

A Phase 1 study represents the first time a biopharmaceutical is used in man. Phase 1 studies focus largely on safety and tolerability of the product, but also may include measurement of efficacy endpoints. The numbers of subjects enrolled in a Phase 1 study are small, usually less than 50 and often under 25. A sponsor may elect to do several Phase 1 studies (i.e., Phase 1a, Phase 1b, etc.) in sequence, each focusing on a particular scientific question. This is often the case with complex and novel biotechnology products. Design is usually with normal and healthy individuals, at least in the first Phase 1 study. Exceptions are products with an excellent safety profile that are intended to treat life-threatening diseases, such as a study of a gene therapy to a treat rapid progression of a cancer or the study of an antiviral agent to treat a chronic infection like HIV. In such cases, actual patients, having exhausted all traditional therapies, are enrolled into a Phase 1 study.

The design of a Phase 1 study may be "open-label," meaning that both patient and investigative staff members know the nature of the treatment (i.e., investigational product or placebo) when it is given, or it may be blinded or double blinded, in which case a placebo (sugar pill) is given to one group of subjects without this knowledge being disclosed. Doses of product may be escalated in Phase I studies, but the scheme is quite conservative and only a few individuals are enrolled in each dosing group. Indeed, standard dosing schemes, such as single-rising dose or multiple-rising dose, are selected for each new biotechnology product. These study designs are shown in Figure 9.2 and Figure 9.3, respectively. In a single-rising dose study, subjects are randomly assigned to groups, perhaps 5 to 10 subjects per group. The lowest dose is given to subjects in the first group and then individuals in the second group are given the next (higher) dose. The process continues until the highest dose is reached, this determined from toxicology study results as the maximum tolerated dose. In a multiple-rising dose design, the dose is constant for any given individual, but the individual returns to the clinical trial site to receive an additional dose or doses. To prevent the possibility that a particular dose might result in acute reactions, only two or three subjects in a group may be dosed with product. This is done hours or even days before the remainder of individuals assigned to this group are dosed in the same manner.

Phase I trial measurements focus on safety endpoints, but measures of efficacy are typically performed whenever possible. Subjects may be kept in a clinic for days or even weeks following treatment so that they can be carefully evaluated at intervals. For example, frequent physical examination of subjects, use of electrocardiograms to identify changes in heartbeat, and regular clinical laboratory testing are hallmarks of Phase I studies of novel biotechnology products. Criteria for ending the treatment or dosing of human subjects whenever an SAE or multiple AEs are identified, "stopping rules," are very important elements of Phase I studies.

Clinical Pharmacology Studies of Biopharmaceuticals in Man

Additional studies are often required to fully understand a biotechnology product before it can enter Phase III trials. These studies are often given creative and complex numbers and letters, such as Phase 1c or Phase IIa, by their sponsors. However numbered, each is designed to specifically support the overall clinical development plan and is best referred to by its purpose (e.g., Pharmacokinetic Study in Normal Adults).

Pharmacokinetic (PK) studies are almost always performed if a new biotechnology product is to be given repeatedly or in significant amounts. In PK studies with human subjects, the product is given in a carefully controlled

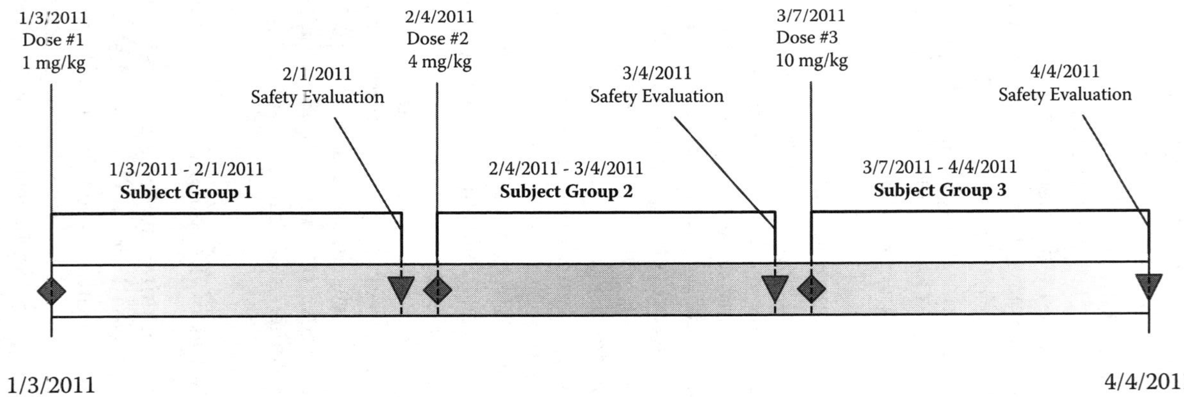

FIGURE 9.2
Design of a single rising dose study. In this study, the dose is raised from 1 to 4 and then to 10 milligrams of product per kilogram of the subject's body weight and a single dose is given to each subject before the next higher dose is given.

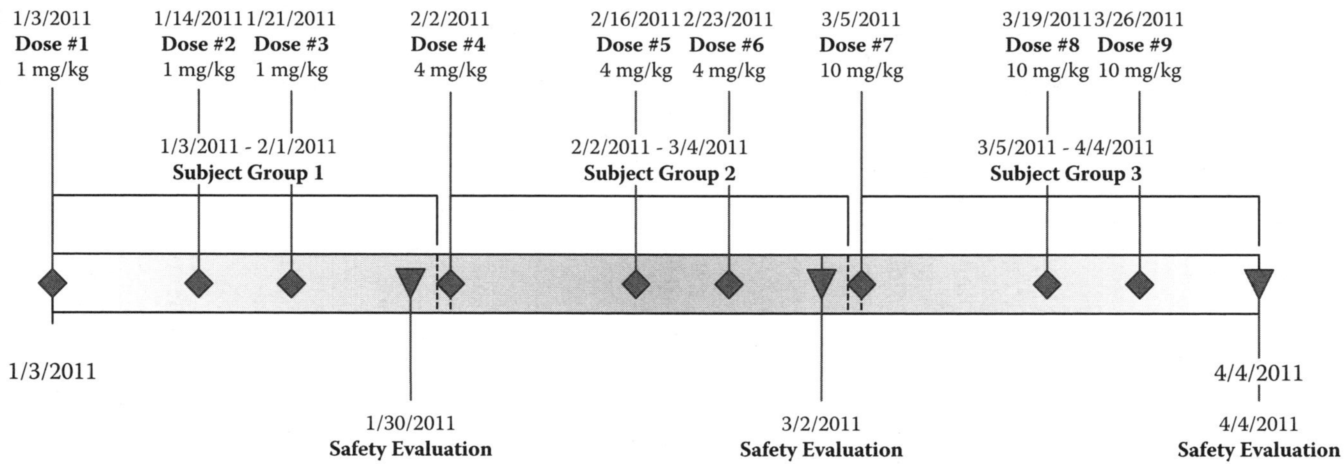

FIGURE 9.3
Design of a multiple rising dose study. In this study, the dose is raised from 1 to 4 and then to 10 milligrams of product per kilogram of the subject's body weight. Multiple doses, three in this design, are spaced by approximately four weeks. The next higher dose is initiated only after the previous dose level has been completed and initial safety evaluation performed after the third dose.

manner and then blood or another body fluid is taken at regular intervals following dosing and these samples are tested to determine the half-life of the product in circulation. While PK studies may be part of the Phase 1 or Phase II investigations, they also are performed as stand-alone studies designed strictly for that purpose.

PK studies have been best developed for drugs, but they are used extensively for studies with monoclonal antibodies and other biopharmaceuticals intended for distribution throughout the body. PK and pharmacodynamic studies are typically performed in animals during nonclinical studies before they are done in man (see Chapter 8).

Mass balance studies are designed to determine where in the human body a new biotechnology product goes and how long it remains in each location. To do these studies, product may be labeled with radioisotopes (having extremely short half lives) and then given. Metabolism, excretion and even localization in organs then can be followed with such tracer agents. For biotechnology products that target a particular tissue, mass balance studies may be performed in conjunction with imaging methods that allow the molecule to be identified in a particular organ. For example, it would be important to understand if a molecule, aimed at cancerous cells in the lung, bound largely to tumor mass and not to critical and unaffected organs, such as heart or kidney.

Food effect studies determine if a particular type or amount of food has an effect on the uptake and effectiveness of a new biopharmaceutical. It is from food-effect studies that we learn whether or not a patient should ingest a product on a full or an empty stomach. While quite important to orally ingested products, such as many drugs, food effect studies also may help to explain pharmacokinetic observations of biopharmaceutical products, such as unexpected patterns of excretion or binding to components of serum.

Additional Phase 1 studies may focus on subpopulations, such as a racial or geographic population, the elderly, adolescents or children. Controlled studies may be performed to determine if a biotechnology product will have greater or lesser effect when another drug is taken. These concomitant medication studies measure the effects of drug–drug interactions. Some classes of products are known to cause very unique types or reactions and these may be studied in more detail with additional pharmacology studies in human subjects.

Phase 2 Clinical Trial: Proof-of-Concept

The second phase of clinical development includes one or more therapeutic exploratory or proof-of-concept studies, referred to as such because they are designed to provide sufficient data to suggest that a biopharmaceutical product may well have the intended effect. Phase 2 studies may be dose ranging and demonstrate the optimal dose to take forward into later studies. With certain other Phase 2 trial designs, the intention is to determine the minimal effective dose, or "threshold effect" of the biopharmaceutical. Another intention at Phase 2 may be to determine the maximum effective tolerated dose, at least within the dosing criteria identified in Phase 1. Phase 2 study designs may examine various endpoints and measurements for those endpoints, searching for ones that will provide the best estimate of drug safety and efficacy in subsequent studies.

Phase 2 studies are often performed at five or more clinical study centers ("multicenter study") because there is a need for more patients (50 to 500 is a typical number) with a single disease and to determine if results vary by study site. A single center cannot often recruit this many qualified individuals. Thus, Phase 2 studies are typically multiarm studies, designed with several "arms" or groups (cohorts) of patients, each receiving a set dose level of the product. Whenever possible, Phase 2 studies are placebo-controlled and double-blinded, meaning that neither the patient nor the investigator and staff know which treatment or placebo a patient has received. If the treatment involves multiple doses of a product, as is often the case, the treatment period will be much longer in Phase 2 than it was in Phase 1, so as to determine a more realistic effect of the biopharmaceutical on both safety and efficacy. Hence, Phase 2 provides a means of determining if enough patients with the condition do exist so that a larger definitive study could be conducted. Thus, another benefit derived from Phase 2 studies is a determination of the best, and worst, clinical sites and investigative teams. Also, the range of subjects' medical conditions enrolled may be limited in Phase 2. Results from Phase 2 studies are seldom definitive because they do not enroll enough patients to absolutely demonstrate safety, tolerability and efficacy. Some argue that a Phase 2 clinical study is actually a "mini-Phase 3," a rehearsal if you will, for the pivotal study and, therefore, the results matter greatly for business development and the decision to move forward. Others suggest that the Phase 2 is often not predictive of outcomes in Phase 3, but is a means of determining the best dosing regimens and a valuable lesson. Whatever the case may be for a given biopharmaceutical, Phase 2 is an important step in the clinical development of any product and, therefore, each study must be carefully designed and executed.

Phase 3 Clinical Trial: Therapeutic Confirmatory

Following successful completion of Phase 2, the sponsor will almost certainly hold a meeting with regulatory authorities to discuss findings and to propose the design of a pivotal or Phase 3 study in man. Discussions between sponsor and agency typically follow this meeting and, within a few weeks, both parties should agree on the design of the all-important Phase 3 or "pivotal" clinical trial for the biopharmaceutical. Phase 3 studies are also referred to as adequate and well-controlled, since they must be just that. They are, in fact, the study on which the product will be registered and labeling claims will be supported.

Phase 3 trials are carefully considered, with significant input from medical experts, statisticians and those who manage and operate the study. Phase 3 studies are always large and multicenter and, today, most are multinational. Some drugs are tested in two Phase 3 trials, both using the same product and indication. The study is "statistically powered," i.e., it includes enough patients so that definitive answers as to safety, tolerability and efficacy of product can be obtained from a single study. Placebo or comparator is typically used and double blinding and other means of preventing bias are always included in the design.

The adaptive design also may be considered, with regulatory agency concurrence, for mid- and late-stage clinical trials of certain products. Adaptive means that changes may be made in the design of a clinical study if such change is guided by examination of data, accumulated at a particular interim milestone. An adaptive design can reduce the duration of a study or decrease the total number of patients required and, because it is based on recent information and experience, it can enhance the value of data that are generated by study completion. While the greatest interest in adaptive design has been with adequate and well-controlled late stage (Phase 3) studies, this approach has worked well with ascending dose or other midstage studies as well. There are caveats, however. Adaptive clinical study designs are prospective and must be carefully considered with regulatory authorities before initiation. As noted above, the interim analysis upon which change is based is itself carefully selected and protocol revisions are previously planned while certain changes may not be acceptable under any circumstances. Nonetheless, given a wide range of acceptable design changes, the adaptive design offers numerous opportunities when properly planned and applied.

As one might imagine, no matter what the size or design, a Phase 3 study can take years to complete and generates millions of data points and huge volumes of source documents as well as IC, CRFs and other documents. These studies are big and expensive. Large Phase 3 studies typically require an operations manual to ensure that all aspects of the study are performed exactly the same way at each center. A manual also serves to resolve problems as they arise, and to facilitate communication and good medical and administrative practices. In contrast, with studies of some rare diseases, the numbers may be low because it is simply impossible to find enough patients to enroll in a reasonable number of geographic locations.

An independent committee, referred to as a Data Safety Monitoring Board (DSMB), is included in the design of most Phase 3 studies. This board of experts is unblinded to the treatment at established intervals. The DSMB may do interim statistical analyses of the data to determine if the product seems to be working and is safe. For example, the DSMB could, early in a study, discover a distressingly large number of serious adverse events; in this case, they may ask that the study be halted because the risks to patients outweigh the possible benefits. In other instances, the DSMB might discover early in the study that the biopharmaceutical is so good that it cannot ethically be withheld from the subjects in the placebo group and so the argument is made that that product be given to all participants.

Phase 4 Clinical Study and REMS

When a biopharmaceutical has been approved for marketing by regulatory authorities, it is not unusual for the agency to ask that extended, "open-label" studies be performed in the postmarket approval period. These studies, sometimes referred to as Therapeutic Use, Phase 4 or postmarketing, are conducted and financed by the sponsor. Biotechnology firms usually welcome the suggestion of Phase 4 studies because it means their drug could be approved without the need for additional Phase 3 trials. This result is called "conditional approval." The firm receives marketing approval and may charge patients for the product, thus generating income, but with the understanding that one or more Phase 4 studies will be conducted by the sponsor and in consideration of regulatory agency guidelines. Product safety and efficacy definitively may be demonstrated during Phase 4 and, if they are not, the regulatory authorities have grounds to pull the market approval and, hence, ask that the product be pulled from market. Thus, the derivation of "conditional." Also considered for Phase 4 are extended testing for drug–drug interactions, effects in special or high-risk populations and additional safety surveillance.

Phase 4 studies are intended to reduce the risk to consumers from newly marketed biopharmaceuticals. The FDA has authority to require for all newly licensed products Risk Evaluation and Mitigation Strategy (REMS), which is intended to reduce risk to patients following market approval under a Biologics License Application or New Drug Application. While REMS does not involve additional Phase 4 studies, it does include specific clinical guidelines to sponsors as based on the clinical use of their product. These include a communication plan and materials for healthcare providers and information and instructions on safe use for prescribers, dispensers (pharmacists) and users of selected products. REMS thus represents a novel approach to clinical information to ensure safe use by healthcare professionals and the general public of already licensed products.

Clinical Trials for New Populations or Indications

Given the expense and complexity of performing any clinical study, it is impossible to expect a firm to test, in their initial pivotal trial, every special population that might benefit from the biopharmaceutical. These are the elderly, infants, children, adolescents, pregnant or lactating women, and certain racial, ethnic or geographic minorities. However, other populations of individuals, notably those with underlying diseases, such as liver or lung disease or impaired kidney function, are also difficult to study in initial Phase 2 and Phase 3 clinical studies. Some would argue that this discriminates against such populations because they have no chance to benefit from the product immediately after market approval. But, it is impossible to study each group of individuals in the pivotal Phase 3, due to resource and time constraints. How does the biotechnology firm go about testing individuals of any population when pursuing a new indication?

The answer is to perform another Phase 3 trial in that new population or with another indication. It is often possible to begin these Phase 3 studies at Phase 2 or after having performed small Phase 1 and Phase 2 studies. Assuming that the product has market approval for an indication and in one population (the first labeling claim), these postmarketing clinical trials may help the sponsor to market a biotechnology product and to bring benefit to patients currently without access to that product. These are not Phase 4 studies. Instead they are Phase 3 studies focused on broadening the indication for the product by testing it in new populations, for new indications or, for example, applying novel methods or routes of administration. However, there is a risk to the sponsor as such studies may uncover previously unknown safety issues, such as side effects of the biopharmaceutical. This can lead to undesirable regulatory action, such as addition of warnings in current labels.

Indeed, biotechnology firms are sometimes encouraged by a number of programs, sponsored by the FDA or other public health agencies, such as the National Institutes of Health (see Chapter 4), to test a product as soon as possible, typically postlicensure, for as many special populations as might benefit from the product. Such studies are typically done postmarket approval, not as Phase 4 studies, but as Phase 2 or small Phase 3 in scope and design. If they successfully demonstrate safety and efficacy in a special population, these studies, if adequate and well designed, are the basis for additional labeling claims for a biopharmaceutical product.

Global Clinical Trials

Many late-stage clinical trials are performed in countries distant from the sponsor's location. Indeed, today it is common to place multicenter clinical trials in numerous countries and to perform specialty studies and even Phase 1 studies in a foreign country. The sponsor often finds such efforts to save significant time and money. However, there are caveats. Cultural and regulatory differences can confound even the best planned global efforts. And, then there are the issues of different medical standards of care and of genetic differences in various populations. Some, but certainly not all, global clinical trials are managed by large CROs, organizations that maintain clinical trial facilities in many countries and, thus, understand the language, customs, regulatory environment, medical practices and population genetics in many countries where they have offices and local national employees. Foreign and multicenter global clinical trials are certainly possible and desirable, but require considerable planning and assistance.

Quality Systems for Clinical Trials: Current Good Clinical Practices

A quality system cGCP is applied throughout the clinical study process, from preparation of a clinical plan to completion of the clinical study report. Why must study integrity and quality be maintained at such high levels for clinical trials? First and foremost, it is the right of each human subject. When a volunteer enrolls in a study, he or she is subjected to a certain degree of risk or potential risk. Because of this, and with no guarantee of benefit, the subject has the right to know that the study will, in the end, provide meaningful scientific results, certainly a correct answer regarding the safety and efficacy of a biopharmaceutical. Hence, maintaining high quality and integrity of the study ensures that a meaningful answer is achieved and that this right of the subjects is fulfilled. Costs or finances are other reasons to complete a study properly. Studies are expensive and few biotechnology firms have the resources to repeat a clinical trial. Indeed, a single clinical study often means the difference between success or failure of a biotechnology firm and this alone is a compelling reason to get it right the first time in clinical research.

Under cGCP, certain systems and procedures and systems are applied to clinical trials to ensure the well-being of subjects, data integrity, overall quality and success of the trial. Some examples include:

- Careful planning before the study and coordination during and after the study.
- Selecting proven clinical sites and investigators.
- Training study staff to follow the protocol and other study documents and to accurately record and transfer data.
- Ensuring quality and integrity of data by using time-proven methods.
- The 100% internal audit of data sets and records.
- "Clean and screen" of all data entries on all documents to examine quality and consistency of data as it is transferred to a database.

These and many other practices regarding integrity of clinical trial data are embodied in the principles and practices of cGCP. The remainder of this section will focus on four clinical trial quality practices that are very important to quality and compliance with cGCP.

Quality and cGCP in Clinical Trial Operations

cGCP is further defined by ICH as "an international ethical and scientific quality standard for designing, conduction, recording, and reporting (clinical) trials that involve the participation of human subjects." The extensive ICH (see Chapter 4) guideline, entitled *Good Clinical Practices,* is the international standard for quality in clinical research. It also is adopted by most countries with a developed regulatory agency and most of these countries have supplemental regulations and guidance for conduct of studies in human subjects. The cGCP in the United States, outlined in Chapter 4, is further defined by several federal regulations, notably 21 CFR, Parts 50, 54, 56, 312, 314, 812 and 814, which collectively provide extensive guidance in this country. The U.S. government has accepted the ICH guidelines and the U.S. and ICH systems are currently harmonized.

Management responsibility, a critical hallmark of any quality system, is clearly identified in cGCP. The ICH guideline and FDA regulations identify responsibilities of the sponsor and the principal investigator. In the United States, cGCPs allow a sponsor to transfer certain responsibilities to an investigator, an institution (e.g., university) or a commercial entity, such as a CRO, but this transfer must be in writing and clearly described. Furthermore, the guidelines state that any responsibilities not transferred in writing to an investigator, institution or CRO are assumed by the sponsor. Thus, cGCP demands management responsibility and vendor and consultant control.

Control of the clinical trial process is clearly mandated by cGCP and this is done in a number of ways. As described earlier, written guidance, e.g., the clinical protocol or an operations manual, directs the clinical trial processes. Standard Operating Procedures (SOPs) are commonly applied for routine tasks performed in support of a clinical trial. Procedures and data are carefully documented on source documents (e.g., medical records), CRFs and electronic databases. These documents also provide for environmental control by establishing a proper environment for both investigational product and clinical processes. For example, cGCPs provide for product identification and traceability and for inspection or testing. The biopharmaceutical cGCPs demand that investigational product be clearly labeled and that the dosage form be clearly identified. The sponsor typically assumes responsibility for delivering a quality biopharmaceutical to the clinical site and then the investigator assumes responsibility for maintaining the integrity of that material and ensuring that each patient receives the correct product (e.g., placebo or active product).

cGCPs apply to virtually every operational aspect of a study. As noted earlier, responsibilities are clearly defined in writing. A case in point is control of a nonconforming study. Clinical practices recognize that, despite the best of intentions and controls, mistakes are made over the course of the study and refer to these situations as noncompliance with the protocol, SOPs, cGCPs and/or applicable regulatory requirements. cGCP guidelines establish the need for self-reporting, for auditing and for careful review by several parties of all documents. Most importantly, noncompliance must be reported and then corrective and preventive action taken. In regards to preventive actions, GCPs allow for changes in processes, as described in study documents, but they also demand that change be controlled, reviewed and approved by responsible individuals.

The performance of clinical trials requires that all study staff have the appropriate education and experience and are properly trained. Earlier we spoke of the sponsor's responsibility to select only qualified investigators and institutions to perform clinical studies. The quality requirement does not stop there, however. The sponsor is ultimately responsible for ensuring proper education, experience and training of individuals in the clinical laboratory, statistical group and at any CRO. All professional staff must fully understand the protocol and IB and know their respective professional roles and responsibilities under the protocol.

Customer concerns and complaints, another hallmark of quality, focus on satisfaction of both human subjects and, in the case of contract studies, the sponsor. As noted in cGCP, "the rights, safety and well-being of trial subjects are the most important consideration and must prevail over interests of science or the study staff, and society." Everyone involved in a clinical trial must, at all times, consider the rights and well-being of each subject. Such consideration does not end with signing the consent form, but continues to the end of the study. Indeed, for some studies of novel biotechnology products, responsibility for the well-being of a subject extends through the lifetime of that person.

Integrity of Clinical Study Data and Documents

Data collection and control is important under cGCPs. The data presented in a clinical summary report accurately reflects information that was recorded in source documents during the study. The objective is for completeness and accuracy, i.e., all data points collected for every patient enrolled. Diligence is taken by the investigator, sponsor and others in handling, analyzing and reporting data.

While 100% complete/0% errors is the goal, a number of seemingly unavoidable problems can occur and data points may be corrupted, questionable or missing from even the best designed and managed clinical study. For example, patients may fail to meet appointments or they may "drop out" of the study all together. This can be tolerated to some degree, but if too many patients leave the study or fail to comply with follow-up visits, it may not be adequate and well controlled and the overall study results are open to question. Also, investigative staff inevitably make errors when entering data into source documents or transferring information from source documents to CRFs or electronic databases. A high error rate can invalidate a study. Serious violations of a protocol may occur if subjects are enrolled in a study without meeting eligibility criteria or completing the informed consent process. In such cases, the data set may be considered incomplete. One patient missing a single dose of product will not invalidate a complete data set, but when several patients each miss a dose or if a few patients each miss several doses, the integrity of that study will, at the very least, demonstrate flaws. These are but a few examples of why cGCPs stress the importance of excellent study management and performance.

Monitoring and Auditing Clinical Trials

Clinical monitoring, not to be confused with the medical safety monitor, is the process of overseeing all aspects of a clinical trial and is a responsibility of the sponsor. It is a requirement of the sponsor under cGCP. Monitoring begins before the first subject is enrolled and ends when the last subject is discharged from the study. Monitoring ensures that the study is performed in accordance with cGCPs, the study documents, notably the protocol, and other regulatory requirements. It is a big job to monitor even a small clinical trial and takes tremendous efforts to properly monitor a Phase III study. But, it is done for every study.

Auditing, a systematic examination of study processes and documents, is an important part of monitoring and carries with it a function of determining whether particular activities are being performed correctly. Auditing normally involves the careful review of clinical trial documents to ensure they are correctly completed according to instructions in the protocol. Auditors, also referred to as clinical research associates, are individuals who perform the audits. The task of auditing a clinical trial is very detail-oriented and analytical. Auditors visit clinical trial sites where they review documents, speak with the investigational staff and, if they have arisen, identify issues or problems. In many cases they assist in resolving those issues by speaking with the principal investigator and sometimes performing staff training. Thus, auditors perform important roles in the overall monitoring process and ensure the integrity of a study and compliance with cGCPs.

Ethical Behavior and the Well-Being of Clinical Trial Subjects

The Declaration of Helsinki, as noted earlier, is the foundation for protection of human volunteers in any study. It holds clinical research to an exceptionally high ethical standard, stating that "... Compliance provides public assurance that the rights, well-being, and confidentiality of trial subjects are protected and that the clinical trial data is [*sic*] credible." This is totally appropriate and cGCPs directly support each principle in the declaration. Over the years, a number of human rights issues have arisen in clinical studies, even as the clinical research community applied cGCPs to thousands of clinical research studies worldwide. The vast majority of human research studies are, however, without breaches of ethical behavior, suggesting that the clinical research community and cGCP are doing an outstanding job of maintaining the principles laid down in the declaration. Unfortunately, more ethical issues may arise in the future. Without further delving into ethical behavior in clinical studies, some examples of common ethical situations faced by biotechnology firms are worth mentioning as a close to this chapter and, perhaps, as word of caution to those entering the field of biopharmaceutical development.

Some years ago, as clinical research expanded to support development of biotechnology products, clinical investigators in private practice or associated with nonprofit institutions accepted stock options in return for providing clinical investigative services for the biotechnology firm. This seemed like a useful model in the beginning, as biotechnology firms essentially deferred compensation, thus saving themselves considerable up-front expenses. However, it was also felt that this practice was a potential conflict of interest and regulatory agencies argued that, at the least, it be fully disclosed by both investigator and sponsor. Others asked how this differed from the accepted practice of employees (of the sponsoring biotechnology firm, including those staff members responsible for clinical monitoring) accepting stock options from their employer. While the issue has not been fully resolved to everyone's satisfaction, the consensus is that clinical investigators must not hold significant interest in the sponsoring entity or, if so, that any interest must be disclosed to regulatory authorities. Today, most regulatory agencies demand full disclosure by outside investigators and the indirect financial remuneration (e.g., stock options) of outside or "independent" investigators is capped in some instances. This situation demonstrates how important it is to ensure high ethical standards when dealing with human subjects and the clinical study process.

In another example, regulatory authorities are authorized to "blacklist" employees of the biopharmaceutical industries, preventing them from working in our industry if there is evidence that they egregiously or repeatedly failed to comply with cGCPs or other regulations. While the practice of individual sanctioning is applied by the FDA to all areas of biotechnology development, it is not uncommonly used to prevent certain clinical investigators, those who repeatedly failed to adhere to cGCPs or those who commit a major infraction, from further participating in studies. Note that even though a blacklisted investigator need not be convicted of a crime, his/her name and affiliation still becomes a matter of public record. While the clinical research community may feel singled out by the practice of blacklisting, it demonstrates how seriously we as a society take the rights of human subjects and the sanctity of clinical trials as a means of ensuring a safe and effective supply of biopharmaceuticals.

Another issue with ethical aspects is the problem of distinguishing clinical research from medical treatment. The distinction between the two is often blurred and this challenges clinical researchers, worldwide, as they seek the best treatment for patients. For example, a new biopharmaceutical product to treat AIDS is taken to market following abbreviated clinical studies. On the one hand, this is good, because it provides access to a seemingly promising treatment. On the other hand, it might also put users at potential risk of using a product that has not been thoroughly tested for safety or efficacy and might have if more extensive research been conducted.

Another example is the "off-label" use of a biopharmaceutical. Sometimes this occurs when it is quietly encouraged by biotechnology firms who wish to increase sales. It also happens when well-meaning, but sometimes poorly informed, physicians treat disease in an effort to save a patient from pain, suffering or even death. An important question is: In such cases, are the patients being enrolled and treated in a clinical research study, but without full informed consent? Does consent alone mean that a patient is being treated ethically?

There is no simple answer to any of these examples, but these types of questions arise repeatedly in biotechnology and they apply to many of today's most exciting biotechnology advances. There is no clear answer for every type of clinical situation. Nonetheless, those of us in the biotechnology industry face ethical issues as they make difficult decisions on how to proceed into clinical studies.

Summary on Clinical Trials

If nothing else, the information provided in this chapter demonstrates that clinical research involves a very large number of individuals with unique skills and technologies, from an understanding of ethical principles to cutting-edge medical practice. An understanding of clinical trials, their phases, design, requirements and potential outcomes, is important to the overall PDP. Indeed, the clinical plan is drafted after the TPP yet before all other sections of the plan because, during the development process of a biopharmaceutical, the clinical studies are the locomotive driving the train and setting the course for other operational activities. Also, unlike other endeavors in biotechnology development, clinical trials consider the rights and well-being of a very important resource, the human volunteer.

Glossary

21 CFR: Part 21 of the United States Code of Federal Regulations, the part in which most food and drug laws are located.

483: See Form 483.

510(k) Premarket Notification Process: A regulatory route by which to seek marketing approval from the FDA for a medical device of low to moderate risk and substantial equivalence to another device.

Absorption Phase: The pharmacokinetic phase during which a biopharmaceutical is absorbed into the body and, presumably, into the blood.

Absorption, Distribution, Metabolism and Excretion (ADME): Measurements of a biopharmaceutical in pharmacokinetic studies.

Accuracy: The measure of an assay to agree with a known true value.

Act: Legislation that begins as a bill before Congress and, once passed by Congress, becomes law.

Active (Pharmaceutical) Ingredient (API): The part of a product that has the desired biological activity, providing the primary therapeutic and biological effect.

Acute Toxicity: An animal safety study that examines the toxicity of a biopharmaceutical following a single dose with short-term follow-up.

Adaptive (Study Design): A study design that allows changes to be made in the protocol at a milestone, if data warrant.

ADE: See Adverse (Drug) Event.

AE: See Adverse (Drug) Event.

Adequate and Well Controlled Study: A scientific study that is carefully designed to test a hypothesis and has proper controls for the intended purpose.

ADME: See Absorption, Distribution, Metabolism and Excretion.

Adulterated: A biopharmaceutical or drug that is putrid, filthy or decomposed or lacking strength, purity or quality, not of cGMP nature or in a deficient container.

Adverse (Drug) Event (AE or ADE): A medical event in a human subject that is undesirable and symptomatic of a physiological change or disease and is due to a particular intervention or treatment, such as use of a biopharmaceutical.

Analyate: Material or product that is being tested.

Analytical Method: Laboratory procedure or test performed on a product to measure an attribute.

Analytical Tool: Laboratory procedure or test performed on a product to measure an attribute.

Analyze: In project management, an assessment of achievement relative to the project plan. May include evaluation of alternatives and resource requirements and usage.

APHIS: Animal and Plant Health Inspection Service of United States Department of Agriculture.

API: See Active Pharmaceutical Ingredient.

Ascending Dose Study: An experimental design in which the dose of investigational product is raised with each subsequent group of volunteers.

Aseptic: Used as a noun it means "without living organisms." As an adjective or adverb, aseptic describes processes that avoid, to a great degree, the inclusion of or contact with microbes.

Assay: Laboratory procedure or test performed on a product to measure an attribute.

Attribute: A positive, desirable, or even necessary characteristic of a product that lends itself to testing.

Audit: A formal review by an auditor of a process, study or product that examines whether actual performance was conducted in accordance with established instructions.

Batch: An amount of product that is produced together as a single entity. Batch usually refers to a defined amount of biopharmaceutical bulk substance.

Batch Production Record (BPR): A document used in manufacturing to both guide a process and to record critical information regarding performance on a particular batch or lot of product.

BDS: See Bulk Substance.

Bias: A predisposition or prejudice in scientific studies or a systemic distortion of a statistical result (Oxford Dictionary, 1997).

BIO: Biotechnology Industry Organization.

Bioavailability: The fraction of biopharmaceutical, of total amount given, available in the blood (or tissue) to have its intended effect.

Bioequivalence: Assessment of the comparative activity and bioavailability of two products after administration to animals or humans.

Biologic or Biological: Historical terms used to describe products that are derived from or represent biological or living sources.

Biologics License Application (BLA): An application made to the FDA for the purpose of gaining marketing approval for a new biological (nontherapeutic and nonpharmaceutical) in the United States. This large document provides complete information on development of the product and its safety and efficacy.

Biomanufacture: Manufacture or production of biological molecules, cells, tissues or other products derived from biotechnology.

Biopharmaceutical: A biological molecule, cell, tissue or other material of biological origin used in the treatment or prevention of disease in humans. Biopharmaceuticals are complex in a molecular sense.

Bioreactor: A closed vessel designed to support the multiplication and growth of eukaryotic cells for the purpose of expanding a cell line or producing a biopharmaceutical.

BIS: Bureau of Industry and Security, U.S. Department of Commerce.

BLA: See Biologics License Application.

Blinded study: A clinical study design in which certain individuals, usually the volunteers and investigative staff, are unaware of the treatment (investigational product, placebo or comparator) given to the volunteer.

BPR: See Batch Production Record.

BS: See Bulk Substance.

Bulk Substance (BS): Biopharmaceutical product that has been produced and purified but has not yet been formulated or aliquoted into the final container. Also referred to as bulk drug substance or BDS.

Campaign: To manufacture more than one product in a facility under an established schedule.

Campaign Manufacture: Manufacture of more than one product in a facility. Each manufacturing area is only used for one product at any given time, so projects are sequential.

CAPA: Corrective and Preventive Action, the process of investigating and correcting a deficiency, deviation or other problem or issue in manufacturing or quality control of product.

Cap: A stopper or other seal that is placed on the container once it has been filled.

Case Report Form (CRF): Paper or electronic form upon which the investigator enters medical information gathered during a clinical trial.

CBER: Center for Biologics Evaluation and Research, United States Food and Drug Administration.

CBP: Customs and Border Protection of the United States Department of Homeland Security.

CDC: See Center for Disease Control and Protection

CDER: Center for Drug Evaluation and Research, United States Food and Drug Administration.

CDRH: Center for Devices and Radiological Health, United States Food and Drug Administration.

Cell: Production cell that replicates and has particular traits. In biomanufacture, often of bacterial, yeast, insect or mammalian origin, but may be derived from almost any species: plant or animal, eukaryotic or prokaryotic.

Cell Bank: A source of live cells, derived from a clone or small number of progenitor cells, that are kept in storage and then used as the seed or source in biomanufacture.

Center for Disease Control and Protection (CDC): A United States federal public health agency under Department of Health and Human Services.

Certificate of Analysis (CoA): A formal document used to identify attributes or traits, quality control tests, specifications and test results for a product or raw material.

CF: See Informed Consent Form (ICF).

cGCPs: Current Good Clinical Practices are regulations promulgated by the U.S. FDA and international bodies that must be followed for conduct of research in human subjects.

cGLPs: Current Good Laboratory Practices are regulations promulgated by the U.S. FDA and international bodies that must be followed for non-clinical safety studies of all biopharmaceuticals.

cGMPs: Current Good Manufacturing Practices are regulations promulgated by the U.S. FDA and international bodies that must be followed for the production (manufacture) and distribution of all biopharmaceuticals.

Change Control: An active process under which proposed changes are introduced, examined and acted upon according to plan and with full knowledge of everyone involved or impacted by the change.

Charter (Team): A mandate and authorization to achieve, as a team, an objective. A charter is bestowed by a higher authority, such as a stakeholder or executive manager.

Chronic Toxicity: A safety study in animals that measures the toxicity of a biopharmaceutical given in multiple doses with follow-up over a long period of time (>6 months).

Classified: A formal designation regarding the level of air quality in a clean area or room.

Clean Room (Area): An area or room in a biomanufacturing facility that is controlled to reduce the chance of microbial or particulate contamination of product.

Clinical (Study) Design: A brief description of a clinical study that describes the scientific approach and hypothesis, as well as ensuring quality elements.

Clinical Summary Report (CSR): A written report that fully summarizes the performance and results of a clinical trial to include tabulated data and statistical analyses.

Clinical Research: See Clinical Trial.

Clinical Study: See Clinical Trial.

Clinical Trial: A designed scientific study in which a principal investigator evaluates an investigational biopharmaceutical product in human volunteers.

C_{MAX}: The maximum amount of biopharmaceutical that is available in blood or tissue following delivery of a given dose.

CMO: See Contract Manufacturing Organization.

CoA: See Certificate of Analysis.

Cohort: Group of human volunteers with common characteristics and treated at the same time although not necessarily in the same manner.

Combination Product: A product that combines two or three of the following: biological, drug and medical device.

Commercial Production: Biomanufacture of a product at the final or commercial scale. Typically happens just before or following marketing approval by regulatory agency.

Common Technical Document (CTD): A format for preparing, organizing and writing market applications and investigational new drug applications in many countries. The eCTD is the electronic version of the CTD.

Comparative Clinical Study: Clinical research in which the investigational product is compared with another product.

Comparator: A control material, typically a product that is licensed for the same indication, is used in a clinical trial to compare against use of investigational product.

Compendium/Compendial: A reference book that provides product, process and test standards and specifications.

Components: Materials that are used in manufacture. Often hardware materials.

Consent Form (CF): See Informed Consent.

Concentration–Effect Relationship: The relationship between the concentration of a biopharmaceutical in the blood (or tissue) and the desired physiological effect it has on an animal or human.

Concept Protocol: A brief design of a nonclinical or clinical study used as the basis for discussions between sponsor, investigator and regulatory authorities and is the foundation for preparing a full study protocol.

Conditional Approval: Regulatory approval for a biopharmaceutical in which the regulatory agency stipulates that certain tasks, often Phase 4 clinical studies or follow-up of patients from Phase 3 studies, must be performed as a condition to that approval.

Conformance: A product or, in a broader sense, a study report or other document that meets specifications and regulations.

Construct (Genetic): A biological material that is or has been derived from genetic engineering of a molecule or a cell. Usually refers to a plasmid.

Container: The vial or other vessel that directly holds a final product.

Contaminant: Particle or chemical that is undesirable and has entered the product stream during manufacture.

Contract Manufacturing Organization (CMO): A manufacturing facility that performs biomanufacturing on a contract basis.

Contract Research Organization (CRO): A corporation or institute that provides contractual support to a biopharmaceutical sponsor in areas of clinical or nonclinical studies or manufacture and control.

Control: A material that is used to ensure performance of an assay. May be a positive or negative control. In project management, using influence to follow the current plan or improve it.

Control Article: The nonactive material that is given to experimental animals as a control and in lieu of active ingredient during a nonclinical study.

Controlled Clinical Research: A clinical trial in which both an investigational product and one or more control substances, such as a placebo or a comparator, are given to patients randomized into groups.

CPMP: Committee for Proprietary Medicinal Products of the EMEA.

CRF: See Case Report Form.

Crimp: Process of sealing or closing a cap onto a container of final product. Often done with metal bands or covers.

Critical Pathway: The pathway in a project that is critical to achieving objectives and schedules. Also called the rate-limiting pathway as it determines the rate at which the product is going forward.

CRO: See Contract Research Organization.

CSR: See Clinical Study Report.

CTD or eCTD: See Common Technical Document.

Cycle or Life cycle (Project): The overall project, from beginning to end, with all elements included.

Decision Points: Precise or particular moments in a project schedule that require consensus on a particular management or technical matter.

Design Control: A formal and documented system of plans and procedures that are used to ensure the quality development of products or processes.

Deviation: When a value or process does not meet established procedures, rules or specifications. Deviations are discovered after the fact and were not planned.

Device: See Medical Device.

Diary: The patient diary is a record kept by each clinical study volunteer to record any medical conditions he/she might encounter following treatment and while not under direct medical supervision.

Distribution Phase: The pharmacokinetic phase during which a biopharmaceutical is distributed throughout the body, normally from blood and to tissues and organs.

Documentation: A formal process of a quality system in which all documents for a product, process or service are carefully and fully managed from beginning to end.

Dose: Single delivery or application of a biopharmaceutical product.

DOT: United States Department of Transportation.

Downstream: The stage of manufacturing in which product, in a crude state, is purified to bulk substance.

Drug: A small molecule with pharmacological effects, usually of organic origin and with a well-characterized chemical structure.

Early Phase Development: Stage of biopharmaceutical development beginning with initiation of development efforts through the end of Phase 1 clinical trials.

EEC: European Economic Community.

Efficacy: Producing the desired effect so that the therapeutic indication of the product is achieved.

EIR: See Establishment Inspection Report.

EMEA: European Agency for Evaluation of Medicinal Products.

Endpoint: A measurable entity, such as weight or blood pressure, in a scientific study.

Enroll: To allow a volunteer to participate in a clinical study after that person has completed the screening and informed consent processes and has been found acceptable to participate.

EPA: United States Environmental Protection Agency.

Establishment Inspection Report: A document prepared by FDA inspectors to note the findings made during an inspection.

Excipient: A material that is added to a biopharmaceutical product and is not an active ingredient, e.g., a carrier or a preservative.

Exclusion criterion: A medical characteristic of a potential volunteer that requires the investigator to disallow that individual from enrollment in a clinical study.

Excretion Phase: The pharmacokinetic phase during which a biopharmaceutical is excreted from the body.

Experimental clinical study: A clinical trial prospectively designed as an experiment with active treatments, controls and other methods of making comparisons between treatment groups.

Expression System: A biological construct consisting of a recombinant gene stably inserted into a living cell.

FAO: Food and Agricultural Organization of the World Health Organization.

FDA: See Food and Drug Administration.

FDP: See Final Product.

Federal Trade Commission (FTC): A U.S. Government agency regulating commercial practices, including advertising, within the United States with the exception of foods, drugs, and biopharmaceuticals.

Feedback: A team member relating any aspect of the project to other team members.

Fermentation: Process of growing bacterial or yeast cells in a closed vessel under defined conditions for the purpose of manufacturing a product.

Fill: To actively place or aliquot a biopharmaceutical into a container.

Final Drug Product (FDP): See Final Product.

Final Product (FP): The biopharmaceutical product once it has been formulated, filled into a container and finished. Also referred to as Final Drug Product or FDP.

Finish: To place a cap onto a container, crimp or otherwise seal the container and, in some cases, package the product.

FOI: See Freedom of Information (Act).

Food and Drug Administration (FDA): An administrative United States government agency under the Department of Health and Human Services (DHHS) and responsible for regulating many products and their development, including food, drug, medical devices and biopharmaceuticals.

Form 483: FDA Inspectional Report, a form given to sponsors immediately postinspection by an inspector and listing deficiencies or deviations identified during the inspection.

Formulation: Addition of various solutions, buffers, excipients or stabilizing materials to a bulk product so as to make a solution or powder that is ready for fill into a container.

FP: See Final Product.

Freedom of Information (Act) (FOI): A law that allows private citizens to petition a government agency, such as the FDA, to release information that is not proprietary or confidential.

FTC: See Federal Trade Commission.

Functional Area: A particular scientific, management or technical activity and suborganization aimed at fulfilling an established purpose. As regards biotechnology operations, seven functional areas are commonly listed: clinical, manufacture, nonclinical, project management, quality assurance, quality control and regulatory affairs.

FWS: Fish and Wildlife Service of the U.S. Department of Interior.

Gantt Chart: A computer-generated rendering of a project using narrative and horizontal bars to show tasks, milestones and their dependencies and relationships.

Genetically Modified Organism (GMO): An organism that has been changed or modified (e.g., additions, deletions to the genetic makeup) by genetic engineering, often referred to as recombinant DNA technology.

GMO: See Genetically Modified Organism.

GMP: See current Good Manufacturing Practices.

Good Tissue Practices (GTP): A regulatory guideline from the FDA for the processing of tissues or cells for human use.

Guideline: A public document written and promulgated by a government agency, such as the FDA, that recommends or suggests practices, both administrative and technical, that would, if practiced, fulfill requirements given under regulations. Guidelines do not have the legal status of regulations, however.

Hallmark of Quality: One of several operational quality criteria that comprise a quality system.

HEPA: High-efficiency particle air is a special filter that removes all but the smallest particles, leaving the exiting air especially clean and >99% free of bacteria and fungi.

Hold (Clinical and Regulatory): A step that is taken either by the FDA or a sponsor to stop or not begin a clinical study of an investigational product.

Hold (Biomanufacture): A step in biomanufacturing where product is kept in a container awaiting further processing.

Host Cell: A live cell that contains a biological molecule or microbe that is not normally found in that cell.

IB: See Investigator's Brochure.

IBC: See Institutional Biosafety Committee.

IC: See Informed Consent.

ICF: See Informed Consent Form.

ICH: See International Conference for Harmonization.

Identity: Individuality of a product and features that distinguish it from all other products. Ability of a product to be of a known and unique nature.

Impurity: Undesirable material, usually macromolecular and submicroscopic, but may be a visible, microscopic or soluble organic or inorganic, in a product or stream. Often inherent to the process itself, such as cell debris.

Inclusion criterion: A medical characteristic of a potential volunteer that is considered a positive trait for enrollment of that individual into a clinical study.

IND: See Investigational New Drug Application.

Indication: A remedy or treatment or preventative that is suggested by the symptoms or disease. For a biopharmaceutical, an indication is a specific medical condition that may be treated or prevented by the product.

Induction: Biomanufacturing step in which a chemical is added to a fermentor to induce or elicit the production of a product by an organism that has been genetically engineered to respond to the chemical.

Informed Consent (IC): The process of informing a volunteer to a clinical trial exactly on the nature of the trial and all possible risks and benefits that person might derive. To enroll, the volunteer must sign the Informed Consent Form.

Informed Consent Form (ICF or CF): This clinical trials document explains to a volunteer the potential risks and benefits of a clinical study. To enroll in a study a volunteer must understand and sign the CF.

In-Process Testing: Testing that occurs on samples taken from the process stream during the manufacturing process.

Input: A component of design review during which the needs of the user are considered and incorporated into the product or process design.

Installation Qualification (IQ): A step in validation of a biomanufacturing facility, utility or equipment in which the installation is demonstrated to be according to specifications.

Institutional Biosafety Committee (IBC): A committee of scientists, ethicists and lay people established by an institution (e.g., a university) to review the engineering, use or transfer of genetically modified organisms and related research and development.

Institutional Review Board (IRB): An institutionally based, peer review committee that reviews all clinical research and protocols at the institution to ensure the proper treatment and well-being of volunteers.

International Conference for Harmonization (ICH): A nonprofit group, supported by regulatory agencies and medical products industries, dedicated to developing and disseminating medical product development guidelines and pathways that are acceptable to regulatory authorities in most countries and to ensuring the quality of those products.

International Standards Organization (ISO): An international organization dedicated to quality through establishing requirements and specifications for products, services and processes. ISO is not a regulatory agency, but develops guidelines and provides ISO certification following review and approval.

Investigational New Drug (IND) Application: Formal application to the U.S. Food and Drug Administration to test in human volunteers a biopharmaceutical that does not have marketing approval (investigational).

Investigational Product (or Drug): A biopharmaceutical product that is tested in human volunteers in clinical trials and under an IND and has not received market approval.

Investigator: Individual leading the scientific and medical portion of a clinical study. The principal investigator has responsibility for the study while subinvestigators assume certain responsibilities under the principal investigator.

Investigator's Brochure (IB): An informative document that identifies for each member of the investigative staff information on the clinical study, the product being tested in the study and possible risks and benefits to volunteers enrolled in the study.

In Vitro Diagnostic (IVD): A laboratory test used to diagnose disease. Regulated by the FDA as a medical device.

IQ: See Installation Qualification.

IRB: See Institutional Review Board.

ISO: See International Standards Organization.

IVD: See *In Vitro* Diagnostic.

Label: The printed identification for a product, usually paper held with adhesive, that is placed onto a container or package of biopharmaceutical.

Labeling: The sum total of printed materials, package insert, package printing, etc., that accompany, or are adherent to, biopharmaceutical containers and packaging. Labeling provides the approved indication, directions for use, dosage, and other critical information provided by the sponsor and approved by the FDA.

Late-Stage Development: The final investigational stage of product development and the activities associated with this stage and with Phase 3 clinical trials.

Leachates: Chemicals that are dissolved from a surface or other solid matrix into the product stream during biomanufacture, thus becoming contaminants.

Limit of Detection (LOD): The minimal amount of analyate that can be accurately detected by a particular assay in a test substrate.

Limit of Quantitation (LOQ): The minimal amount of analyate that can be reasonably measured, in a quantitative sense, by a particular assay.

Linearity: For a quantitative measurement by an assay, the ability, within a given range of analyte in sample, to obtain test results that are directly proportional to the concentration of the analyte.

LOD: See Limit of Detection.

LOQ: See Limit of Quantitation.

Lot: Defined amount of manufactured final product that constitutes a legally defined entity. A lot has unique character, quality and source and is a specifically identified amount, labeled as such.

Lyophilize: A process in which a biological material in solution is subjected to freezing and drying, simultaneously, to preserve the cells or molecules.

Marketed Product: A biopharmaceutical that has been approved for sale in that country.

Marketing Application: Application to a regulatory agency to market a product in that country. See also Biologics License Application or New Drug Application.

Market Approval: Permission from the Food and Drug Administration to market a biopharmaceutical in the United States for the indication and at the dosage given in the approved labeling.

Master Cell Bank (MCB): The ultimate source of any seed. A bank of cells, usually derived from a single clone or source, that is kept as a unique resource for later expansion or use.

Master File (MF): A regulatory document under which a sponsor may file confidential information with the FDA. Investigational use of a product is not allowed under a Master File, as it is under an IND. Drug Master Files (DMF) or Biologic Master Files (BMF) are used in biopharmaceutical development.

Matrix: The medium in which a product is disbursed or suspended to include those of natural origin, e.g., serum, or of synthetic origin, e.g., phosphate-buffered saline.

Maximal Efficacy: The greatest effect or response that is given by a biopharmaceutical (and in the absence of toxicity).

MCB: See Master Cell Bank.

MDR: See Medical Device Reporting.

Measurement/Measure: The act of or a system for determining, through laboratory, clinical or nonclinical testing and evaluation, the quantity or quality of a particular endpoint or process. Use of an assay or instrument or scientific and technical skills to determine an unknown parameter. It may be qualitative or quantitative. Usually determined in a stated unit or capacity.

Medical Device Reporting (MDR): An FDA regulation aimed at ensuring manufacturers report defects in, or adverse events associated with, medical devices.

Medical Device (Device): An object that may be any one of many classes of physical or engineered products that achieves its intended primary action in a manner other than pharmacological, biological or metabolic means. Medical devices are used to diagnose, prevent, monitor and treat disease.

Medical Monitor: Also referred to as the safety monitor, a medical professional assigned to review AE or SAE or other matters relating to safety of volunteers.

Method (Analytical): A test or analytical procedure that is used in a laboratory to measure the quality.

Metrics: In project management, measurement of progress against established milestones, schedules, budgets or other resources.

MF: See Master File.

Middle-Stage Development: Product development that occurs before, during and immediately after Phase 2 clinical trials.

Milestone: A readily identified interim event or set of events; a major waypoint in a project that demonstrates achievement of a planned outcome.

Misbranding: Labeled or branded falsely or in a misleading manner or without supporting scientific data and in violation of FDA regulations. Not completely or legibly labeled or not accurately reflecting the truth.

Monitor: In a clinical trial, the Medical Monitor is a safety monitor, a medical professional assigned to review AE or SAE or other matters relating to safety or volunteers. A monitor is also a sponsor's monitor, an individual that reviews activities and progress of a clinical study at the study site.

Multiarm Clinical Study: A clinical study design that includes several treatment groups, each group receiving a different treatment or dose.

Multicenter Clinical Study: A trial that is performed in more than one medical center yet under the same protocol and for the same purpose. Multiple centers are used because insufficient volunteers can be recruited at one center.

NAI: See No Action Indicated.

National Regulatory Authority (NRA): A regulatory body, such as the FDA, appointed by a national government in the area of food and drug regulation.

NDA: See New Drug Application.

Neat: Test article used full strength or undiluted in a nonclinical study.

New Drug Application (NDA): An application made to the FDA for the purpose of gaining marketing approval for a new drug (pharmaceutical) in the United States. An NDA also applies to certain therapeutic biopharmaceuticals receiving review at CDER. This large document provides complete information on development of the product and its safety and efficacy.

NIH: National Institutes of Health of the U.S. Department of Health and Human Services.

NIOSH: National Institute for Occupational Safety and Health, CDC.

No Action Indicated (NAI): No action indicated refers to findings from an inspection report or EIR in which the FDA states that no findings in an inspection warrant further investigation or action.

Nonclinical: Studies, both *in vitro* and *in vivo*, that are performed outside of man to define in the laboratory and animals the pharmacology or toxicity of a biopharmaceutical.

Nonconformance: A product or, in a broader sense, a study report or other document that fails to meet specifications following quality control testing or quality review.

NRA: See National Regulatory Authority.

NRC: Nuclear Regulatory Commission, Department of Energy.

OAI: See Official Action Indicated.

OBA: Office of Biotechnology Activities, Office of the Director, NIH.

Observational Clinical Study: A study that observes patients for distribution and incidence of disease, in the absence of specific treatments and interventions. May be an epidemiological study.

Off-Track: A project is not on schedule or budget.

Official Action Indicated (OAI): Refers to findings on an EIR in which the FDA recommends that action be taken immediately by a manufacturer and followed up by the FDA regarding deficiencies or deviations noted during an inspection. It is the most serious of the three EIR finding categories.

On-Track: A project is on schedule or budget.

Open Label Study: A type of clinical study in which the investigator, and in some cases the patient, is aware of the treatment regimen (placebo or investigational product). It is not a blinded study.

Operational Area: As regards biotechnology operations, a functional area is one of seven commonly listed (clinical, manufacture, nonclinical, project management, quality assurance, quality control, and regulatory affairs) or other developmental specialty.

Operational Management: Managing technology development under a product development plan in an operational area.

Operational Qualification (OQ): A stage of manufacturing facility validation in which the operation of a utility or piece of equipment is shown to meet specifications. OQ is performed prior to process qualification.

OQ: See Operational Qualification.

OSHA: Occupational Health and Safety Administration, U.S. Department of Labor.

Outcome: Broad results or visible effects that form the basis for a study hypothesis. In clinical studies, these are often medical items.

Output: A component of design control in which the process, service or product design, based on input and review by professionals, is proposed and documented as the process and/or the product, in whole or in part.

Package Insert: The printed, extended instructions and information, approved both by the manufacturer and a regulatory agency, for a product and enclosed in the package. The package insert is an important part of labeling that states indication, directions for use, warnings, etc. (See Labeling).

Package: The inner (surrounds a primary product container) or outer (surrounds multiple inner packages) material that is used to protect a product from damage. Often cardboard or plastic packaging material.

Parenteral: A route of delivery given beneath or through the epidermal layer and is not oral, mucosal or topical.

Parenteral Product: A product that will be given beneath the skin or injected.

Pathway: In project management, a well-defined course of action or sequence of events.

Particle: A microscopic or visible piece of material, usually an undesirable contaminant in a product or stream.

Patients: Individuals with a preexisting medical condition enrolled in a clinical trial for the purpose of testing a therapeutic product for that condition.

PCB: Production Cell Bank. See Working Cell Bank.

PDP: See Product Development Plan.

PDS: Product Development Strategy. See Product Development Plan.

PERT: The Program Evaluation and Review Technique, actually an illustration of a project schedule to depict interrelationships of various tasks and milestones in a project.

PhRMA: Pharmaceutical Research and Manufacturers of America.

Pharmaceutical: A small molecule drug. See Drug.

Pharmacodynamics: The study of how a biopharmaceutical interacts with various tissues, fluids or organs to achieve a therapeutic effect.

Pharmacokinetics: The study of how, when and where a biopharmaceutical gains access (e.g., absorption), is distributed, is metabolized or is excreted by the body.

Pharmacology: The study of pharmacological agents (drugs or biopharmaceuticals) and their mechanisms of action and effects on organisms.

Pharmacopeia: A reference book providing product, process and test standards and specifications.

Phase 1: The first clinical phase and the early phase of product development.

Phase 2: The second clinical phase and the midphase of product development.

Phase 3: The third clinical phase and the last phase of product development prior to market approval.

Phase 4: Any development activities that occur following market approval of a product.

Phased Manufacture: Production of product over time using phases of development, going from simple systems to more complex, from producing small to large batches, from small to larger clinical studies, etc. Phases numbered as 1, 2, 3 or 4 or as early, mid- and late development phases.

PI: See Principal Investigator.

Pilot Production: Earliest production of a new product in the biomanufacturing cycle. Usually done at small scale and in an experimental mode.

Pivotal: A clinical study, usually in Phase 3, designed to demonstrate or confirm beyond reasonable doubt the safety and efficacy of a biopharmaceutical.

Placebo: A "sugar pill," any substance that is known to be safe and not cause a direct physiological or therapeutic effect and is given to volunteers assigned to a controlled clinical trial.

PMA: See Premarket Approval.

Portfolio: A collection of projects or programs with common technologies or goals.

Potency: The direct or indirect, surrogate measurement of a product's biological or therapeutic effect, generally in a quantitative manner, and established as a quality control test to evaluate BS or FD.

PQ: See Process Qualification.

Precision: The ability of an assay to repeatedly produce the same or very similar result on repeated testing when variables are held constant.

Preclinical: Research and early development activities that occur prior to Phase 1 clinical studies. Most often used in reference to research activities.

Premarket Approval (PMA): The document and process used as one route to gain marketing approval for a medical device in the higher risk classes.

Preventive Action: An activity that prevents a problem or issue from occurring or recurring.

Principal Investigator (PI): The medical professional, usually a physician, who is the responsible individual at a clinical study site and for a particular clinical trial.

Process: As a verb, process means to actively produce a product using defined technical skills. As a noun, a process refers to a defined portion of biomanufacture.

Process Qualification (PQ): A stage of manufacturing validation in which a process or part of a larger process is qualified by actual performance against specifications.

Product: A thing, substance or material that is produced by manufacture during biotechnology operations. A result of planning and labors.

Product Development Plan (PDP) or Strategy (PDS): A document developed early in the life cycle of a product's development cycle that provides a roadmap and specifications needed to conduct rational, compliant and resource-effective biopharmaceutical development from early through late phases.

Product Life Cycle: The manufacturing cycle involved in product development, beginning with planning and continuing through all phases of development and stages of manufacture.

Product Stream: See Stream.

Production: The act of biomanufacturing.

Production Cell Bank (PCB): See Working Cell Bank.

Program: A group of related projects, often coordinated and sharing a common objective.

Project: A distinct and planned enterprise that is identified by a unique objective or goal, a beginning, an end and a schedule. A unique effort, temporary in nature and defined by objectives, tasks, processes, budget and a schedule, notably start and completion.

Project Champion: Individual serving as a stakeholder or on the project team who has a strong personal and professional interest in achieving the objective or product. Champions argue and strongly support the objective.

Project Management: The function of planning, organizing and managing resources and schedules to bear on performance of a defined project.

Project Management Plan: A written plan that outlines how a project will be managed using modern project management processes and tools, both technical and social.

Project Manager: A professional appointed to manage and lead a project team and the processes, tasks, budget, schedule and other activities that fall within the scope of the project.

Project Schedule: A calendar with dates that demonstrates a project from beginning to completion, with all major tasks and processes having start and finish dates.

Protocol: An instructive document that directs the performance of a study and schedule of events, such as a clinical, nonclinical or validation study.

Purification: Process used in biomanufacturing to remove contaminants and impurities while leaving desired product.

Purity: The amount of product in relation to impurities that might exist in a product. Freedom from chemical or biological contamination or impurities.

QSR: See Quality Systems Regulation.

Qualification: To ensure that something, such as a test or process, is suitable for use. Qualification is typically less stringent than validation and the process is often applied to critical laboratory tests or manufacturing processes in early to midphase development.

Quality: The degree of excellence of a thing; general excellence (Oxford Dictionary, 1997).

Quality Agreement: A contract between two parties, generally a contractor and client, that identifies quality aspects of a relationship, especially regarding compliance issues and conformance with a quality system.

Quality Assurance: Function of planning, managing, operating and ensuring the performance of a quality system.

Quality Assurance Unit: The functional area at a biotechnology firm responsible for all aspects of Quality Assurance.

Quality by Design (QbD): A planning process in which a product and its biomanufacture are carefully described along with attributes and specifications.

Quality Control: Laboratory, test or metrology function to ensure quality of a product.

Quality Manual: A written document that proscribes the quality policies and, in general, criteria, operations and organization for quality systems at a biotechnology firm.

Quality Plan: A written document that expands on the Quality Manual, describing the quality systems and plans to operate those systems at a biotechnology firm. The Quality Plan is referenced in or is part of each Product Development Plan.

Quality System: A designated set of components, connected in a logical fashion and focused upon quality of a product or service. Quality systems for biopharmaceutical development are codified in regulations, guiding quality operations of a particular activity, such as manufacture (e.g., cGMP), nonclinical studies (e.g. GLP) or clinical studies (e.g., GCP).

Quality Systems Regulations (QSR): An FDA guideline, based on a regulation, and focused on quality design and manufacture of medical devices.

RAC: See Recombinant Advisory Committee.

Randomize: The act of randomly placing a subject or animal into one or another treatment group in a controlled and blinded clinical or nonclinical study.

Range: The value defined by the upper and lower limits of a specification or concentration of an analyate.

Raw Materials: Items that are used to manufacture a product. Generally refer to materials, such as solutions, reagents, chemicals or biological substances.

Recombinant Advisory Committee (RAC): A committee established by the director, NIH, to review and approve certain studies, *in vitro* or in animals or man, dealing with genetic engineering or transfer or use of genetically modified organisms.

Recruiter: Individual employed by a clinical investigator or sponsor for the purpose of identifying individuals who might wish to volunteer for a clinical trial.

Reference Standard: A well-characterized and known product or material against which the attribute of a test material may be compared.

Regulation: A rule that has the force of law. Interpretation of the law by an administrative government agency in the executive branch and intended to carry out the intent of the law.

Regulatory Intelligence: The process of finding and analyzing publically available regulatory information.

Release: The action of allowing a product or study to be provided to the user, customer or client.

Release Testing: The panel of tests that are performed on a product before it can be released.

Requirement: An attribute that a product possesses, usually defined in scientific terms.

Research Seed: A microbial seed (e.g., clone) or cell line derived from a research laboratory.

Robust: A vigorous, strong and sturdy manufacturing process or test method in biopharmaceutical development. Overall reproducibility of a process or of test results when operational conditions are held within established ranges.

Root Cause: The ultimate or original cause for an issue or problem that occurs. It can be clearly described in technical or scientific language and, hence, is preventable in the future.

Run: A single and clearly identifiable manufacturing process. A single run produced as an individual lot or batch or product.

SAE: See Serious Adverse Event.

Scale up: Increase in amount of biomanufacturing for a product so as to increase the total amount of product in a single batch or lot of BS or FP.

Seed: A defined cell or viral particle from which other cells or particles may be derived.

Serious Adverse Event (SAE): An adverse event that is serious by medical diagnosis or is life threatening or causes death of the volunteer.

Sociotechnical Skills: Practiced ability to integrate sociological, or people management, expertise together with a technical knowledge and capability so as to lead a project team.

SOP: See Standard Operating Procedure.

Source Document: Any record, data or other piece of information that is closest to the source. Data that are initial, original or raw and on a written or electronic document.

Sparge: Move a gas into or through a liquid in a vessel such as a fermentor.

Specification: A stated value or range of values that is specific to a product attribute and quality control test. Specifications are specific, strict and fully defined criteria upon which a product is found either suitable or unsuitable for use.

Specificity: The degree to which a measurement made by an analytical test is due to the actual analyate of interest and not to other materials in the test matrix.

Sponsor: The entity (institute, individual, corporation) that is ultimately responsible, in a scientific, business and legal sense and to regulatory agencies, the public and the users, for a product and its development.

Stability: For a product, the trait of maintaining purity, potency and strength over time and in a given environment. Property of a biopharmaceutical or product not to degrade or break down.

Stability-Indicating: A test that is capable of identifying when a product has lost or is losing purity, potency or strength.

Stability Protocol: A document that designs and plans a stability testing program for a specific product and applies various assays under an established schedule.

Stage: A major division of a biomanufacturing scheme, such as upstream processing.

Stakeholder: An individual who, by status or dominant position, is influential to a project team and, while not always serving directly on the team, has a vested interest in success of the team's efforts. Project teams serve, in part, to meet expectations of a stakeholder.

Standard Operating Procedure (SOP): An instructive document providing exact or detailed technical or administrative procedures. An SOP may include forms to capture data during performance of that procedure.

Steady State: Pharmacokinetic phase during which the concentration of a biopharmaceutical is maintained at a given level.

Step: A small but important part of any stage of biomanufacture.

Sterile Fill: A process to fill final containers with product and in the absence of microbial contamination. Aseptic methods are used throughout.

Stop criterion: A medical situation that arises and, by definition, leads to cessation in the enrollment and treatment of volunteers in a clinical trial.

Stream: Product stream or main bulk of product, in process, during biomanufacture. The material that is under production at a single time in biomanufacture.

Strength: Measure of active ingredient in a product. Typically using an analytical method that does not measure biological activity but instead measures the amount of chemical or biological substance present.

Study Director: Individual responsible for overall design, performance and reporting of a nonclinical safety study.

Subacute Toxicity: A safety study that evaluates toxicity of a biopharmaceutical, given in multiple doses, in animals over a brief period (e.g., 30–60 days).

Subchronic Toxicity: A safety study that evaluates toxicity of a biopharmaceutical, given in multiple doses, in animals over a moderate period (e.g., 3–6 months).

Subject: Individual without underlying disease who volunteers for a clinical trial.

Synchronize (Project): Integrate and bring together the various parts under a schedule and series of events.

System Suitability: Ability of an analytical test to achieve the objectives of the assay. All components of the test are suitable for the intended purpose.

$T_{1/2}$: The time elapsed from when a biopharmaceutical reaches C_{MAX} until it reaches ½ the value of C_{MAX}.

T_{MAX}: The elapsed time from when a biopharmaceutical is given until the maximum concentration is seen in blood (or tissue).

Tags: Molecular identifiers genetically engineered into a molecule. They may be used for identification or affinity purification of that molecule.

Tangential Flow Filtration (TFF): Type of biomanufacturing preparative filtration that allows filtration through a selective membrane as flow of liquid sweeps the membrane surface to prevent clogging.

Targeted Product Profile (TPP): Targeted Product Profile is a written document that prospectively identifies the attributes and intended therapeutic indications for a biopharmaceutical product. It is written in the format of product labeling, but is a planning tool, not a means of reporting results.

Task: A piece of work included in a project that is exactly defined in technical terms and itself has a beginning and end. An individual and defined piece of work, administrative or technical, in a project.

Team (Project): A group of professionals from different functional areas working together toward a common objective, each of them bringing a specific expertise.

Team Dynamics: The sociological and psychological energies and motions that affect behavior and change for a project team.

Team Leader: An influential individual serving on the project team, but not often serving as the project manager. He/she may be a chief scientist or key executive on the team.

Test: Analytical method or laboratory procedure performed on a product to measure an attribute.

Test Article: Test product, or the biopharmaceutical in formulation, as given to animals in a nonclinical study.

TFF: See Tangential Flow Filtration.

Timeline: A visualization of tasks or processes and milestones set against a schedule.

Tolerated: Ability of an organism to be subjected to a biopharmaceutical over a period of time without experiencing adverse effects or harm due to that product.

Toxicology: The study of toxic effects of chemicals, biologicals or ionizing radiation on a living organism.

Toxic Substances Control Act (TSCA): A U.S. law dealing with testing prior to use or release to the environment of chemical substances.

TPP: See Targeted Product Profile.

Track: As a noun, it is the pathway of a project. The verb, to track, means to monitor the processes and tasks within a project to ensure completion on budget and schedule.

Tracking: The project management process of reviewing all aspects of the project to ensure tasks are completed on schedule and budget.

Trait: Distinguishing feature or characteristic of a product. A specific chemical or biological feature that can be measured with an analytical test.

Transfection: Placing a foreign or recombinant gene into a mammalian or other cell derived from animals or plants.

Transgenic: Organism, such as a plant or animal, that retains one or more genes of another organism.

Treatment group: In a scientific study design, a group of humans or animals that all receive the same treatment, such as either investigational product or placebo.

TSCA: See Toxic Substances Control Act.

United States Pharmacopeia (USP): A compendium or reference volume that provides information on biopharmaceutical raw materials, products, processes, tests and formulations.

Upstream Manufacture: Biopharmaceutical production that yields the product as a crude or unpurified material. The early stages of biomanufacturing from cell bank to crude cell paste.

USDA: United States Department of Agriculture.

USP: See United States Pharmacopeia.

VAI: See Voluntary Action Indicated.

Validate: To provide strong evidence, usually through experimentation, that a piece of equipment, facility, utility or process performs exactly as intended and within established specifications.

Validation: A process in which a test or process is demonstrated to perform exactly as intended and planned and meets established specifications.

Variance: A measurement, outcome or part of a process or study that does not meet established procedures, rules or specifications, but is known or planned before it occurs in fact.

Vector: A live organism or construct of DNA (e.g., plasmid) that contains DNA or RNA of another organism, usually through recombinant technology.

Vehicle: A material, such as saline, in a formulation that serves to enhance transfer, absorption, or distribution of a biopharmaceutical.

Vendor: An entity that provides a material or service to a client.

Verification: To demonstrate, with documented evidence, that something, such as a piece of equipment, is what it is purported to be. Specific act of verifying and documenting that a compendial test performs as intended in a quality control laboratory.

Viable Particle: Living contaminant (typically bacterial, fungal or yeast) found in a product or manufacturing stream. An undesirable occupant of a manufacturing area.

Virtual Team: A project team that is separated by space and time yet still functions as an effective group of individuals working together toward a common objective.

Volume of Distribution: The distribution of a biopharmaceutical, in quantitative terms, between blood and other tissues of the body following dosing. It measures the volume in which a drug would be uniformly distributed at any point in time.

Voluntary Action Indicated (VAI): Refers to findings on an EIR in which the FDA recommends that action be taken by a manufacturer to correct minor deficiencies or deviations noted during an inspection.

Volunteer: Any individual who requests to be enrolled in a clinical trial.

Water for Injection (WFI): Highly purified water free of microbes, contaminants or impurities and of a quality that can be injected into humans.

WCB: See Working Cell Bank.

Well Characterized: A product or material for which there is a significant amount of scientific information, often chemical, biological and physical, that provides a high degree of understanding on the nature and functional properties of the product.

WFI: See Water for Injection.

WHO: World Health Organization, United Nations.

WI: See Work Instruction.

Withdrawal: The instance of a volunteer in a clinical study leaving that study on his or her own initiative or at the request of the principal investigator.

Work Breakdown Structure: A tool used in project management to identify the various pieces of a project and place them in a logical sequence of events or hierarchy. It is the basis for project planning. A visualization or narrative outline of project tasks as mapped to component parts.

Working Cell Bank (WCB): Derived from an MCB, this is the source of cells for production. Also referred to as a Production Cell Bank (PCB).

Work Instruction (WI): A document used in biomanufacturing to both guide a process and to record critical information regarding a particular batch or lot of product. Also called a Batch Production Record (BPR).

Yield: The amount of product that results from a step or stage of biomanufacture. Often presented as a percentage, the starting amount divided by amount obtained at the end.

Additional Reading

Chapter 1

APA. 2010. *Handbook of nonprescription drugs*. American Pharmacists Association, Washington, D.C.

Billups, N.F. 2010. *American drug index*. 54th ed. St. Louis, MO: Wolters Kluwer.

Dorland's illustrated medical dictionary. 2008. New York: Elsevier.

FDA. 2010. *Approved drug products with therapeutic equivalence evaluation*. U.S. Food and Drug Administration, Rockville, MD.

FDA. 2010. *National drug code directory (NDC)*. U.S. Food and Drug Administration, Rockville, MD.

Hetferon, K.L., ed. 2009. *Biopharmaceuticals in plants. Toward the next century of medicine*. Boca Raton, FL: CRC Press.

Merck & Co. 2010. *The Merck index*. Whitehouse Station, NJ: Merck & Co.

Mosby's drug consult for health professions. 2010. St. Louis, MO: Elsevier Mosby.

PDR Drug interactions and side effects. 2010. Florence, KY: Thompson Healthcare.

Physician's desk reference (PDR). 2010. Florence, KY: Thompson Healthcare.

Physician's desk Reference (PDR) generics. 2010. Florence, KY: Thompson Healthcare.

Physician's desk reference (PDR) for nonprescription drugs, dietary supplements and herbs. Definitions guide to over the counter (OTC) medicines. 2008. Florence, KY: Thompson Healthcare.

Rathone, A.S., and R. Mhatre. 2009. *Quality by design for biopharmaceuticals: Principles and case studies*. New York: John Wiley & Sons.

Sobti, R.C. 2008. *Essentials of biotechnology*. Boca Raton, FL: CRC Press.

Walsh. G. 2003. *Biopharmaceuticals: Biochemistry & biotechnology*. New York: John Wiley & Sons, Inc.

Chapter 2. Project Management

Fetterman, J.E. et al. 2008. *A framework for pharmaceutical risk management*. Food Drug Law Institute (FDLI), Washington, D.C.

Harpum, P. 2010. *Portfolio, program and project management in the pharmaceutical and biotechnology industries*. New York: John Wiley & Sons.

PMI. 2008. *A guide to the project management body of knowledge (PMBoK Guide)*. 2008. Newtown Square, PA: Project Management Institute.

Chapters 3 and 4. Regulatory Affairs and Regulatory Compliance

Adams, et al., eds. 2008. *Food and drug law & regulation*. Food and Drug Law Institute (FDLI), Washington, D.C.

Danzis, S.D. and E.J. Flannery, eds. 2010. *In vitro diagnostics: The complete regulatory guide*. Food and Drug Law Institute, Washington, D.C.

Kahan, J.S. 2008. *Medical device development: Regulation and law*. Waltham, MA: Parexel International Corp.

Klincewicz, S.L. et al. 2009. *Global pharmacovigilance laws and regulations: The essential reference*. Food and Drug Law Institute, Washington, D.C.

Mathieu, M. 2004., *Biologics development: A regulatory overview*, 3rd ed. Waltham, MA: Parexel International Corp.

Pines, W.L. 2003. *How to work with the FDA*. Food and Drug Law Institute, Washington, D.C.

Pisano, D.J. and D. Mantus, eds. 2008. *FDA regulatory affairs: A guide for prescription drugs, medical devices, and biologics*. Boca Raton, FL: CRC Press.

RAPS. 2008. *Fundamentals of EU regulatory affairs*, 4th ed. Regulatory Affairs Professionals, Rockville, MD.

RAPS. (2010). *Fundamentals of international regulatory affairs*. Regulatory Affairs Professionals, Rockville, MD.

RAPS. 2009. *Fundamentals of U.S. regulatory affairs*, 6th ed. Regulatory Affairs Professionals Society, Rockville, MD.

RAPS. 2010. *U.S. regulatory acronyms & definitions pocket guide*, 4th ed. Regulatory Affairs Professionals, Rockville, MD.

Chapter 5. Quality Assurance

Avis, K.E., et al., eds. 1998. *Biotechnology: Quality assurance and validation*. Boca Raton, FL: Interpharma/CRC.

Avo, K.E., ed. 1998. *Quality assurance and validation*. Boca Raton, FL: CRC Press.

Sarker, D.K. 2008. *Quality systems and control for pharmaceuticals*. New York: John Wiley & Sons.

Chapter 6. Biomanufacturing

Avis, K.E., ed. 1996. *Biotechnology and biopharmaceutical manufacturing, processing and preservation*. Boca Raton, FL: CRC Press.

Berry, I.R., ed. 2003. *Pharmaceutical process validation*. New York: Marcel Dekker.

Doble, M. 2004. *Biotransformations and bioprocess*. Boca Raton, FL: CRC Press.

Haider, S.I. 2009. *Biotechnology. A comprehensive training guide for the biotechnology industry*. Boca Raton, FL: CRC Press.
Haider, S.I. 2010. *Cleaning validation manual. A comprehensive guide for the pharmaceutical and biotechnology industries*. Boca Raton, FL: CRC Press.
Jamel, F., and S. Hershenson. 2010. *Formulation and process development. Strategies for manufacturing biopharmaceuticals*. New York: John Wiley & Sons.
Ozturk, S. 2005. *Cell culture technology for pharmaceutical and cell-based therapies*. Boca Raton, FL: CRC Press.
Sofer, G. 2005. *Process validation in manufacturing of biopharmaceuticals*. Boca Raton, FL: CRC Press.

Chapter 7. Quality Control

Rodriguez-Diaz, R., T. Wehr, and S. Tuck. 2005. *Analytical techniques for biopharmaceutical development*. Boca Raton, FL: Taylor & Francis/CRC Press.
Schalkhammer, T.G.M. 2002. *Analytical biotechnology*. Basel, Switzerland: Birkhauser Verlag.

Chapter 8. Nonclinical

Cavagnaro, J. A. 2008. *Preclinical safety evaluation of biopharmaceuticals* New York: John Wiley & Sons.
Choy, W.N. 2007. *Genetic toxicology and cancer risk assessment*. New York: Marcel Dekker.
Descotes, J. 2004. *Principles and methods of immunotoxicology*. New York: Elsevier.
Meibohm, B. 2007. *Pharmacokinetics and pharmacodynamics of biotechnology drugs. Principles and case studies on drug development*. New York: Wiley-VCH.
Timbrell, J.A. 2007. *Introduction to toxicology*, 3rd ed. Boca Raton, FL: Taylor & Francis, CRC Press.
Weinberg, S. 2003. *Good laboratory practice regulations*, 4th ed. New York: Marcel Dekker.

Chapter 9. Clinical

Chin, R., C. Yoonsik, and B.Y. Lee. 2008. *Principles and practices of clinical trials medicine*. New York: Elsevier.
Cook, T.D., and D.L. DeMets. 2007. *Introduction to statistical methods for clinical trials*. Boca Raton, FL: CRC Press.
Hackshaw, A.K. 2009. *A concise guide to clinical trials*. New York: Wiley-Blackwell.

Rozovsky, F.A. and R.K. Adams. 2003. *Clinical trials & human research: A practical guide to regulatory compliance*. New York: John Wiley & Sons.

Stone, J. 2006. *Conducting clinical research: A practical guide for physicians, nurses, study coordinators and investigators*. Cumberland, MD: Mountainside MD Press.

Practical Problems and Questions

Overview

This addendum to the book provides an opportunity for readers to consider situations that occur in biotechnology operations and to develop a Targeted Product Profile (TPP) and Product Development Plan (PDP) for a biopharmaceutical product directed at treating or preventing a disease. In the first section, situations are posed in which biopharmaceutical products are to be developed as a project by a biopharmaceutical firm. A small amount of technical background is provided and students are encouraged to make and state assumptions regarding the research results, intended population, indication, etc., for each product. In the second section, questions are posed regarding development of the products. Each question pertains to any of the six products.

Products and Projects

Problem #1. The product to be developed is a monoclonal antibody. Your firm, ABC Biologicals, Inc., is developing a proprietary monoclonal antibody to treat cutaneous T-cell lymphoma (CTCL), a cancer of white blood cells. This biopharmaceutical product is a "humanized" monoclonal antibody that targets and binds specifically to a cell surface molecule, CD545. The molecule, developed in your laboratory, has been characterized in research and was shown, in a transgenic mouse model, to bind to cancer cells, leading to their death. To be effective in mice, the product is given by subcutaneous injection. Your firm now wishes to develop the monoclonal antibody product and apply for market approval.

Problem #2. Your firm, EbolVac, Inc., has discovered and patented a surface protein of the Ebola virus, an organism that causes hemorrhagic fever in man. Ebola virus is considered a threat agent for bioterrorism and so governments are the potential customers. A gene for an Ebola virus surface protein has been cloned and expressed in a host system as a 65 kD glycoprotein. The product can be handled safely in the laboratory. Using a model of Ebola virus in nonhuman primates kept in a high containment facility, your firm has demonstrated proof of principle by protecting monkeys using a nasal spray of the vaccine. You now wish to develop the product for human use and obtain market approval.

Problem #3. Your firm, GenTherLine, holds an exclusive license for a novel gene therapy product composed of "naked" plasmid DNA. The gene therapy is intended to resolve nevoid basal cell carcinoma, a cancer of the skin. The naked plasmid DNA is a vector that includes the gene for PTCH. It must be delivered by a micro-needle injection apparatus that exposes the DNA to the cells of the basal epidermis. Unfortunately, there is not an animal model in which the mutation causes this disease. The PTCH gene, once it enters the nucleus of a cultured cancerous basal epidermal cell, replaces the mutant gene that causes this disease, but the concept has not been attempted in man. You wish to develop the product and obtain market approval.

Problem #4. Your firm, MalarTher, has discovered a 35kD protein, MER24, that interrupts the life cycle of malaria parasites by blocking their ability to further infect red blood cells. It acts directly on the merozoites stage in the erythrocytic or asexual cycle of erythrocyte infection, mimicking a red blood cell receptor. It is considered a potential therapeutic molecule to be given intravenously to individuals infected with *Plasmodium falciparum*. You now wish to develop the product and obtain market approval.

Problem #5. Your firm, StemTechnolUS, develops human therapeutic products from human pluripotent stem cells. You have discovered and cloned a stem cell-derived cell line that differentiates to neurons and is suitable for repair of uncomplicated spinal cord injuries from blunt trauma. Cells would be given by surgical intervention. You now wish to develop the product and obtain market approval.

Problem #6. Your firm, TheraGentCure, has the patent for a gene therapy construct. It is a retrovirus, specifically a lentivirus, that carries the gene for a tissue inhibitor of a matrix metalloproteinase (MMP) that inhibits neuroblastoma tumor cell growth. It would be used by direct injection into tumor mass or into blood vessels feeding the mass in an attempt to slow the growth of the tumor and perhaps cure the cancer. You now wish to develop the product and obtain market approval.

Questions by Chapter and Functional Area

Section 1. Background and Targeted Product Profile

Discuss the class of biopharmaceutical represented by this product.

Provide a name for the product.

What competitive products are on the market, if any, and how have they been developed and marketed? Discuss market advantages your product should possess.

Describe the disease or condition (indication) and the population that is subject to the disease.

Develop a TPP for this product with all elements of biopharmaceutical labeling.

Define exactly the indication and population your product would diagnose, treat or prevent as considered for the first market approval. Discuss the rationale for choosing this indication and the intended population.

How, in general, will this TPP impact the plans for each functional area?

Provide a product design, to include input, design and output based on elements of the TPP.

Section 2. Project Management

Identify the project and its purpose and provide an overview of the project from the standpoint of the project manager. Include a purpose, scope and technical and management objectives.

Provide the composition of the project team enlisted for this project. Identify team members, their affiliations and locations, authorities and responsibilities, and the rational for these choices. Identify stakeholders to the project and define their roles apart from the team.

Describe the team communications plan to include types of communication and frequency of meetings.

Describe metrics that would be used to measure progress of the project.

Describe the methods the project manager will apply for allocating, tracking and managing resources allocated to the project.

Identify the 10 major risks this project is likely to encounter. Outline the risk assessment and risk management plans for the project, focusing on risk mitigation early in the project. Describe the roles for the project manager in risk management.

Describe your plan for identifying actual problems or issues and resolving them as a team.

Establish a Work Breakdown Structure both as a narrative explanation and as an illustration (e.g., Gantt or PERT chart).

Section 3 and Section 4. Regulatory Affairs and Regulatory Compliance

Is there regulatory precedent for treatment of this indication and population and, if so, are the predicate products approved by the FDA or any other regulatory agency for marketing? How might this history of predicate products impact the regulatory development of your product? Is there precedent for this type of product (e.g., molecular or cellular nature) having received market approval or having been tested as an investigational drug? If so, what were the outcomes and how might this precedent impact the development and regulatory activities for your product? What do you think is the status of your product in the eyes of the FDA?

Each product falls under one or another office at the FDA. Describe which office or division at the FDA is likely to review your product. Explain how your product will fall into the scheme of FDA organization. Where will you submit applications within the FDA organization?

Will market approval be sought, initially, in the United States alone and/or in other countries. Describe your plans to submit marketing applications in each chosen country.

What types of investigational use and marketing applications will be submitted to the FDA over the life of the product?

For this product, outline the elements of the IND, in IND or in CTD format, as intended for the FDA.

Outline briefly an FDA marketing application in an acceptable format. Highlight the key elements that must be achieved during development so that application will be complete. Will user fees be necessary and, if so, how much are the fees in today's dollars?

What other types of applications might be submitted to the FDA in an effort to facilitate development, increase market share or exclusivity, or speed the process of approval?

What FDA guidance documents are most important for your firm to consider for this product and indication.?

What is the nature and timing of meetings or teleconferences that might be held with the FDA during the development process and prior to receiving market approval?

As designed, could your product be considered in any way a combination product and explain the rationale for this conclusion? If it is a combination product, then describe the impact on the regulatory plan and development pathway.

Describe the major risk and benefits for the product as they might be perceived by the FDA. Discuss the factors that will enter into a regulatory review of the risk-to-benefit ratio for this product.

Describe postmarketing activities that might be required by regulatory agencies for this type of product or indication.

What non-FDA regulatory hurdles must be considered for this type of product? Do these pose major obstacles to product development and, if so, how will they be addressed and resolved?

Section 5. Quality Systems and Quality Assurance

Outline the contents of a quality manual that will serve your product's development operation. Cite specifically each quality system that will be included and explain why and when in development it will be applied. Include a brief quality policy and a brief statement regarding management responsibility as it would be later signed by upper management.

Describe the need to design your product and explain how that will be accomplished. Provide product-related specifics regarding quality by design and design control.

Review the elements of a design program specific for your product, relating user needs, product attributes and technical elements of design. Is it possible that your product must be reworked in the research laboratory before it begins development? If so or if not so, explain exactly why this is the case.

Choosing from the list of hallmarks of quality, choose what you consider to be the six most important in regards to the quality of your product. Do not choose hallmarks of design. Explain why each was chosen and specifically how it will be applied to your product during both development and marketing phases.

Describe the fully functional quality assurance unit that will support development of your product and in your narrative specifically identify examples of how the unit will function in the areas of audit, investigations and change control.

Describe the five most likely problems you will experience as you establish quality systems and a quality assurance unit for this particular product and project.

Section 6. Biomanufacturing

Describe the biomanufacturing schemes that have been used to produce biopharmaceuticals of a similar molecular or cellular nature.

Identify regulatory guidelines that apply to the manufacture of such products.

Discuss major risks associated with manufacturing this type of product.

Design a biomanufacturing plan for your product to include objectives, input, process, equipment and facility considerations, output and review. Provide the process flow to include premanufacturing preparation of constructs, upstream and downstream processing, holds, and formulation, fill, finish and formulation and raw material requirements.

Outline how and at what phase of development the biomanufacturing process would be increased in scale. Provide plans to increase facility size or to use contract manufacturing operations.

Provide general plans to validate the process, facility, equipment and utilities.

Section 7. Quality Control

Describe quality control test schemes that have been used to produce biopharmaceuticals of a similar molecular or cellular nature. Identify regulatory guidelines that apply to the quality control of such products.

Define and justify the attributes of bulk substance (BS) as they are based on its known nature and manufacturing scheme for BS. Once the attributes are listed, draft a Certificate of Analysis, adding analytical methods and specifications to the attributes and considering testing more than one parameter of key attributes.

Define and justify the attributes of final product (FP) as they are based on its known nature and manufacturing scheme for FP. Once the attributes are listed, draft a Certificate of Analysis, adding analytical methods and specifications to the attributes. Consider testing more than one parameter of key attributes.

Describe in-process samples that will be taken during the manufacturing process and identify attributes, tests and possible specifications for each sample.

Briefly describe each test chosen for testing BS and FP, for both release and stability and assays for in-process samples. Explain why each test was chosen based on performance, intended use and meaningful results. Describe what is known about the specificity, accuracy, precision, range and robustness of each assay when applied to this or to other classes of product. If an assay (e.g., potency) is to be developed just to test this product, justify the need to develop the assay and present ideas on its nature. Discuss the nature or need for control reagents and reference standards for these assays.

Develop stability protocols for BS and for FP and explain why each was chosen for the purpose of indicating stability. Describe any tests that are used to measure stability that were not applied to product release.

Consider attributes, analytical tests and specifications for critical raw materials used in the process, working closely with the manufacturing plan. Prepare a draft Certificate of Analysis for what you consider to be the five most critical raw materials.

Identify which analytical tests will be qualified, verified and/or validated and mention the most likely point in the development cycle for each activity. Highlight critical assays that might require special attention in development, qualification or validation.

Section 8. Nonclinical Studies

Describe the nonclinical studies that are typically performed for this class of product and for products used with this indication. Consider both regulatory guidelines and precedent.

What additional work must be performed to complete a pharmacokinetic and pharmacodynamic profile for this product if it is given by the route and doses currently suggested in the clinical studies plan. Outline the studies that must be performed and describe how and where they might be done. Consider major design criteria for these studies.

Describe the nonclinical toxicology studies that have been performed with other products in this class of product. Consider both regulatory guidelines and precedent.

Identify the nonclinical toxicology studies that should be performed to ensure safety of this product. For each study, provide a brief purpose and design and recommend when they might be performed by phase of development.

Describe how a quality system, cGLP, will be ensured for the planned studies.

Section 9. Clinical Studies

Outline, in general, the clinical program that will be performed during the course of development.

Given the nature of the product and the indication, outline the pivotal or Phase 3 clinical study and describe the objectives and indication and population in which your product will be tested. Describe the patient or subject population, approximate size and scope and critical elements of study design.

Outline the Phase 2 clinical study upon which you will demonstrate proof of principle and from which you intend to derive information upon which to base the Phase 3 clinical study.

Outline the Phase 1 clinical study or studies that will be the foundation for Phase 2 study design.

Present requirements for choice of an investigational site and a principal investigator for each phase of clinical development you outlined and in relation to your product and indication.

Discuss elements of current Good Clinical Practices (cGCPs) that would be required for each phase of clinical development you have outlined.

Describe any ethical considerations that might impact any one of your clinical trials given the nature of the product, the study design and the intended patient population.

Index

D

E

F

G

H